Current Topics in Microbiology and Immunology

177

Cover illustration:
Region of adherent layer that supports haemopoiesis in vitro. The shown region has spontaneously detached from substratum, as indicated by the exposed flattened undersides of several cell profiles. A mixture of adipocytes, developing granulocytes, macrophages as well as their precursors including stem cells can be visualised "underneath" the stromal layer of blanket cells. Magnification x 3000.

Hematopoietic Stem Cells

Animal Models and Human Transplantation

Edited by C. Müller-Sieburg, B. Torok-Storb, J. Visser and R. Storb

With 18 Figures and 14 Tables

Springer-Verlag
Berlin Heidelberg New York
London Paris Tokyo
Hong Kong Barcelona
Budapest

CHRISTA E. MÜLLER-SIEBURG, Ph.D.
Medical Biology Institute
11077 North Torrey Pines Road
La Jolla, CA 92037
USA

BEVERLY TOROK-STORB, Ph.D.
Fred Hutchinson Cancer Research Center
1124 Columbia Street
Seattle, WA 98104
USA

JAN W. M. VISSER, Ph.D.
Radiobiologisch Instituut—TNO
151 Lange Kleiweg
2288 GJ Rijswijk
The Netherlands

RAINER STORB, Ph.D.
Fred Hutchinson Cancer Research Center
1124 Columbia Street
Seattle, WA 98104
USA

ISBN 3-540-54531-X Springer-Verlag Berlin Heidelberg New York
ISBN 0-387-54531-X Springer-Verlag New York Berlin Heidelberg

Library of Congress Catalog Card Number 15-12910
Printed in Germany

Typesetting: Thomson Press (India) Ltd, New Delhi; Offsetprinting: Saladruck, Berlin; Bookbinding: Lüderitz & Bauer, Berlin.
23/3020-5 4 3 2 1 0 – Printed on acid free paper.

Preface

The idea for this volume was conceived during a discussion in the hallway at a conference in early 1990. "What is the *best* way to detect and define pluripotent hematopoietic stem cells?" was the question posed by Dr. Fritz Melchers. After discussing the pros and cons of the available assays for quite some time, it became apparent that this topic required a wider expertise and merited a larger forum. Thus, we decided to extend the discussion and to compile the results in this volume. Much to our delight, many of the pioneers of recent experimental and theoretical developments in stem cell research agreed to contribute their expertise to answer the question. These authors review both past findings and present insights, thus providing an overview of the evolution that has been and is occurring in the field of stem cell research.

In the light of recent trailblazing developments in both experimental models and in clinical application it is indeed time to reevaluate our knowledge about stem cells. Transplantation of hematopoietic stem cells has become more and more prevalent as a curative therapy in a variety of acquired and genetic diseases, including cancer, radiation accident, as an agent for gene therapy, and perhaps even as treatment for autoimmune diseases. Stem cells are now derived not only from bone marrow but also from peripheral blood, cord blood, and fetal liver, greatly increasing their availability for human transplantation and in some cases (fetal tissues) obliterating the need to match donors and hosts.

The developments in clinical application have gone hand in hand with progress in experimental stem cell research. The demonstration that murine embryonic stem cells can differentiate in vitro into hematopoietic cells has made the process of stem cell development amenable to in vitro analysis and manipulation. Recent experimental findings using in vivo models have changed our concepts and views about the identity and characteristics of stem cells. Several assays that were believed to detect stem cells have now been shown conclusively to identify lineage-restricted precursors both in the myeloid and the lymphoid differentiation pathways. Therefore, it is necessary to reevaluate many of the concepts about differentiation and replication of stem cells that were based on these assays. This is particularly true for the theories that were developed using the spleen colony assay. Furthermore, mounting evidence indicates functional heterogeneity in the stem cell compartment itself, although the cellular basis of this functional difference remains to be elucidated.

While several well-defined clonal assays for restricted precursors are available, currently, long-term repopulation assays after transplantation are the only valid detection systems for the most primitive pluripotent stem cells. In fact, stem cells are *defined* by their capacity to give rise to all hematopoietic lineages for a long time, reflecting extensive differentiation and self-renewal capacity of these cells. Several contributors to this volume have explored in detail the most commonly used in vivo detection systems for stem cells. Each of these assays has drawbacks; they require large numbers of animals, take a long time to complete, and tend to be tedious to quantitate. Thus, short-term, clonal in vitro assays are urgently needed, especially in the human system. The recent explosion in our understanding of the role of the microenvironment in regulating hematopoiesis promises rapid progress in finding such an in vitro stem cell assay. Culture methods based on the maintenance of hematopoiesis by stromal cells and by newly discovered cytokines and negative regulators are currently being explored.

In summary, the field of stem cell research is at an exciting juncture. New techniques enable us to enrich stem cells, to examine them at the molecular level, and to identify regulatory signals. This volume takes inventory of our present body of knowledge and outlines future directions. We thank Dr. Fritz Melchers for initiating this book by his persistent questions and for guiding us through the jungle of being an editor, Marga Botsch at Springer-Verlag for her patience and for being there when we needed her, Danette Stroh at MBI for cheerfully sending out numerous letters and Faxes and for fending endless telephone calls. In particular, we would like to thank the contributors to this volume for producing wonderful manuscripts on short notice.

CHRISTA E. MÜLLER-SIEBURG and JAN W. M. VISSER

List of Contents

Microenvironments for Hematopoiesis

Human Stem Cell Transplantation

Gene Transfer

List of Contributors

(Their addresses can be found at the beginning of the respective chapters.)

Introduction

Closely guided by principles developed in experiments with animals, the modern phase of clinical marrow transplantation began in the late 1960s. Initially used only as a last resort, marrow grafts have become therapeutically effective in selected patients with hematological diseases. In many transplant centers worldwide, thousands of marrow grafts are carried out every year to treat congenital defects of the hematopoietic and immune systems (e.g., thalassemia major and severe combined immune deficiency disease), acquired marrow dysfunction such as aplastic anemia, and malignant hematological diseases.

The longest survival time after transplant is now more than 20 years. The hematopoietic system remains entirely of marrow donor origin while marrow stromal cells are host derived. This indicates that the original marrow inoculum contained pluripotent stem cells with self-renewal capacity. Because of the self-renewal capacity of the blood forming tissue, the same donor may be used repeatedly for marrow harvests. In at least one instance, a marrow recipient treated for aplastic anemia served as the marrow donor 10 years later when the original marrow donor developed malignant lymphoma.

In distinction to the clonal reconstitution found in mice using retroviral marking, studies in humans are consistent with polyclonal reconstitution of marrow after transplant. This was determined in recipients of female marrow using restriction fragment length polymorphisms which can distinguish the active from the inactive X chromosome. The discrepancy between mono- or oligoclonal reconstitution in mice versus polyclonal reconstitution in humans raises several questions concerning the dogma associated with the two models. Is the difference due to qualitative interspecies variation; to the fact that successful retroviral marking requires stem cells that are dividing, which provides an initial advantage to the marked cell; because only the progeny of the marked cell is detected while the contribution of nonmarked cells remains unnoticed;

or because there is a more than 3-log size difference between a mouse and a human, so it would take many more stem cells to repopulate a human hematopoietic system. Understanding the mechanisms which underlie these differences becomes important as we begin to modify transplant protocols for broader applications.

One limitation in the use of marrow grafts has been that the ideal donor, a genotypically HLA-identical sibling, is available for only 35% of patients. Over the last decade marrow grafting has been extended by the use of phenotypically HLA-identical, unrelated and HLA-nonidentical, family member donors. However, more acute and chronic graft-versus-host disease has been seen with alternative donors, adversely influencing survival. The use of fetal cells or of T-cell depleted marrow grafts has only partially solved the problems seen with alternative donors. With T-cell depleted marrow grafts, two principal problems have been encountered: increases in graft failure and recurrence of underlying disease. Exactly how T-cell depletion compromises hematopoietic reconstitution has not been fully explained. Little is known about what critical signals T cells may provide during initial engraftment; therefore, it remains to be seen whether this T-cell function can be replaced by hematopoietic growth factors or immune response modifiers.

An alternative to allogeneic marrow transplants has been the use of the patient's own marrow cells. Presently, autologous marrow grafts are restricted to patients with malignant diseases. In patients with leukemia, the marrow is stored during a chemotherapy-induced remission, while in patients with solid tumors, the marrow is often free of metastatic disease. The presence of tumor in marrow has prompted the development of procedures to remove malignant cells. These include the use of chemical agents, e.g., hydroperoxycyclophosphamide, monoclonal antibodies coupled to complement or toxins, or positive selection of stem cells with the help of a monoclonal antibody directed at CD34. An alternative way of purging marrow consists of the use of marrow cells maintained in long-term in vitro culture. Often after purging techniques have been applied, the kinetics of engraftment can be slowed. In such instances, hematopoietic growth factors may be used to hasten recovery. To circumvent the need for purging, peripheral blood stem cells have been used in lieu of marrow for autologous transplantation. The presence of circulating hematopoietic stem cells in experimental animals has been recognized since the early 1960s. Their long-term repopulating potential has been established in allografted dogs using blood genetic markers, although such evidence in humans is still lacking. Retrovirally marked blood stem cells are needed to assess their long-term repopulating potential. Blood stem cells have been used as an adjunct to allogeneic marrow in patients with aplastic anemia to increase the rate of engraftment, although at the price of an increase in chronic graft-versus-host disease. More efficient techniques of stem cell separation from the peripheral blood are needed for allogeneic grafts.

Finally, the ability of pluripotent stem cells not only to self-renew, but also the differentiate into committed hematopoietic progenitors has made them potential targets for somatic gene therapy. Many hereditary hematopoietic disorders can be cured by marrow grafting. This and recent advances in gene cloning and transfer have made it possible to contemplate the replacement of missing or defective genes in hematopoietic tissues. Also, the transfer of genes that confer resistance to antineoplastic agents might increase the therapeutic index of chemotherapy by making a patient's marrow more resistant to anticancer drugs. Gene transfer may also enable us to address questions on stem cell commitment and proliferation in humans. Retroviral vectors have been the preferred vehicle for attempts at infecting pluripotent stem cells in various animal species. Best results have generally been seen in mice. Nevertheless, even in this model, expression of the transferred genes is often poor and the frequency of cells expressing the gene of interest may decrease over time. Successful in vitro gene transduction into fetal hematopoietic cells resulted in intermittent neomycin-resistant hematopoietic colonies in sheep for more than 2 years. Also, successful transduction of long-term repopulating marrow cells in adult dogs has been achieved. Neomycin resistance has been shown in marrow, peripheral blood and lymph node lymphocytes, and granulocytes for over 2 years after transplant. However, much work still needs to be done to increase the relatively low levels of gene expression in vivo through improvements in retroviral vectors and packaging cell lines so that the highest possible virus titers can be generated in the absence of detectable helper virus. As our understanding of gene transfer techniques increases and our knowledge of gene regulation in the course of cell differentiation improves, manipulation of genes for the treatment of hematopoietic disorders may one day become clinical reality.

RAINER STORB and BEVERLY TOROK-STORB

This book is dedicated to
Cheryl A. Whitlock, Ph.D., M.D. (1953–1990).

Friend, teacher, inspiration.

Experimental Models for Pluripotent Stem Cells

Careful Maintenance of Undifferentiated Mouse Embryonic Stem Cells is Necessary for Their Capacity to Differentiate to Hematopoietic Lineages In Vitro

U. CHEN

1 Introduction

It is well established that pluripotent stem cells derived from adult bone marrow give rise to all cells in the myeloid, erythroid, and lymphoid lineages. The origin of the stem cells and their lineage commitment to lymphocytes from mouse embryos is a controversial issue. Stem cells from yolk sac have been shown to give rise to erythrocytes, megakaryotic cells, and myeloid cells (LABASTIE et al. 1984). B lymphocytes can be found when yolk sac stem cells are transferred into animals, but not in the differentiation condition in vitro (MELCHERS 1977; MELCHERS and ABRAMCZUK 1980; PAIGE et al. 1979; C. Paige, personal communication). However, stem cells obtained from the fetal liver have been demonstrated to give rise to myeloid, erythroid, and B-lymphoid cells in vitro (OWEN et al. 1975; MELCHERS 1977; OGAWA et al. 1988; ROLINK et al. 1991). T-lymphoid cells require a thymic microenvironment and can be demonstrated in vivo (see review by VON BOEHMER 1990).

Pluripotent stem cells have been difficult to study since they are rare cells found in the heterogeneous bone marrow and fetal liver. Great progress has been made in obtaining subpopulations of bone marrow from adult and fetal liver that are highly enriched for pluripotent stem cells. Yet, very little is known

Basel Institute & Immunology, Grenzacherstr. 487, CH-4058 Basel, Switzerland

Current Topics in Microbiology and Immunology, Vol. 177

about how stem cells commit to differentiation and how stem cells self-maintain. This lack of knowledge is at least partially due to the inability to maintain stem cells in vitro. Culture systems that would allow the growth of very primitive stem cells, such as mouse embryonic stem cells and their differentiated progeny, the primitive hematopoietic stem cells, would be invaluable in addressing these issues.

I describe here the growth condition important for the maintenance of totipotent embryonic stem cells and the criteria for the status of "undifferentiation." I also describe the derivation of pluripotent hematopoietic stem cells from totipotent embryonic stem (ES) cells in vitro and discuss the implication and tools for the analysis of hematopoietic stem cell differentiation.

Mouse ES cells are cell lines cloned from the inner cell mass (ICM) of blastocysts and retain their totipotency in vivo and in vitro (GOSSLER et al. 1986; DOETSCHMAN et al. 1985, 1987; MANSOUR et al. 1988; THOMPSON et al. 1989; ZIJLSTRA et al. 1989; SCHWARTZBERG et al. 1989; KOLLER and SMITHIES 1989; DECHIARA et al. 1990). In vitro, the differentiation of ES cells has been carried through to the generation of simple and cystic embryoid bodies (MARTIN and EVANS 1975; MARTIN 1977; ROBERTSON 1987). The development of blood islands which contain erythryopoietic cells in the cystic embryoid bodies has been reported (GOSSLER et al. 1986; DOETSCHMAN et al.1985; ROBERTSON 1987). Recently, the mechanism of developmental regulation of the β-globin gene has been studied in the in vitro ES cell model in three different culture conditions (LINDENBAUM and GROSVELD 1990; WILES and KELLER 1991; SCHMITT et al. 1991).

The culture conditions containing methylcellulose, developed by Wiles and Keller, a modification of the protocol ISCOVE et al. (1989), can also be used for studying myeloid lineage differentiation. However, it is not suitable for studying lymphoid lineage development since the culture conditions do not support lymphopoiesis (M.V. Wiles, personal communication). It is possible that in that system the pluripotent stem cells differentiated to the stage equivalent found in the yolk sac, or that their culture system did not allow lymphocyte commitment, or both. On the other hand, a liquid culture system has been described (CHEN 1991) where pluripotent hematopoietic stem cells can develop from ES cells; and all known hematopoietic lineages including lymphoid (T and B) lineages are identified. This suggests that this liquid culture system favors the differentiation of stem cells that resemble those existing in fetal liver (Fig. 1).

Under the microscope, the in vitro differentiation of ES cells could be visualized by observing the cells grown in three dimensions. The first to appear are simple embryoid bodies within the cluster of cells, followed by the cystic embryoid bodies in which a complicated inner structure develops. The ES fetuses that contained pulsating cardiac muscle, yolk sac blood islands, and small lymphoid-like cells in the cup-shaped structures were then selected for immortalization of hematopoietic cell lines. It is the cup-shaped structure containing lymphoid-like cells inside the ES fetuses, rather than the blood islands, that are more important in locating the pluripotent stem cells (see below).

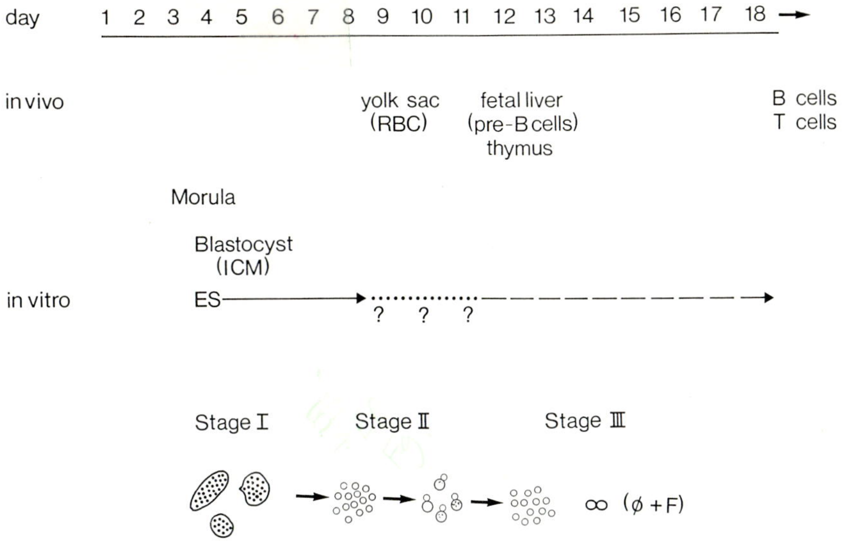

Fig. 1. Mouse embryogenesis. It requires 18 days of gestation before the birth of a mouse. Embryonic stem cells were established from the ICM of blastocysts, the stage after morula formation and equivalent to day 3.5 of gestation. ES cells are known to form germ-line transmitted chimeric mice when implanted in an animal and to differentiate to the equivalent of the yolk sac stage (day 8.5) in culture, where erythroid (*RBC*) and myeloid lineages could be identified. Lymphocytes (both T and B) are found in the fetal liver or thymus which appear around day 11–12 of gestation. The question was whether one could find an optimal culture condition for the differentiation of totipotent ES cells to pluripotent hematopoietic precursor cells (*question marks*), to induce the generation of different lineages, such as erythroid, megakaryotic, myeloid, and lymphoid. ES cells grow two-dimensionally on feeder cells, primary embryonic fibroblasts or STO cells (stage I). They appear like "pancakes" or "islands" on feeder cells. For controlled differentiation, the ES cells were removed from feeder cells and transferred to new dishes to become small clusters of cells and were allowed to grow in a suspension culture (stage II). Under optimal conditions, these cells will grow three-dimensionally, and cavity-like structures will develop (simple embryoid bodies), and complicated tissues inside the embryoid body (cystic embryoid bodies) will appear. ES fetuses which possess cup-shaped structures containing lymphoid-like cells were selected for establishment of cell lines (∞) by retrovirus infection ($\varnothing$) and/or growth factors (*F*) (stage III)

2 Maintenance and Characterization of Undifferentiated ES Cells (Stage I)

In order to maintain undifferentiated ES cells, it is critical to examine their "undifferentiation" status and totipotency regularly by several criteria: (a) morphologic appearence; (b) immunotyping; (c) karyotyping; and (d) potency to differentiate in vivo and in vitro. The method of karyotyping has been described by ROBERTSON (1987). The most important criteria of totipotency of ES cells is their ability to form chimeric animals (see ROBERTSON 1987; SCHWARTZBERG et al. 1989; GOSSLER et al. 1986) and to differentiate in culture to lympho-hematopoietic cells. The ES cell lines were monitored for lack of differentiation in two ways: morphologic observation and immunotyping.

Table 1. Immunotyping of ES cells and their derivated hematopoietic cells

Cell type	Antibody staining				
	SSEA-1	Ig	CD3	Mac-1	AA4
ES cells (undifferentiated)	+	–	–	–	–
Hematopoietic cells (differentiated from ES cells)	–	+	+	+	+

Morphologically, ES cells grow two-dimensionally on top of non-dividing feeder cells, which are either mouse fibroblast STO cells (WARE and AXELRAD 1972) or primary embryonic fibroblasts. The ES cell clusters are of "pancake"- or "island"-like morphology. The edge of the individual cells is not clearly distinguishable by light microscopy. Once the cells begin to grow three-dimensionally and the edge of the cluster begin to be distinct, the culture has already moved toward "chaotic" differentiation. The stage of undifferentiation can also be identified by immunotyping, ES cells express a stage-specific embryonic antigen (SSEA-1), recognized by the monoclonal antibody MC-480 (SOLTER and KNOWLES 1978). If the cells have started to differentiate, the expression of SSEA-1 is lost (Table 1). This method provides a simple and quick way to identify undifferentiated ES cells (CHEN 1991).

ES cells are fast-growing cells. The doubling time is 8–10 h as estimated by a BrdU/Hoechst two-parameter flow cytometric method (CHEN and SEYSCHAB 1989). It is critical to maintain a neutral pH and sufficient nutrients in the medium, and thus regular feeding and frequent passage of cells are required. ES cells were used for fewer than 20 passages. ES cells, feeder cells, and all reagents must be free from mycoplasma. Contamination by mycoplasma will severely affect their ability to differentiate in vivo and in vitro. In my experience, ES cells cultured with leukemia inhibitory factor (LIF) alone and for passages of over 20 generations lose their capacity to differentiate into lymphohematopoietic cells. Thus, fetal fibroblasts are routinely used to maintain the undifferentiated stage of ES cells, and multiple aliquots of early passages are kept frozen in liquid nitrogen.

3 Differentiation of ES Cells to ES Fetuses In Vitro (Stage II)

The culture condition for efficient and controlled differentiation of ES cells to ES fetuses, which contain all known hematopoietic cells, is a liquid suspension culture. It should allow slow but free movement of small clusters of cells under optimal nutrient conditions (high glucose and serum in medium with low carbondioxide in the air supply). It is important that a neutral pH is maintained and a rich nutrient is supplied, hence daily feeding is required. This culture condition, that has been described in detail (CHEN 1991), seems to provide a favorable microenvironment for the differentiation of ES cells to the stage where ES fetuses containing hematopoietic cells can be found.

In brief, mouse ES cells derived from strains 129 and C57B1/6 were removed from feeder cells and and/or growth factors and allowed to differentiate in the suspension culture. The ES cells differentiated three-dimensionally, and cavities were formed to become simple, and then cystic, embryoid bodies within 11–30 days. The ES fetuses were selected from cystic embryoid bodies. "ES fetuses" are characterized by the presence of pulsating cardiac muscle, yolk sac blood islands, and most importantly, the cup-shaped structures containing small lymphoid-like cells. Five to ten ES fetuses were transferred to a new Petri dish for retrovirus infection.

4 Establishment of Lymphoid and Myeloid Cell Lines from ES Fetuses (Stage III)

Hematopoietic cell lines were established by retrovirus infection using viruses such as Abelson murine leukemia virus (A-MuLv) (ROSENBERG and BALTIMORE 1976; ROSENBERG et al. 1975) and RIM viruses (CLYNES et al. 1988). A virus stock of 10^5–10^6 PFU/ml was used to infect ES fetuses in the presence of polybrene for a period of 24 h. After infection, ES fetuses were fed regularly, and cells spontaneously detached from the bodies. Most of the hematopoietic cells attached lightly to the surface of the Petri dishes. Thus, cell passages were performed by using a rubber policeman. It takes 4 weeks to 6 months to establish cell lines. It is also possible to enrich normal untransformed hematopoietic cells from ES fetuses by culturing them in a combination of growth factors in the presence of stroma cells (U. Chen, unpublished). However, this approach is less efficient than viral transformation, and greater care needs to be taken in order to maintain these normal cells. It is strongly recommended to select ES fetuses from complicated cystic embryoid bodies for the establishment of cell lines, especially when the virus stocks and growth factors are limiting.

5 Morphologic Characterization of Cell Lines

Cytocentrifuge preparations were evaluated following staining with Giemsa. A diversity of cell types was found after 4 weeks to 6 months of culture. Cells with morphology characteristic of monocytes and macrophages were observed, and many of them were found to cluster around erythrocytes. Megakaryocytes were also found to be associated with these clusters. Lymphoid-like cells with little cytoplasm and nuclei staining dark blue were observed either in clusters or interspersed between myeloid cells.

6 Expression of Cell Surface Antigens as Characterized by Immunofluorescent Staining

Immunofluorescent labeling was carried out on cells cultured for 3 days on coverslips. Table 1 summarizes the pattern of labeling obtained using antibodies specific for IgM, CD3 (HAVRAN et al. 1987), Mac-1 (SPRINGER et al. 1979), and AA4.1 antigen (MCKEARN et al. 1984). Macrophages and lymphoid cells of both B and T lineages were shown to establish, while fotipotent ES cells (which stained with anti-SSEA-1 antibody) were lost.

7 In Vivo Characterization of Hematopoietic Stem Cells

By adoptive transfer experiments, hematopoietic cell lines established by A-MuLV or RIM virus infection can further differentiate and repopulate the immune system of sublethally irradiated mice. Mature B lineage cells of at least three phenotypes and T lymphocytes can be identified in the spleens (CHEN et al. 1992). These in vivo repopulation experiments strongly indicate that substantial numbers of hematopoietic cells can be found in culture and that some of them can mature to lymphocytes. Thus, the liquid culture conditions in conjunction with retroviral infection supports the maintenance and differentiation of all hematopoietic cells.

8 Outlook

The ES-D3 and CCE ES cell lines, which were used in this study were established from mouse strain 129. ES cell lines from other mouse strains have also been established, such as the (C57B1/6 X DBA/2) F_1 strain (K. Burki, Sandoz, Basel Switzerland) and the C57B1/6 strain (the ES-B6 cell line, R. Kelmer, Max Planck Institute of Immunobiology, Freiburg, FRG, also used in this study), and other cell lines (K. Burki, Sandoz, Basel Switzerland; G. Keller, M. Wiles, National Jewish Center, Denver, USA). ES cell lines from both C57B1/6 and DBA/2 strain from Keller and Wiles did not differentiate into hematopoietic cells as the CCE lines, and those from K. Burki were not tested for their capacity to differentiate in vitro. However, the ES-B6 cell line has similar totipotent capacity as the ES-D3 line in vitro (CHEN 1991). Traditionally, feeder cells have been used for the establishment and maintenance of ES cell lines. It is now possible to use LIF-containing medium in gelatin-coated Petri dishes to establish ES cells from any strain of mouse (E. Robertson, personal communication). Thus, it can be expected that many more ES cell lines will be available for analysis.

In this communication, I have described the growth behavior, the criteria of the undifferentiated stage of ES cells, the differentiation condition, immortalization methods, and the characterization of ES derived hematopoietic cells. This stage of maturation in vitro appears to be equivalent to day 11–14 fetal liver of mouse embryogenesis. Mixed populations of myeloid, erythroid, and lymphoid cells were identified in culture. Thus, the cultures resemble the differentiation events found in adult bone marrow and fetal liver. This system has many advantages for studying the development of the lympho-hematopoietic system. It supplies unlimited numbers of totipotent ES cells and both transformed and normal pluripotent hematopoietic cells. Due to the fact that fetus-like embryoid bodies were formed in vitro in a controlled nutrient environment, steps and elements controlling the growth requirements for hematopoietic lineages, especially the lymphoid lineage commitment, may be studied.

It is intriguing that it is possible to infect ES fetuses with retroviruses containing oncogenes. These retroviruses, such as Abelson viruses, are known to immortalize pre-B and other hematopoietic cells from in vivo sources. The infected ES fetuses appeared to increase the cell mass and life span up to 6 months in culture, and the established cell lines from these ES fetuses contained all known lineages of hematopoietic cells. This process amplifies the number of hematopoietic cells and it may enable the molecular analysis and gene manipulation of stem cells at this stage.

It seems that factors required for lineage commitment are provided from within the culture, but the frequency of committed cells might be increased if proper growth factors and stromal cells could be included in the culture. For studying lineage commitment, several available tools for measuring stage-specific antigen expression in differentiation are critical, for example: (a) antibodies against cell surface markers which appear in early development, such as Thy-1 (DENNERT et al. 1980; MÜLLER-SIEBURG et al. 1986), Sca-1 (SPANGRUDE et al. 1988), AA4 (MCKEARN et al. 1984), G-5-2 (STRASSER 1988), Joros (PALACIOS et al. 1990), and J11-D (BRUCE et al. 1981); (b) polymerase chain reaction (PCR) technique for Ig/TCR gene rearrangement and RNA expression at the single cell level (N.N. Iscove, personal communication); (c) mutant ES cells missing specific genes which are important for embryogenesis and lineage commitment. Examples are uromorulin (R. Kemler, personal communication), which is important for early embryogenesis. GATA-1 (PEVNY et al. 1990; M.C. Simon and S.Orkin, personal communication), which is located on the X chromosome and is essential for erythroid lineage development, and mutants which deleted both alleles of B- and T-cell-specific genes, such as Octamer-binding protein-2 and RAG-2 (L. Staudt and F. Alt, personal communication).

It is possible to obtain beating cardiac muscle, blood islands, and lymphoid cell positive clusters (the so-called ES fetuses) from ES lines derived from several mouse strains. The ES-D3 cell line has been shown to be germ-line transmissible (GOSSLER et al. 1986; DOETSCHMAN et al. 1987), while the ES-B6 cell line has not been able to develop into germ-line chimeras in vivo. I have, nevertheless, obtained ES fetuses, established hematopoietic cell lines, and found T and B cells. Thus, the ability of the ES-D3 line to differentiate in vitro to ES fetuses and further to lymphoid and myeloid lineages seems to be independent of their ability to form chimeras in vivo. These cultures resemble the situation in bone marrow and fetal liver since stem cells are found as a minority in a heterogeneous mixture of cells. It is inconceivable that with the help of growth factors, stroma cells, and/or retroviruses, pure population stem cells of unlimited quantity may be grown for multiple studies.

Acknowledgements. I acknowledge the generous gifts of embryonic stem cells and kind help in establishing the ES cell technology from Rolf Kemler, Elizabeth Robertson, and Colin Stewart. I thank Steve Bauer for helpful discussion and Fiona McConnell for critical reading during the preparation of this manuscript, and I thank Nicole Schoepflin for preparing the manuscript. I particularly acknowledge the contribution, encouragement and support of Fritz Melchers, who introduced me to the field of developmental immunology. The Basel Institute of Immunology was founded and is supported by Hoffmann-La Roche Ltd., Basel, Switzerland.

References

Bruce J, Symington FW, McKearn IJ, Sprent J (1981) A monoclonal antibody discriminating between subsets of T and B cells. J Immunol 127: 2496–2501

Chen U (1991) Differentiation of mouse embryonic stem cells to lympho-hematopoietic lineages in vitro. Dev Immunol (in press)

Chen U, Kosco M, Staerz U (1992) Establishment and characterization of lymphoid and myeloid cell lines from mouse ES fetuses. Proc Natl Acad Sci USA

Chen U, Seyschab H (1989) Analysis of cell cycle progression of mouse ES cells by using BrdU/Hoechst flow cytometry. In: 7th International Congress of Immunology. Fisher, Stuttgart (abstract 29-2)

Clynes R, Wax J, Stanton LW, Smith-Gill S, Potter M, Marcu KB (1988) Rapid induction of IgM-secreting murine plasmacytomas by pristane and an immunoglobulin heavy-chain promoter/enhancer-driven *c-myc*/*v*-Ha-*ras* retrovirus. Proc Natl Acad Sci USA 85: 6067–6071

DeChiara TM, Efstratiadis A, Robertson EJ (1990) A growth-deficiency phenotype in heterozygous mice carrying an insulin-like growth factor II gene disrupted by targeting. Nature 345: 78–80

Dennert G, Hyman R, Lesley J, Trowbridge IS (1980) Effects of cytotoxic monoclonal antibody specific for T200 glycoprotein on functional lymphoid cell populations. Cell Immunol 53: 350–364

Doetschman T, Eistetter H, Katz M, Schmidt W, Kemler R (1985) The in vitro development of blastocyst-derived embryonic stem cell lines: formation of visceral yolk sac, blood islands and myocardium. J Embryol Exp Morphol 87: 27–45

Doetschman T, Gregg RG, Maeda N, Hooper ML, Melton DW, Thompson S, Smithies O (1987) Targetted correction of a mutant HPRT gene in mouse embryonic stem cells. Nature 330: 576–578

Havran WL, Poenie M, Kimura J, Tsien R, Weiss A, Allison JP (1987) Expression and function of the CD3-antigen receptor on murine $CD4^+8^+$ thymocytes. Nature 330: 170–173

Iscove NN, Shaw AR, Keller G (1989) Net increase of pluripotential hematopoietic precursors in suspension culture in response to IL-1 and IL-3[1]. J Immunol 142: 2332–2337

Gossler A, Doetschman T, Korn R, Serfling E, Kemler R (1986) Transgenesis by means of blastocyst-derived embryonic stem cell lines. Proc Natl Acad Sci USA 83: 9065–9069

Koller BH, Smithies O (1989) Inactivating the β_2-microglobulin locus in mouse embryonic stem cells by homologous recombination. Proc Natl Acad Sci USA 86: 8932–8935

Labastie M-C, Thiery J-P, Le Douarin NM (1984) Mouse Yolk sac and intraembryonic tissues produce factors able to elicit differentiation of erythroid burst-forming units and colony-forming units, respectively. Proc Natl Acad Sci USA 81: 1453–1456

Lindenbaum MH, Grosveld FG (1990) An in vitro globin gene switching model based on differentiated embryonic stem cells. Gene Dev 4: 2075–2085

Mansour SL, Thomas KR, Capecchi MR (1988) Disruption of the protooncogene *int*-2 in mouse embryo-derived stem cells: a general strategy for tergeting mutations to non-selectable genes. Nature 336: 348–352

Martin GR (1981) Isolation of a pluripotent cell line from early mouse embryos cultured in medium conditioned by terato-carcinoma stem cells. Proc Natl Acad Sci 78: 7634–7638

Martin GR, Evans MJ (1975) Differentiation of clonal lines of teratocarcinoma cells: formation of embryoid bodies in vitro. Proc Natl Acad Sci USA 72: 1441–1445

McKearn JP, Baum C, Davie JM (1984) Cell surface antigens expressed by subsets of pre-B cells. J Immunol 132: 332–339

Melchers F (1977) B lymphocyte development in fetal liver. I. Development of reactivities to B cell mitogens "in vivo" and "in vitro". Eur J Immunol 7: 476–481

Melchers F, Abramczuk J (1980) Murine embryonic blood between day 10 and 13 of gestation as a source of immature precursor B cells. Eur J Immunol 10: 763–767

Müller-Sieburg C, Whitlock CA, Weissman IL (1986) Isolation of two early B lymphocyte progenitors from mouse marrow: a committed pre-pre B cell and a clonogenic Thy-1^{lo} hematopoietic stem cell. Cell 44: 653–662

Ogawa M, Nishikawa S, Ikuta K, Yamamura F, Naito M, Takahashi K, Nishkawa S-I (1988) B cell ontogeny in murine embryo studied by a culture system with the monolayer of a stromal cell clone, ST2: B cell progenitor develops first in the embryonal body rather than in the yolk sac. EMBO J 7: 1337–1343

Owen JJT, Raff MC, Cooper MD (1975) Studies on the generation of B lymphocytes in the mouse embryo. Eur J Immunol 5: 468–473

Paige CJ, Kincade PW, Moore MAS, Lee G (1979) The fate of fetal and adult B-cell progenitors grafted into immunodeficient CBA/N mice. J Exp Med 150: 548–563

Palacios R, Samaridis J, Thorpe D, Leu T (1990) Identification and characterization of Pro-T lymphocytes and lineage-uncommited lymphocyte precursors from mice with three novel surface markers. J Exp Med 172: 219–230

Pevny L, Simon MC, Robertson E, Klein WH, Tsai S-F, D'Agati V, Orkin SH, Costantini F (1991) A targeted mutation in the gene for transcription factor GATA-1 block erythroid differentiation in chimeric mice. Nature 349: 257–260

Robertson EJ (1987) Teratocarcinomas & embryonic stem cells, a practical approach. IRL Press, Oxford

Rolink A, Kudo A, Karasuyama K, Kikuchi Y, Melchers F (1991) Long-term proliferating early pre B cell lines and clones with the potential to develop to surface Ig-positive, mitogen reactive B cells in vitro and in vivo. EMBO J 10: 327–336

Rosenberg N, Baltimore D (1976) A quantitative assay for transformation of bone marrow cells by Abelson murine leukemia virus. J Exp Med 143: 1453–1463

Rosenberg N, Baltimore D, Scher CD (1975) In vitro transformation of lymphoid cells by Abelson murine leukemia virus. Proc Natl Acad Sci USA 72: 1932–1936

Schmitt RM, Bruyns E, Snodgrass HR (1991) Hematopoietic development of embryonic stem cells in vitro: cytokine and receptor gene expression

Schwartzberg PL, Goff SP, Robertson EJ (1989) Germ-line transmission of a *c-abl* mutation introduced by targeted gene disruption in ES cells. Science 246: 799–803

Solter D, Knowles BB (1978) Monoclonal antibody defining a stage-specific mouse embryonic antigen (SSEA-1). Proc Natl Acad Sci USA 75: 5565–5569

Spangrude GJ, Heimfeld S, Weissman IL (1988) Purification and characterization of mouse hematopoietic stem cells. Science 241: 58–62

Springer T, Galfrè G, Secher DS, Milstein C (1979) Mac-1: a macrophage differentiation antigen identified by monoclonal antibody. Eur J Immunol 9: 301–306

Strasser A (1988) PB76: a novel surface glycoprotein preferentially expressed on mouse pre-B cells and plasma cells detected by the monoclonal antibody G-5-2. Eur J Immunol 18: 1803–1810

Thompson S, Clarke AR, Pow AM, Hooper ML, Melton DW (1989) Germ line transmission and expression of a corrected HPRT gene produced by gene targeting in embryonic stem cells. Cell 56: 313–321

von Boehmer H (1990) Developmental biology of T cells in T cell-receptor transgenic mice. Annu Rev Immunol 8: 531–556

Ware LM, Axelrad AA (1972) Inherited resistance to N- and B-tropic murine leukemia viruses in vitro: evidence that congenic mouse strains SIM and SIM.R differ at the Fv-1 locus. Virology 50: 339–348

Wiles MV, Keller G (1991) Multiple hematopoietic lineages from embryonic stem (ES) cells in culture. Development 111: 259–267

Zijlstra M, Li E, Sajjadi F, Subramani S, Janish R (1989) Germ-line transmission of a disrupted β_2-microglobulin gene produced by homologous recombination in embryonic stem cells. Nature 342: 435–438

Evaluating Functional Abilities of Primitive Hematopoietic Stem Cell Populations

D. E. HARRISON

1 Introduction

1.1 Primitive Hematopoietic Stem Cells

As outlined in Table 1, all renewable lymphoid cells and myeloid cells are continuously descended from the primitive hematopoietic stem cell (PHSC). It is characterized by maximal self-renewal and maximal differentiating ability. It appears to be the most primitive somatic cell type functioning in mammals and thus of considerable interest in developmental studies. Furthermore, success of clinical marrow transplantation depends on PHSC function. The PHSC is essential in every clinical treatment requiring long-term production of myeloid or lymphoid cells and will be vital in human gene transplantation via autologous marrow cells, a putative therapeutic measure for heritable disorders.

The Jackson Laboratory, Bar Harbor, ME 04609, USA

Current Topics in Microbiology and Immunology, Vol. 177

Table 1. Primitive Hemopoietic Stem Cell Differentiation

		P. B cell	—	B cells
	LSP? —			
		P. T. cell	—	T cells
PHSC —				
		P. erythrocyte	—	Erythrocytes
	MSP —	P. mega	—	Platelets
		P. granulocyte	—	Granulocytes
		P. monocyte	—	Monocytes/macrophages

The PHSC is the precursor of both the immune and myeloid cell lineages, differentiating into the widest variety of cell populations, and repopulating both over the longest times. From PHSC are descended all the more differentiated lymphoid and myeloid precursors, as well as mast cells, natural killer (NK) cells, and null cells. The question mark indicates that there is no definitive evidence showing an important role in repopulation for the lymphoid-specific precursors (LSP), the common precursor of B and T cells, but not myeloid cells. The myeloid analogue of the LSP is the myeloid-specific precursors (MSP) and is usually measured as the macroscopic spleen colony of Till and McCulloch (1961)

The PHSC is quantified in assays that often require 3–6 months or more, but which are necessary because short-term assays measure only the more mature precursors. Functions of mature precursors do not necessarily predict functions of PHSC. This chapter will outline assays that directly test the defining characteristics of PHSC, their long-term function and maximal differentiating ability.

1.2 Mouse Models

The mouse is the best species for measuring long-term function of transplanted cells accurately and conveniently. Therefore PHSC can be assayed best in murine model systems. Fortunately, insights from these systems should be relevant to human PHSC, as murine and human myeloid and lymphoid cells are very similar functionally. Using mice congenic for quantitative markers, PHSC functions can be accurately defined.

Mice are valuable model systems for several reasons. Not only are they far better defined in terms of genotype and environment that humans, but their lives pass 30 times faster, allowing experiments to be performed in a few years that would require a researcher's entire life span in human beings. The numbers of genetically identical mice required to obtain statistically valid data are readily available, and convenient to work with. Finally, a wide variety of hematopoietic mutations and congenic lines with useful markers have long been available (Harrison 1979); their value has recently been illustrated in defining the importance of the *c-kit* (dominant spotting or *W*) receptor and its ligand (*Sl*) in hematopoietic development (Reith et al. 1990; Nocka et al. 1990; Witte 1990).

1.3 PHSC Are Not Measured by Colony-Forming Cell Assays

For many years, the spleen colony-forming unit (CFU-S) measured by the macroscopic spleen colony assay of TILL and MCCULLOCH (1961) was considered the stem cell, although many colonies contain only one myeloid cell type and may be derived from more differentiated precursors (WU et al. 1967; MAGLI et al. 1982). Furthermore, it has long been known that both lymphoid and myeloid cells can be descended from a single precursor (ABRAMSON et al. 1977; MINTZ et al. 1984; DICK et al. 1985; KELLER et al. 1985; LEMISCHKA et al. 1986; JORDAN and LEMISCHKA 1990). Recently, we showed that this is the usual differentiation pathway, because after engraftment all or most lymphocytes and erythrocytes are eventually descended from a few common precursors (HARRISON et al. 1988, 1989, 1990). Spleen colony assays and other measures of differentiated precursors do not measure both myeloid and lymphoid cells and thus cannot measure PHSC (see also R. Jones, this volume).

Assays of colony-forming cells often fail to predict long-term repopulation, the PHSC function. Repopulating abilities of transplanted marrow (HARRISON et al. 1978; ROSS et al. 1982; HARRISON and ASTLE 1982), blood (MICKLEM et al. 1975), or marrow cultures (HARRISON et al. 1987a), and life-sparing abilities of spleen cells (KRETCHMAR and CONOVER 1970) are all less than expected from CFU-S numbers. Assays of CFU-S, and low and high proliferative-potential in vitro colony-forming cells fail to predict long-term restoration of hematopoiesis by transplanted marrow (BERTONCELLO et al. 1988; JONES et al. 1989). Finally, estimates of PHSC number are 10–100 times lower than predicted from CFU-S concentrations (BOGGS et al. 1982; MICKLEM et al. 1987; HARRISON et al. 1988, 1989; NAKANO et al. 1989). Owing to the lack of a colony-forming assay for PHSC, the only definitive measure is to assess their long-term function in vivo (BERTONCELLO et al. 1988). The long-term repopulation assays described in this chapter are thus required to measure PHSC.

2 Materials and Methods

2.1 Mice

Different mouse strains may have genetic disparities affecting histocompatibility and hybrid or allogeneic resistance. Since these may influence PHSC development, we used C57BL/6J (B6) and congenic B6-*Hbb*^{d}/*Hbb*^{d}, *Gpi-1*^{a}/*Gpi-1*^{a} (B6-*Hbb*^{d} *Gpi-1*^{a}) male donors with B6 recipients. Recipients were lethally irradiated and marrow cell suspensions were prepared from femurs and tibias of each donor as detailed previously (HARRISON et al. 1989; HARRISON and ASTLE 1992).

2.2 Recipient Cell Recovery

Contributions due to recovery of recipient cells were measured in groups of irradiated recipients given only 4×10^5 young B6-*Hbb*^{d} *Gpi-1*^{a} marrow cells. Percentages of host-type lymphocytes (P_1) and erythrocytes (P_e) in these recipients were very low, 2.2 ± 2.2 (12) and 4.0 ± 5.0 (12), respectively, giving data as means $\pm$ standard deviation (SD) (number of means of groups averaging 5.3 mice each and averaging the means for tests at each time point from 80 to 304 days). Thus host recovery did not contribute significant numbers of differentiated cells (HARRISON et al. 1989).

2.3 Percentages of B6-Type P_l and P_e

Lymphocytes from the blood were separated by density gradient centrifugration as detailed previously (HARRISON et al. 1989; HARRISON and ASTLE 1992), and the percentages of B6-type glucosephosphate isomerase (Gpi) were determined by densitometry after electrophoresis. B6 mice have the *Gpi-1*^{b} allele at the *Gpi-1* locus; in electrophoresis this type forms a more rapidly migrating band from positive to negative than does the *Gpi-1*^{a} allele. Erythrocytes were taken from the same blood samples and percentages of B6-type hemaglobin (Hb) were determined by densitometry after electrophoresis. B6 mice have *Hbb*^{s} (*s*, single) hemoglobin, which forms an electrophoretic band migrating farthest from negative to positive on the standard cellulose acetate gels used, while *Hbb*^{d} *(d,* diffuse) hemoglobin has the *Hbb-b1*^{d} and *Hbb-b2*^{d} bands that are intermediate and slowest migrating, respectively. Details of the Gpi and hemoglobin electrophoretic techniques have been published previously (HARRISON et al. 1989; HARRISON and ASTLE 1992). Blood was sampled at intervals of more than 100 days, and each blood sample was tested for both lymphocyte and erythrocyte percentages of B6 type. Results are given as sample means estimating P_l and P_e.

2.4 Mathematical Model: Estimation of Numbers and Concentrations for Multilineage Precursors (from HARRISON and ASTLE 1992)

2.4.1 Mathematical Model for Estimating PHSC Numbers and Duration of Function: General Explanation

PHSC numbers and functional life spans are estimated using a mathematical model. This has the advantage that concentrations are estimated for the entire cell population. Usually a group 15–20 recipients are each given equal portions from a mixture of marrow cells of two distinguishable types, and the percentage of differentiated cells of each type is measured. Numbers of PHSC are estimated from a "binomial formula" in which the number depends inversely on the variability in the percentage of donor cell types between the different recipients (NESBITT 1971; STONE 1984; STONE and HARRISON 1990).

The reason for this dependence is illustrated using a bowl of randomly mixed black marbles and white marbles to represent the marrow cell pool used for repopulation. The PHSC given to each recipient from the pool are represented as samples of marbles taken randomly from the bowl. If each sample contains many marbles, then all will have similar percentages of each marble type, and variability will be low. But if each sample contains few marbles, the percentage of each type will vary greatly from sample to sample.

Estimates are not exact because each precursor may not contribute equally to the differentiated cell populations (STONE 1984). However, our estimates represent the most interesting cells, the precursors from which most erythrocytes and lymphocytes are descended (MICKLEM et al.1987). In the mathematical model, we pioneered the use of covariance (HARRISON et al. 1988, 1989, 1990), rather than variance to avoid bias from random experimental error or any other uncorrelated source of error. When variances are used (NESBITT 1971; MICKLEM et al. 1987; HARRISON et al. 1987b), inherently inaccurate estimates of measurement error are needed to correct for the amount of variance due to error rather than to limited precursor numbers.

2.4.2 Estimation of Numbers and Concentrations for Multilineage Lymphoid and Erythroid Precursors: Detailed Description of the Model

The mathematical model for studying precursor populations is based on the simple binomial neutral-marker model (NESBITT 1971; STONE 1984; STONE and HARRISON 1990) associating increasing variance with decreasing PHSC numbers, and using variance instead of covariance. For each individual in a group numbering n recipients getting identical aliquots from the same marrow cell pool, we measure the B6 percentages for lymphocytes and for erythrocytes (P_l and P_e) at the same sampling time after transplantation. We calculate:

— Mean = the average of all $2n$ percentages which is the average of the two sample means estimating P_l and P_e, the percentages of B6-type lymphocytes and erythrocytes.
— SD_L, SD_E = the standard deviations of the two sets of n measurements.
— r = lymphocyte:erythrocyte Pearson correlation coefficient (L:Er); high correlations suggests that the sampled lymphocytes and erythrocytes are descended from the same precursors.
— Covariance = $r \times SD_L \times SD_E$.

Estimates of PHSC number and concentration are calculated from the binomial formula as in Eqs. 1 and 2 below. If in each individual there were just n such PHSC that contributed equal fractions $1/n$ of lymphocytes and of erythrocytes, then equivalent PHSC number would be an estimate of *n* and equivalent PHSC concentration an estimate of the concentration of such PHSC.

$$\text{Equivalent PHSC number} = (\text{mean})\ (100 - \text{mean})/\text{covariance}, \quad (1)$$

$$\text{Equivalent PHSC concentration} = \text{equivalent PHSC number/number of injected cells.} \quad (2)$$

The reciprocal of equivalent PHSC number is an estimate of the parameter introduced by STONE (1984) and labeled π_c. If the genetic marking (B6 or not) does not effect clonal development, as should be the case using congenic lines, π_c is the probability that a randomly picked lymphocyte and a randomly picked erythrocyte are descended from the same transplanted PHSC. Equivalent PHSC number and concentration clearly have useful interpretations in terms of the fundamental probability parameter π_c, even when the assumption that each PHSC contributes equally is unrealistic. They should then be regarded as experimentally ascertainable, inverse measures of the common PHSC clonality of the lymphocyte and erythrocyte compartments. Furthermore, if each precursor does not contribute equally, estimates of the equivalent PHSC number and concentration reflect those precursors contributing the most, and these are especially interesting.

2.4.3 Estimation of Numbers and Concentrations for Continuously Functioning Precursors

The methods described in the preceding paragraphs are also applicable if the two measurements are B6 percentages of either lymphocytes or erythrocytes at dufferent sampling times. As long as these times are far enough apart so that circulating cells have been renewed during that interval, estimations of common continuously functioning precursors will still be labelled as PHSC. In these cases, *r* is the correlation coefficient relating the measures on the same recipients at the two times; SD1 and SD2, the standard deviations of measurements at those times, are substituted for SD_L and SD_E. High correlations suggest that the differentiated cells are descended from the same precursors over the time interval used.

3 Results and Discussion

3.1 Competitive Repopulation Assay (from HARRISON and ASTLE 1992)

3.1.1 Description of Assay

Table 2 illustrates a competative repopulation experiment using four donors. Mixing cells from each different donor with alioquots from the same congenic

Table 2. Long-term repopulating ability directly measured as competitive repopulation

Donor A (genotype 1) + Competitor pool (genotype 2)—Repopulated hosts (#1–8)
Donor B (genotype 1) + Competitor pool (genotype 2)—Repopulated hosts (#9–16)
Donor C (genotype 1) + Competitor pool (genotype 2)—Repopulated hosts (#17–24)
Donor D (genotype 1) + Competitor pool (genotype 2)—Repopulated hosts (#25–32)

Cells from the donors (A–D) are each mixed with a standard dose of genetically distinguishable marrow (Competitor pool). Mixtures are injected into stem cell-deficient hosts, and precursors of each type compete to produce differentiated cells. The percentage of differentiated cells in the hosts that are descended from each donor type measures its repopulating activity. Blood cells are assayed, testing hosts repeatedly over many months. To provide distinguishable cells, C57BL/6J (B6) and congenic B6-*Hbb*^{d} *Gpi-1*^{a} mice are used as donors and competitors. These are genetically the same except at loci determining hemoglobin and Gpi, and closely linked loci. This avoids complications such as graft versus host reactions or hybrid resistance. Lymphocytes from the blood are separated by density gradient centrifugation, and percentages of donor and competitor Gpi types are determined by densitometry after electrophoresis. Erythrocytes are taken from the same blood samples, and percentages of donor and competitor hemoglobin types are determined by densitometry after electrophoresis. (From HARRISON and ASTLE 1992)

competitor pool provides an extremely sensitive measure for donor-repopulating abilities.

3.1.2 Dose-Response Curve

Figure 1 shows the experiment in Table 2, except that donors A, B, C, and D are replaced with 20%, 40%, 60%, and 80% donor cells to test the accuracy of the assay.

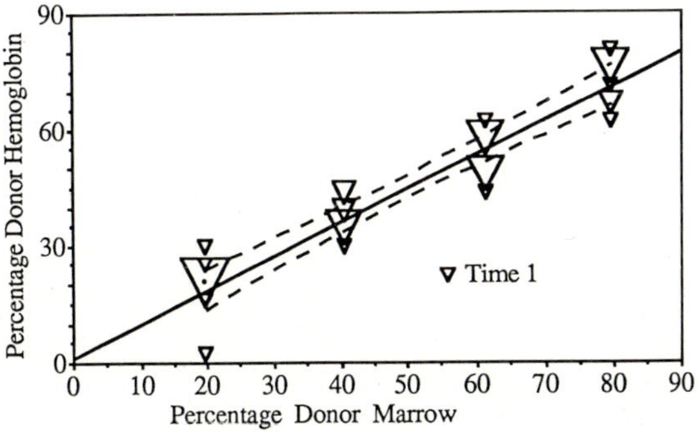

Fig. 1. Mixtures containing 15 million total marrow cells were injected into each lethally irradiated B6 recipient, with eight per dose. Doses contained 20%, 40%, 60%, or 80% B6 cells and the balance B6-*Hbb*^{d}*Gpi-1*^{a} cells. Control recipients of 3 million B6-*Hbb*^{d}*Gpi-1*^{a} cells had only that hemoglobin type when percentages of B6 hemoglobin in the recipients were analyzed after 3 months (*Time 1*). Confluent points are enlarged. Analyzed as a simple regression using Statview SE (Abacus Concepts, Inc., Berkeley, CA), the adjusted R^2 is 0.91, and *dotted lines* show 99% confidence bands for the true mean of Y. The 99% confidence limit of Y was less than ±5% over the donor cell range of 20%–80%. The mean was 45.0%, and its 99% confidence limits were ±3.1%. Thus small differences between group means in percentage donor-type hemoglobin are likely to be meaningful in competitive repopulation. (From HARRISON and ASTLE 1992)

3.1.3 Advantages of the Competitive Repopulation Assay

Competitive repopulation is very sensitive in detecting differences in poliferative ability. If one donor's cells have slightly longer cell cycles and many cycles are required for repopulation, a large defect will develop in comparison to the intermixed competitor cells. PHSC defects, such as those found in previously transplanted marrow (HARRISON et al. 1990) or marrow from long-term cultures (HARRISON et al. 1987a), become most apparent hen measurements are made several months after transplantation. This shows the need to study long-term repopulation. This is done readily by sampling the recipients for as long a time period as required, up to and over 1 year if necessary.

In some studies, long-term precursor function has been tested by "life-sparing" assays which measure the ability to rescue lethally irradiated recipients from death. Competitive repopulation is superior to the life-sparing assay for three reasons. First, competitive repopulation measures in a group of eight recipients are accurate over a ten fold range of repopulating abilities (Fig. 1), while measuring an effective dose saving 50% of recipients (ED_{50}) as accurately requires many more recipients. Second, competitive repopulation is much more reproducible. Life-sparing measures are greatly affected by types and numbers of micro-organisms in the mouse colony, things that are almost impossible to control entirely. Third, competitive repopulation is easily followed over the entire life span of the recipient and thus is ideal for long-term measures, while host cells often contribute substantially to hematopoiesis after the first 1–2 months in radio-protection assays.

3.1.4 Examples of Results Obtained with Competitive Repopulation Assays

Four examples are summarized.

Transplantation Effects. Competitive repopulation assays first showed that transplantation has strong deleterious effects on the PHSC (HARRISON et al. 1978; ROSS et al. 1982). Detectable effects occur even when large numbers of marrow cells (up to 10^8) are transplanted. Three to twelve months after transplantation of 10^7 marrow cells, erythroid repopulation is only about 10% as effective as in fresh marrow cells (HARRISON and ASTLE 1982). Deleterious effects of transplantation are not removed by using W/W^v genetically anemic recipients that are repopulated without irradiation (GARDNER et al. 1988). This suggests that irradiation damage to recipients is not responsible for effects of transplantation. Unfortunately in these experiments donors and host were not completely matched which could have affected the outcome. This emphasizes the need to use congenic animals, particularly *W*-anemic recipients (HARRISON and ASTLE 1991).

Repopulating Abilities Decline in Marrow Cultures. Competitive repopulation assays found no evidence that long-term marrow cultures (LTMC; also called

Dexter cultures) support PHSC proliferation in vitro (HARRISON et al. 1987a). Cultured adherent cells cosistently showed less capacity to repopulate irradiated recipients than fresh marrow cells. Cultured nonadherent cells had much less capacity than adherent cells, although 9- and 12-day CFU-S concentrations were similar in adherent and nonadherent cells. In every experiment, the long-term repopulating ability of the marrow was substantially reduced after transfer into tissue culture (HARRISON et al. 1987a). Thus, primitive stem cells fail to proliferate in such cultures enough to maintain their numbers despite extensive production of CFU-S.

A Single Treatment with 5-fluorouracil Does Not Affect Repopulation. Functional abilities of cells from 5-fluorouracil (5-FU)-treated marrow donors were assessed by competitive repopulation (LERNER and HARRISON 1990). Bone marrow from mice treated with 5-FU 1, 5, and 8 days prior to transplantation produced circulating lymphoid and erythroid cells as well as equal fractions of untreated fresh marrow when surveyed approximately 90 days after transplantation. Normal reconstitutive ability is thus maintained despite a tenfold reduction in marrow cell numbers when donor mice had been treated with 5-FU 5 days prior to transplantation. To evaluate PHSC function more rigorously, a second round of repopulation was stimulated, subsequent to the initial 90-day screening, by giving recipients a sublethal (500 R) dose of irradiation. After 3–4 months, percentages of donor and competitor lymphocytes and erythrocytes were the same as pre-irradiation values. Thus, the radioresistance of 5-FU-treated PHSC is comparable to that of fresh marrow. These results indicate that PHSC are unaffected after a single treatment with 5-FU, although they are concentrated ten fold (LERNER and HARRISON 1990).

Antigenic Effects on Long-Term Repopulating Abilities. Competitive repopulation was also used to define effects of immune responses against particular antigens on erythroid repopulation. We analyzed these effects for 16 independent non-*H-2* antigenic loci (HARRISON 1987). Antigens specified by the *Thy-1*, *H-1*, *H-24*, *Ly-1*, *H-37*, and *H-17* loci affected repopulating abilities slightly, if at all. Repopulating abilities were moderately reduced by responses against antigens specified by four others, the *H-15*, *H-16*, *Ea-2*, and *Ly-2*, *Ly-3* loci, and the differences between the B6 and B10 genotypes, although marrow of these types cured *W*-anemic recipients. Immune responses against antigens specified by the *H-8* locus strongly reduced repopulating abilities, although this type of marrow also cured *W*-anemic recipients. Antigens specified by the *H-3*,*H-12*, and *H-25* loci prevented repopulation or cures. Comparing these results with skin graft survival times suggests that the antigens specified by the *H-17* and *H-24* loci are strongly immunogenic on skin but not marrow, while those specified by *H-8* and *H-12* are strongly immunogenic on marrow but not skin (HARRISON 1987). Antigens specified by the *H-24* locus were so specific that *W*-anemic mice repopulated by PHSC from such donors still rejected their skin (HARRISON and MORBRAATEN 1984). Such tissue specificity leads to hopes that PHSC-specific antigens may be found.

3.2 Concentrations and Functions of PHSC Populations

3.2.1 Illustrations of the Mathematical Model

All recipients received aliquots from a mixture of two genetically distinguishable marrow cell types, and percentages of B6-type lymphocytes and erythrocytes, P_l and P_e, were measured repeatedly at intervals of about 100 days to follow long-term changes in these populations. In the first study (Table 3), 16 recipients were each given 4×10^5 B6 marrow cells, and 4×10^5B6-*Hbb*^{d}*Gpi-1*^{a} marrow cells. Excellent correlations between percentages of B6-type lymphocytes and erythrocytes in the same recipient ($r = 0.73$–0.94) strongly suggest that they are descended from the same precursors. The same excellent correlations were found in the same recipients at different sampling times 101–136 days apart ($r = 0.61$–0.97 in lymphocytes and 0.86–0.93 in erythrocytes), strongly suggesting that differentiated cells are continuously descended from the same precursors.

PHSC concentrations are estimated in two ways: as common precursors of lymphocytes and erythrocytes in one blood sample, or as precursors producing differentiated cells during two samplings 101–136 days apart. Numbers of PHSC appear highest in the earliest samples, when variances and covariances are lowest. However, these high estimates may reflect some contribution to differentiated cell populations present 107 days after transplantation from precursors more differentiated than true PHSC. These may be exhausted by 243 days and later, leaving only descendants of PHSC, in concentrations of about 1 per 10^5 marrow cells.

Table 4 describes an independent experiment with a similar design, except that marrow doses for each recipient totaled 8, 20, or 80×10^5. The basic feature of the simple binomial neutral-marker model (NESBITT 1971; STONE 1984; STONE and HARRISON 1990) associating increasing variance with decreasing PHSC numbers is supported. SDs are highest in recipients of 8×10^5 cells, declining to less than one-half of their value as the dose was increased to 80×10^5 cells.

After 251 and 393 days, and at all cell doses, excellent correlations between percentages of B6-type lymphocytes and erythrocytes in the same recipients ($r = 0.84$–0.87) strongly suggest that they are descended from the same precursors. Excellent correlations between percentages of B6 type in the same recipients at sampling times 132 days apart ($r = 0.78$–0.96) strongly suggest that differentiated cells are continuously descended from the same precursors. PHSC concentrations were again about one per 10^5 marrow cells, whether estimated as common precursors of lymphocytes and erythrocytes, or as precursors producing differentiated cells during two samplings 132 days apart.

3.2.2 All Myeloid and Lymphoid Cells are Descended from PHSC

Tables 3 and 4 demonstrate that after engraftment all or most lymphocytes and erythrocytes are eventually descended from a few common precursors. In many independent experiments, we have consistently shown this (HARRISON et al.

Table 3. PHSC frequencies estimated using the mathematical model

Sampling times (days)	Calculated statistics	(%) B6 type
107	P_l, P_e	49, 45
	SD_L, SD_E	12, 16)
	L:E *r*	0.73(16
	Conc	2.1
243	P_l, P_e	54, 47
	SD_L, SD_E	14, 21
	L:E *r*	0.94(16)
	Conc	1.1
362	P_l, P_e	59, 50
	SD_L, SD_E	19, 23
	L:E *r*	0.83(16)
	Conc	0.8
463	P_l, P_e	60, 50
	SD_L, SD_E	21, 23
	L:E *r*	0.85(9)
	Conc	0.8
	Lymphoid cells	
107–243	L1:L2 *r*	0.61
	Conc	2.9
243–362	L2:L3 *r*	0.91
	Conc	1.3
362–463	L3:L4 *r*	0.97
	Conc	0.8
	Erythroid cells	
107–243	E1:E2 *r*	0.86
	Conc	1.0
243–362	E2:E3 *r*	0.91
	Conc	0.7
362–463	E3:E4 *r*	0.93
	Conc	0.6

Marrow doses totalled 8×10^5, half B6 cells, and half B6-*Hbb*^{d}*Gpi-1*^{a} cells. Recipients were bled and tested 107, 243, 362, and 463 days after they were given the marrow cell mixtures. P_l and P_e are the percentages of B6-type lymphocytes and erythrocytes, respectively, and SD_L, SD_E are their sample standard deviations. In the *upper portion* of the table, *r* values (L:E *r*) are Pearson's product moment correlation coefficients of P_l and P_e, and numbers of recipients tested are given in *parentheses.* In the *lower portion* of the table, *r* values are based on P_l and P_l (or P_e and P_e) for sampling times 1–2, 2–3, and 3–4. For each correlation, the equivalent PHSC number was calculated as P (100-P)/covariance, where P = over-all mean percentage of B6 type, and covariance = (*r*) x ($SD_L \times SD_E$). *Conc* is the equivalent PHSC concentration = equivalent PHSC number/number of injected cells. Estimations become more meaning-ful as *r* approaches 1.00 and the SDs approach the same value. These data were previously published in a different form (HARRISON et al. 1990)

Table 4. Multipotential PHSC concentrations and continuous function

Time in recipient (days)	Calculated statistics	B6 type in recipients of (× 10⁵ cells)		
		8 (%)	20 (%)	80 (%)
251	P_l, P_e	55, 50	52, 46	50, 47
	SD_L, SD_E	15, 21	11, 14	7, 6
	L:E *r*	0.87(18)	0.86(17)	0.84(20)
	Conc	1.0	1.0	0.9
383	P_l, P_e	53, 49	51, 47	52, 48
	SD_L, SD_E	19, 20	14, 17	8, 7
	L:E *r*	0.84(16)	0.85(16)	0.84(16)
	Conc	1.0	0.6	0.7
252–383	L3:L4 *r*	0.80	0.89	0.78
	Conc	1.4	1.0	0.7
251–383	E3:E4 *r*	0.96	0.88	0.89
	Conc	0.8	0.6	0.9

Marrow doses totalled 8, 20, and 80 × 10^5 cells , half from B6, and half from the B6-*Hbb*^{d}*Gpi-1*^{a} marrow pools. Recipients were bled and tested 251 and 383 days after they were given the marrow cell mixtures. P_1 and P_e are the percentages of B6-type lymphocytes and erythrocytes, respectively, and SD_L, SD_E are their sample standard deviations. In the *upper portion* of the table, *r* values (L:E *r*) are Pearson's product moment correlation coefficients of P_l and P_e, and numbers of recipients tested are given in *parentheses*. In the *lower portion* of the table, *r* values are based on P_l and P_l (or P_e and P_e) for sampling times 251 and 383 days. For each correlation, the equivalent PHSC number was calculated as P (100-P)/covariance, where P = overall mean percentage of B6 type, and covariance = (*r*) × (SD_L × SD_E). *Conc* is the equivalent PHSC concentration = equivalent PHSC number/number of injected cells. Estimations become more meaningful as *r* approaches 1.00 and the SDs approach the same value.

1988, 1989, 1990). Many previous workers used markers of individual clones to demonstrate that both lymphoid and myeloid cells can be descended from a single precursor (Abramson et al. 1977; Mintz et al. 1984; Dick et al. 1985; Keller et al. 1985; Lemischka et al. 1986; Jordan and Lemischka 1990). However these studies were not quantitative. They did not determine whether few, many, or all differentiated cells actually are descended from this pluripotent precursor (see G. Keller and I.R. Lemischka, this volume).

3.3 Observing Recipients for Long Time Periods Is Critically Important

3.3.1 Different Precursor Cells Are Responsible for Short and Long-Term Repopulation

After transplantation, two fundamentally different types of precursor cell function occur. Short-lived precursors are responsible for most of the initial repopulation of myeloid and lymphoid systems. However, later than 3–4 months after transplantation, most or all differentiated cells are descended from a few long-lived stem cells, and these repopulate the recipient for the rest of its life. Only these few stem cells functioning for long periods show true PHSC behavior; these can only be distinguished by direct measures of long-term function.

Initially after transplantation, the circulating cells are descended from differentiated precursors soon to be exhausted. These differentiated precursors are present in much larger numbers in the transplanted marrow than are PHSC, and they can rapidly produce large numbers of a single differentiated cell type: erythrocytes, granulocytes, macrophages, platelets, T cells, B cells, and so on. However, differentiated precursors are replaced by descendants of the transplanted PHSC.

3.3.2 PHSC Function Continuously

This behavior of true PHSC can only be detected 3–4 months after marrow transplantation or later. Using groups of recipients given equal aliquots from the same mixture of genetically different marrow cells, there are high correlations (r usually > 0.8) between the donor-type differentiated cells in the same recipient even when tested hundreds of days apart. For example, in Table 4, percentages of donor type in each recipient were closely correlated in samples taken after 251 and 383 days, with r values ranging from 0.78–0.96. These high correlations are found between samples hundreds of days apart in the same individuals in many experiments (HARRISON et al. 1988, 1989, 1990). Thus, most differentiated cells are descended from the same PHSC during much of the recipient's life span.

Further evidence that PHSC function continuously is tne fact that percentages of newly produced erythrocytes (reticulocytes) of each donor type varied little when individual recipients were followed for 165–295 days after transplantation (HARRISON et al.1988). Furthermore, estimated PHSC numbers increase in direct relation to the numbers of marrow cells transplanted over a wide range, 2 to 80×10^5 marrow cells (HARRISON and ASTLE 1992). If only a few PHSC were active at a time, estimated numbers would not increase once the number of active cells was exceeded.

Finally, our data also contradict the model in which a constant proportion of donor PHSC are sequentially active. This model predicts that variability within recipients should be higher in recipients of smaller amounts of marrow. In fact, variability within recipients is the same, and very small, at marrow doses from 8 to 200×10^5 cells (HARRISON et al. 1988).

3.3.3 Similar Results from Mathematical Models and Clonal Markers

When we first published these results (HARRISON et al. 1988), they appeared to contradict studies in which stem cell clones were identified by unique retroviral insertions. Uniquely identified clones appeared and disappeared, suggesting that PHSC are sequentially activated and inactivated (DICK et al. 1985; KELLER et al. 1985; LEMISCHKA et al. 1986). However, these retroviral markers were used in short-term experiments lasting only a few weeks or months.

The requirement for long-term studies is demonstrated by 6–12-month studies on stem cell clones identified by unique retroviral insertions (JORDAN and LEMISCHKA et al. 1990a; KELLER and SNODGRASS 1990). These confirm interpretations from the mathematical model (HARRISON et al. 1988), showing that differentiated cells are usually descended from the same stem cell over many months. The same few clones produce differentiated cells for at least 12 months, although many more clones are found early after transplantation, but disappear during the first 3–4 months. Reports that PHSC are sequentially activated and inactivated probably result from recipients not followed for long enough time periods. When descendents of PHSC replace more mature precursors, the latter would appear to be inactivated.

3.3.4 Untransplanted PHSC

PHSC may also function continuously in chimeric mice made by combining embryos from two different strains so that they would have two distinct stem cell populations with each producing a different hemoglobin type (HARRISON et al. 1987b). The original interpretation was that large numbers of PHSC were active simultaneously; however, even small numbers of active PHSC may give the same result if the same PHSC remain active over the entire observation period, in this case, over 700 days.

3.4 Further Examples of Results Obtained from Competitive Repopulation Used with the Mathematical Model

3.4.1 Defining Effects of Transplantation

These effects may be caused by reductions in PHSC numbers or by reductions in repopulating abilities per PHSC. Tests of these hypotheses provide a final example of using the mathematical model (HARRISON et al. 1990). Equivalent PHSC numbers were estimated in recipients of mixtures of genetically different cells using the binomial model with covariance. Percentages of lymphocyte and erythrocyte types were closely correlated, as were percentages of either type sampled at intervals of several months. This suggests that the same PHSC produced lymphoid and myeloid cells, and that most circulating cells were descended from the same PHSC over hundreds of days. Equivalent PHSC concentrations were about one per 10^5 fresh marrow cells, and were about two fold lower using previously transplanted marrow. However, such marrow repopulated only one-seventh to one-eighth as well as fresh marrow. Apparently transplantation not only reduces PHSC concentrations, but also reduces the repopulating ability per PHSC. This may result from excessive stimuli to differentiate that overbalance the stimuli for PHSC to replenish themselves (see P.J. Quesenberry, this volume).

3.4.2 Defining Effects of Age

Effects of age on numbers and functional abilities was estimated for common PHSC from which lymphocytes and erythrocytes are descended (HARRISON et al. 1989). Blood samples from recipients of aged or young marrow plus competitor were drawn at successive sampling times about 100 days apart. Percentages of lymphocytes and erythrocytes types measured in each recipient were closely correlated, especially after 6 months and later. Close correlations were also found in cells sampled at successive 100-day intervals. Apparently, a few PHSC or their direct descendents produced most of the blood lymphocytes and erythrocytes, and this production continued for many months. Concentrations of these PHSC were about one per 10^5 marrow cells from young donors, and were about two fold higher in old donors. Old marrow produced correspondingly higher percentages of lymphocytes and erythrocytes, but these declined with time. To explain the temporal trends, we suggest that excess concentrations of precursors less primitive than PHSC are present in old marrow, and their contribution to the differentiated cell population gradually declines. Possibly such percursors, as well as true PHSC, proliferate in old donors to compensate for deficiencies that develop with age (HARRISON et al. 1989).

3.5 Future Studies of PHSC Populations

3.5.1 A Comparison of Mathematical Models and Clonal Markers

Competitive repopulation is best used in conjunction with clonal markers for PHSC analysis because the advantages of each technique compensate for the weaknesses of the other. Competitive repopulation is a technique that "looks at the forest, rather than the trees" since the results of long-term repopulation and differentiation are directly measured for the entire population of narrow cells tested. It is ideal for evaluating average functional abilities of different precursor populations. This is vital in developing procedures to enrich PHSC or to stimulate their proliferation. However, competitive repopulation and mathematical models give information about populations, not individual cells.

Unique retroviral insertion can quantitatively follow individual precursors and determine how much each contributes to the differentiated cell population. However, this is a technique that "looks at the trees, rather than the forest." It cannot test populations, only individual cells. Using this with the complimentry competitive repopulation technique should give definitive information about PHSC.

3.5.2 Are PHSC a Distinct Class of Cells with Maximal Repopulating Abilities?

The hypothesis that PHSC are part of a continuum with differing repopulating abilities depending on their proliferative history (HELLMAN et al. 1978) predicts that cells with varying degrees of repopulating ability contribute to differentiated cell populations, and that the assumption of the mathematical model that each PHSC contributes equally is false. On the other hand, if there is a class of "true" PHSC, they should contribute equally to differentiated cell populations. These hypotheses can only be distinguished by simultaneous studies of populations and individually marked clones.

3.5.3 Physical Separation of PHSC and Other Precursors

A final and popular hypothesis is that marrow contains much larger numbers of potential PHSC than the one per 10^5 that actually are found after transplantation. In this model, the recipient would provide 10–100 sites into which potential PHSC would seed, for every one that supported long-term PHSC function. This leads to assays designed to detect minimal levels of repopulating ability (SZILVASSY et al. 1989), rather than assays designed to rigorously measure levels of repopulating ability.

This hypothesis can be tested by enriching PHSC so much that cells with greater and lesser repopulating abilities would be separated if they had different physical properties. Enrichment techniques using defined antigenic markers have enriched short-term functioning cells at least 1000-fold (SPANGRUDE et al. 1988; JORDAN et al. 1990). These enrichment procedures should be evaluated by long-term competitive repopulation as well as by shorter-term assays to determine if PHSC are enriched in the same proportion as multilineage precursors with limited proliferation capacity. If these cell types can be separated, the hypothesis would be disproved; if they cannot, the hypothesis would be supported.

Acknowledgements. I thank C.M. Astle and C. Lerner for their work in collecting and helping to analyze the data herein, M. Stone for helping to design the mathematical model, and B. Stork for dependable technical assistance. This work was supported in part by National Institutes of Health Grants DK-25687, AG-00594, and AG-06232.

References

Abramson S, Miller RG, Phillips RA (1977) The identification in adult bone marrow of pluripotent and restricted stem cells of the myeloid and lymphoid systems. J Exp Med 145: 1567–1579

Bertoncello I, Hodgson GS, Bradley TR (1988) Multiparameter analysis of transplantable hemopoietic stem cells. II. Stem cells of long-term bone marrow-reconstituted recipients. Exp Hematol 16: 245–249

Boggs DR, Boggs SS, Saxe DF, Cress KA, Confield DR (1982) Hematopoietic stem cells with high proliferative potential. J Clin Invest 70: 242–253

Dick JE, Magli MC, Huszar D, Phillipis RA, Bernstein A (1985) Introduction of the selectable

gene into primitive stem cells capable of long-term reconstitution of the hematopoietic system of W/Wv mice. Cell 42: 71–79
Gardner RV, Astle CM, Harrison DE (1988) The decrease in long-term repopulating capacity seen after transplantation is not the result of irradiation-induced stromal injury. Exp Hematol 16: 49–54
Harrison DE (1979) Use of genetic anemias in mice as tools for hematological research. In: Lajtha LG (ed) Cellular dynamics of hematopoiesis, vol 8, no 2. Clinics in Haematology. Saunders, London, pp 239–262
Harrison DE (1987) A semi-quantitative measure of immune responses against erythropoietic stem cell antigens. Immunogenetics 26: 123–129
Harrison DE, Astle CM (1982) Loss of stem cell repopulating ability with transplantation: effects of donor age, cell number and transplant procedure. J Exp Med 156: 1767–1779
Harrison DE, Astle CM (1991) Lymphoid repopulation delayed in B6 congenic W-anemic mice, a new unirradiated recipient. Exp Hematol 19: 374–377
Harrison DE, Astle CM (1992) Primitive hemopoietic stem cells (PHSC): Direct assay of most productive populations by competitive repopulatión with simple binomial, correlation and covariance calculations. Exp Hematol (in press)
Harrison DE, Mobraaten LE (1984) Skin graft rejection in mice repopulated with marrow of the skin donor type = a skin gene in a congenic line. Immunogenetics 19: 503–509
Harrison DE, Lerner C, Hoppe P, Carlson GA, Alling D (1987a) Large numbers of primitive stem cells are active simultaneously in aggregated embryo chimeric mice. Blood 69: 773–777
Harrison DE, Lerner CP, Spooncer E (1987b) Erythropoietic repopulating ability of stem cells from long-term marrow culture. Blood 69: 1021–1025
Harrison DE, Astle CM, Delaittre JA (1978) Loss of proliferative capacity in immunohemopoietic stem cells caused by serial transplantation rather than aging. J Exp Med 147: 1526–1531
Harrison DE, Astle CM, Lerner C (1984) Ultimate erythropoietic repopulating abilities of fetal, young adult and old adult cells compared using repeated irradiation. J Exp Med 160: 759–771
Harrison DE, Astle CM, Lerner C (1988) Number and continuous proliferation pattern of transplanted primitive immunohemopoietic stem cells. Proc Natl Acad Sci USA 85: 822–826
Harrison DE, Astle CM, Stone M (1989) Numbers and functions of transplantable primitive immunohemopoietic stem cells: effects of age. J Immunol 142: 3833–3840
Harrison DE, Stone M, Astle CM (1990) Effects of transplantation on primitive immunohematopoietic stem cells (PSC). J Exp Med 172: 431–437
Hellman S, Botnick LE, Hannon EC, Vigneulle RM (1978) Proliferative capacity of murine hematopoietic stem cells. Proc Natl Acad Sci USA 75: 490–494
Jones RJ, Celano P, Sharkis SJ, Sensenbrenner LL (1989) Two phases of engraftment established by serial bone marrow transplantation in mice. Blood 73: 397–401
Jordan CT, Lemischka IR (1990) Clonal and systemic analysis of long-term hematopoiesis in the mouse. Genes Dev 4: 220–232
Jordan CT, McKearn JP, Lemischka IR (1990) Cellular and developmental properties of fetal hematopoietic stem cells. Cell 61: 953–963
Keller G, Snodgrass R (1990) Life span of multipotential hematopoietic stem cells in vivo. J Exp Med 171: 1407–1418
Keller G, Paige C, Gilboa E, Wagner EF (1985) Expression of a foreign gene in myeloid and lymphoid cells derived from multipotent hematopoietic precursors. Nature 318: 159–154
Kretchmar AL, Conover WR (1970) A difference between spleen-derived and bone marrow-derived colony-forming units in ability to protect lethally irradiated mice. Blood 36: 772–776
Lemischka IR, Raulet D, Mulligan RC (1986) Developmental potential and dynamic behavior of hematopoietic stem cells. Cell 45: 917–927
Lerner CP, Harrison DE (1990) 5-fluorouracil spares hemopoietic stem cells responsible for long term repopulation. Exp Hematol 18: 114–118
Magli MC, Iscove NN, Odartchenko N (1982) Transient nature of early haematopoietic spleen colonies. Nature 295: 527–529
Micklem HS, Anderson N, Ross E (1975) Limited potential of circulating haemopoietic stem cells. Nature 256: 41–43
Micklem HS, Lennon JE, Ansell JD, Gray RA (1987) Numbers and dispersion of repopulating hematopoietic cell clones in radiation chimeras as functions of cell does. Exp Hematol 15: 251–257
Mintz B, Anthony K, Litwen S (1984) Monoclonal derivation of mouse myeloid and lymphoid lineages afrom totipotent hematopoietic stem cells experimentally engrafted in fetal hosts. Proc Natl Acad Sci USA 81: 7835–7839

Nakano T, Waki N, Asai H, Kitamura Y (1989) Effect of 5-fluorouracil on "primitive" hematopoietic stem cells that reconstitute whole erythropoiesis of genetically anemic W/Wv mice. Blood 73: 425–430

Nesbitt MN (1971) X-Chromosome inactivation mosaicism in the mouse. Dev Biol 26: 252

Nocka K, Tan JC, Chiu E, Chu TY, Ray P, Traktman P, Besmer P (1990) Molecular bases of dominant negative and loss of function mutations at the murine c-kit/white spotting locus: W37, Wv, W41, and W. EMBOJ 9: 1805–1813

Reith AD, Rottapel R, Giddens E, Brady C, Forrester L, Bernstein A (1990) W mutant mice with mild or severe developmental defects contain distinct point mutations in the kinase domain of the c-kit receptor. Genes Dev 4: 390–400

Ross EAM, Anderson N, Micklem HS (1982) Serial depletion and regeneration of the murine hematopoietic system. Implications for hematopoietic organization and the study of cellular aging. J Exp Med 155: 432–434

Spangrude GJ, Heimfeld S, Weissman IL (1988) Purification and characterization of mouse hematopoietic stem cells. Science 241: 58–62

Stone M (1984) Variance-covariance modeling with chromosome markers. J Theor Biol 107: 275

Stone M, Harrison DE (1990) A stratified binomial marker model for bone-marrow repopulation experiments. J Theor Biol 144: 267–273

Szilvassy SJ, Fraser CC, Eeaves CJ, Landsdorp PM, Eaves AC, Humphries RK (1989) Retrovirus-mediated gene transfer to purified hemopoietic stem cells with long term lymphomyelopoietic repopulating ability. Proc Natl Acad Sci USA 86: 8798–8802

Till JE, McCulloch EA (1961) A direct measurement of the radiation sensitivity of normal mouse bone marrow cells. Radiat Res 14: 213–222

Witte ON (1990) Steel locus defines new multipotent growth factor. Cell 63: 5–6

Wu AM, Till JE, Siminovitch L, McCulloch EA (1967) A cytological study of the capacity for differentiation of normal hemopoietic colony-forming cells. J Cell Physiol 69: 177–184

The Pre-Spleen Colony-Forming Unit Assay: Measurement of Spleen Colony-Forming Unit Regeneration

G. J. SPANGRUDE

1 Historical Background and Definition

Very soon after the description of the spleen colony-forming unit (CFU-S) assay by TILL and MCCULLOCH in 1961, the question of whether splenic colonies truly represented the proliferation of primitive hematopoietic stem cells became an important issue. Since the criteria for defining hematopoietic stem cells at that time included multilineage potential as well as a capacity for self-renewal, it naturally followed that one should be able to demonstrate the presence of new colony-forming cells within spleen colonies. This hypothesis was tested using a double transplant experiment, either by excising individual spleen colonies (SIMINOVITCH et al. 1963) or by pooling the entire spleen from a CFU-S experiment (SIMINOVITCH et al. 1964) for transplant into secondary irradiated recipients. The results revealed a broad heterogeneity in the ability of splenic colonies to initiate new splenic colonies, with the additional observation that allowing a longer period of time to elapse prior to harvesting spleen colonies for secondary transplant resulted in an increase in the number of colonies containing secondary CFU-S. Many more colonies harvested 14 days after injection contained CFU-S compared to colonies harvested after 8 days. While a reasonable interpretation at the time invoked an extended time frame necessary for the self-renewal process within spleen colonies (LEWIS and TROBAUGH 1964), later work revealed that the majority of early (day 8) CFU-S disappear a few days later and that

Laboratory of Persistent Viral Diseases, National Institute of Allergy and Infectious Diseases, National Institutes of Health, Rocky Mountain Laboratories, Hamilton, MT 59840, USA

Current Topics in Microbiology and Immunology, Vol. 177

at least half of the late (day 11–13) colonies arise from a separate group of hematopoietic cells (MAGLI et al. 1982; PRIESTLEY and WOLF 1985; WOLF and PRIESTLEY 1986). These intrinsic differences in CFU-S hinted at the limited utility of the CFU-S assay for detecting hematopoietic stem cells, since only after a secondary transplant could one be assured of the self-renewing potential inherent in the original spleen colony.

From these historical origins, the pre-CFU-S assay evolved as a double transplantation technique which measures the regeneration of CFU-S in the hematopoietic tissues of an experimental animal after lethal irradiation and bone marrow reconstitution. The bone marrow and/or spleen cells from the repopulated animal are collected about 2 weeks after the primary transplant and are intravenously transferred into a second irradiated animal at an appropriate concentration to result in the formation of discrete splenic colonies several weeks later. The primary recipient animal provides the environmental support in which primitive hematopoietic stem cells among the transplanted inoculum can develop, while the secondary recipient serves to read out the formation of spleen colonies by the differentiated progeny of the hematopoietic stem cells in the original population. Both of these in vivo phases of the experiment could conceivably occur in vitro (VAN DER SLUIJS et al. 1990), thus simplifying the assay tremendously, however the in vitro culture systems need further refinement before this transition can be made. Some success has been reported in using in vitro colony formation in place of the second transplant recipient (BERTONCELLO et al. 1985; PLOEMACHER and BRONS 1989), and stromal cell lines or cytokines derived from them may eventually be identified, which can provide the unknown conditions present in the primary transplant recipient that contribute to the maintenance and development of primitive hematopoietic stem cells (SUTHERLAND et al. 1989; ISCOVE and YAN 1990).

It is important to note a distinction between the pre-CFU-S assay as described above and the concept of a pre-CFU-S cell population. In the present context, the pre-CFU-S assay as discussed in this manuscript has been considered as an experimental assay which is sensitive to primitive hematopoietic stem cells. A population of pre-CFU-S cells has been described which lacks CFU-S activity but which possesses very potent repopulating capacity and has the potential to differentiate into cells that are CFU-S (HODGSON and BRADLEY 1979; PLOEMACHER and BRONS 1989; JONES et al. 1990). Populations of pre-CFU-S cells are very potent in many aspects of reconstituting hematopoiesis in irradiated animals, including platelet regeneration, reconstitution of in vitro and in vivo colony-forming potential, and long-term repopulation. However, these pre-CFU-S populations are not the only cells which are able to read out in the pre-CFU-S assay, as populations of cells can be isolated that are highly enriched for CFU-S and are also able to function in double transplantation assays (SPANGRUDE and JOHNSON 1990). In this manuscript, the term "pre-CFU-S assay" has been confined to describing a measurement of the ability of any population of cells to form spleen colonies after a secondary transplant, regardless of other activities inherent within that population.

2 Refinement

Hodgson and colleagues (HODGSON and BRADLEY 1979, 1984; HODGSON et al. 1982) extended the concept of analyzing splenic colonies for the ability to form secondary colonies as a means to identify hematopoietic stem cells. By evaluating many hematopoietic parameters within irradiated and reconstituted animals, including marrow and spleen cellularity, peripheral blood platelet counts, and the capacity of marrow cells as well as splenic cells to initiate colony formation in vivo and in vitro, the concept of a precursor for CFU-S began to evolve (HODGSON and BRADLEY 1979, 1984). The pre-CFU-S assay, as we now understand it, is a measure of the ability of a population of cells to differentiate into CFU-S and thus measures a more primitive population of cells than that detected by the simple CFU-S assay.

By extending the approach of SIMINOVITCH et al. (1964) to include the analysis of marrow in addition to spleen in secondary transplantation assays, HODGSON et al. (1982) demonstrated a number of important characteristics inherent in the hematopoietic stem cell compartment. Utilizing 5-fluorouracil as a mechanism to deplete many actively cycling cells in the hematopoietic compartment, these workers revealed the existence of a quiescent, very primitive population of cells which was highly enriched in secondary CFU-S activity, in spite of a relative depletion of primary CFU-S activity (HODGSON et al. 1982). Histologic investigation of the remaining primary CFU-S activity in 5-fluorouracil-depleted marrow revealed the presence of microscopic colonies, primarily consisting of megakaryocytes (HODGSON et al. 1982). These observations supported the concept of a pre-CFU-S activity, not necessarily associated with primary macroscopic spleen colonies. Further, the pre-CFU-S activity exhibited a preference for colonizing the irradiated marrow cavity rather than the spleen, leading to the suggestion that the most primitive hematopoietic stem cells do not colonize the spleen at all (BARNES et al. 1968; HODGSON et al. 1982; VAN ZANT 1984; BERTONCELLO et al. 1985; WOLF and PRIESTLEY 1986). Subsequently, LORD et al. (1989) have argued that the marrow-homing preference of very primitive hematopoietic stem cells does not reflect an obligatory association with marrow stromal elements for their survival and differentiation, since marrow ablation by treatment with strontium-89, a bone-seeking radionuclide, resulted in no decrease in hemopoietic activity in the spleen. However, the work of HODGSON and colleagues clearly underscores the importance of the regenerating marrow cavity as an environment for hematopoietic stem cell differentiation.

PLOEMACHER and BRONS (1989) have confirmed and extended the pre-CFU-S concept by enriching from normal marrow populations of hematopoietic stem cells which lack macroscopic CFU-S activity. In agreement with the work of HODGSON and colleagues, these cells tend to generate microscopic spleen colonies consisting of megakaryocytes and/or granulocytes (PLOEMACHER and BRONS 1988). Further, JONES et al. (1990) have demonstrated long-term repopulation of irradiated animals in the apparent absence of macroscopic CFU-S activity in 13-day assays. However, in those experiments the active population of cells was not isolated, and thus the potential for regulatory effects mediated by other bone marrow cells as an explanation for the lack of primary CFU-S activity cannot be excluded.

3 Utility of the Pre-CFU-S Assay

Table 1 shows an example of data obtained using the double transplantation technique to measure the regeneration of CFU-S in the marrow of irradiated animals. Bone marrow donors were cogenic with the primary and secondary recipients at the CD45 gene locus, so that the contribution by the donor cells to long-term hematopoiesis in the secondary animals could be detected by staining peripheral blood cells with monoclonal antibodies specific for the donor and host alleles of CD45 (SPANGRUDE and JOHNSON 1990). As Table 1 shows, 13 days after a primary transplant of 10^5 bone marrow cells containing 15 CFU-S (Table 1, line 2), the recipient animals yielded 1.6×10^6 cells/femur and about 5 CFU-S/femur, which equals 0.05 CFU-S/femur per 10^3 injected cells. Two populations of cells enriched from mouse bone marrow on the basis

Table 1. Analysis of the repopulating activity of several bone marrow subpopulations

Cells	Transferred[a]	CFU-S[b]	Primary animals[c]		Secondary animals[f]	
			Cells/ femur[d]	CFU-S/ femur per 10^3 [e]	Dead/total[g]	% donor[h]
1.	No cells	0	3.9×10^5	0	10/10	—
2.	10^5 bone marrow	15	1.6×10^6	0.05	ND	ND
3.	10^3 Rh-123^{lo}	23	5.2×10^6	201	0/9	92
4.	10^3 Rh-123^{hi}	105	3.2×10^6	4.7	4/9	0
5.	10^4 Ly-6A/E^{neg}	25	9.1×10^4	0.04	5/5	—

[a] Transferred cell populations were enriched from normal bone marrow as described (SPANGRUDE and JOHNSON 1990). Rh-123^{lo} and Rh-123^{hi} indicate subsets of Thy-1^{lo} Lin^{neg}Ly-6A/E^{+} bone marrow cells discriminated by Rh-123 staining. Ly-6A/E^{neg} indicates Thy-$1^{lo}$$Lin^{neg}$Ly-6A/$E^{neg}$ bone marrow cells. All cell transfers were by intravenous injection
[b] Number of day 13 CFU-S transferred into the primary animals
[c] Primary and secondary animals were lethally irradiated (11 Gy, ^{137}Cs)
[d] Number of cells present per femur of primary recipient animals 13 days after the cell transfer
[e] Number of day 13 CFU-S present per femur of primary recipient animals 13 days after the cell transfer, per 10^3 cells injected
[f] Irradiated secondary animals were transplanted with 0.1 femoral equivalents of bone marrow from the primary animals 13 days after the primary transfer
[g] Survival of lethally irradiated secondary animals 30 days after transplant of 0.1 femoral equivalents from primary animals
[h] Eight weeks after the secondary transplant, the percentage of cells in the peripheral blood of the secondary animals which were derived from the original transferred population was assessed by staining for allelic expression of the CD45 molecule (SPANGRUDE et al. 1988)
ND, not determined

of Thy-1 and Ly-6A/E antigen expression in the absence of lineage antigens (SPANGRUDE et al. 1988), and differing in their retention of the mitochondrial dye rhodamine-123 (Rh-123^{lo} and Rh-123^{hi}, Table 1, lines 3 and 4) required 100–1000-fold fewer cells to produce spleen colonies in primary recipient animals. In spite of a fourfold lower number of CFU-S transplanted by the Rh-123^{lo} population, 40-fold more CFU-S were generated in the marrow of the primary recipient animals relative to the Rh-123^{hi} population. The hematopoietic repopulating potential of the Rh-123^{lo} cells was reflected by an analysis of the secondary transplant recipient animals, which contained >90% donor-derived cells in peripheral blood samples, in contrast to the undetectable levels observed in the Rh-123^{hi}-recipient animals which survived the transplant. The separation of radioprotective activity from long-term repopulating potential apparent in this experiment has been noted previously (BRECHER et al. 1988; JONES et al. 1989; SPANGRUDE and SCOLLAY 1990). A population of cells characterized by Thy-1 antigen expression in the absence of Ly-6A/E or lineage antigens (Table 1, line 5) was equivalent in CFU-S-repopulating activity relative to normal bone marrow, in spite of the enrichment of primary day 13 CFU-S in the transferred population of cells. Secondary transplantation of marrow from irradiated recipients of these cells resulted in a failure to radioprotect recipient animals. Thus, quantitaitive measurements of the CFU-S-generating ability of hematopoietic stem cell populations can be obtained by this technique, and the pre-CFU-S activity correlates with long-term repopulating potential. Since the assay is not clonal, no indication of the actual frequency of pre-CFU-S cells is obtainable. However, the assay is quantitative and thus allows comparisons to be made in determining the primitive hematopoietic stem cell activity contained in various cell populations.

4 Disadvantages

Although one can, utilizing the pre-CFU-S assay, obtain a quantitative estimate of secondary colony forming potential, the pre-CFU-S assay is not a clonal readout of the initiating hematopoietic stem cell. The quantitative aspect of the assay allows some estimate of the long-term hematopoietic activity contained in the original cell suspension and allows comparisons to be made between cell populations. However, no distinction can be made between a population that contains a rare subset of cells with high repopulating potential and one that is highly enriched with cells of a lower repopulating potential per cell. Such a distinction might be observed using a competitive repopulation assay. Further, unless genetic markers such as chromosomal differences or allelic cell surface markers between donor and host are employed, the secondary colonies which form may be derived from cells in the primary host which survived the irradiation conditioning. These problems limit, but do not obviate, the utility of the pre -CFU-S assay as a means to monitor enrichment of hematopoietic stem cells from mouse hematopoietic tissues. Unfortunately, the inability to perform in vivo colony-forming assays using human hematopoietic cells prevents this assay from being applied in that system. Perhaps further refinements in the chimeric human-mouse model system (McCUNE et al. 1988) will allow the development of a similar assay for human hematopoietic stem cells.

A further disadvantage of the pre-CFU-S assay involves technical difficulties. The assay has a long-term readout, requiring 24–26 days to complete. One can modulate this time frame to some extent by shortening the residence time in the primary animal, with the recognition that this will decrease the amplification of the secondary colony-forming cells. The use of marrow cells, rather than spleen, from the primary host would be much more important in experiments utilizing shorter residence periods in the primary animals, since the spleen may not contain as high a content of primitive cells prior to day 11 or so (SIMINOVITCH et al. 1963; BERTONCELLO et al. 1985). However, the 26-day time frame of the assay is much shorter than long-term repopulation assays, which can require several months to complete (BRECHER et al. 1988). The correlation between pre-CFU-S activity and long-term repopulating potential (Table 1) argues that the pre-CFU-S assay measures a very primitive cell population in a relatively short period of time.

A further technical problem in these studies is an inability to predict in advance the repopulating potential of a given preparation of cells. The readout of CFU-S in the secondary animal has a narrow useful range for accurate quantitation, since fewer than two or three splenic colonies per recipient animal results in a statistical limitation in data interpretation when small numbers of recipient animals are analyzed, while greater than ten colonies exceeds the linear range of the readout assay. As a result, the secondary transplant must involve either multiple recipient groups in five fold titrations, or multiple experiments to establish the correct dilution range for the secondary transplant following primary transfer of any particular cell population. Fortunately, the results of secondary transfers of marrow from primary recipients of cell populations enriched by flow cytometry are generally very consistent from experiment to experiment, allowing fairly confident prediction of the CFU-S content of the secondary transplant (BERTONCELLO et al. 1988; PLOEMACHER and BRONS 1989; SPANGRUDE and JOHNSON 1990). This problem can also be overcome to some extent by substituting in vitro colony formation for the secondary transplant; however, this readout may not be as accurate in reflecting the content of primitive hematopoietic stem cells (BERTONCELLO et al. 1988).

5 Self-Renewal: Is It Real?

What does the pre-CFU-S assay tell us about hematopoietic stem cell populations? Although originally intended to analyze the self-renewal potential of hematopoietic stem cell populations (SIMINOVITCH et al. 1963), the pre-CFU-S assay is really only an indirect measurement of the precursors of splenic colony-forming cells. As such, the assay cannot be claimed to be a measure of self-renewal of hematopoietic stem cells, since the CFU-S readout is not a specific characteristic of primitive hematopoietic stem cells. Whether it is appropriate to attempt to measure self-renewal of hematopoietic stem cells is questionable, since the emerging evidence of the heterogeneous nature of the hematopoietic stem cell compartment precludes the requirement for self-renewal of hematopoietic stem cells at a clonal level (HELLMAN et al. 1978; HODGSON et al. 1982; SPANGRUDE and JOHNSON 1990). In a manner analogous to the mammalian female reproductive system, the hematopoietic stem cell compartment may consist of a reserve of primitive cells which are sufficient in number to maintain hematopoiesis for several normal life spans. As each primitive cell is called into service, subsequent proliferation and differentiation result in a net increase in assayable multipotent precursors, with a concomitant loss of proliferative potential (ROSENDAAL et al. 1979). In this scheme of hematopoiesis, self-renewal of hematopoietic stem cells does not need to occur at a clonal level, and the hematopoetic stem cell compartment is continually replenished from an ample reserve of primitive cells. This hypothesis predicts that there should be an upper limit on the number of late-stage progenitors (such as day 13 CFU-S, day 8 CFU-S, or in vitro colony-forming cells) that can be generated by the most primitive cell from the reserve supply. The pre-CFU-S assay will be able to reflect this upper limit, but will not be able to distinguish between progeny of a fewer number of more primitive cells versus a larger number of more differentiated cells. Adaptation of the assay to a clonal amplification of primitive hematopoietic stem cells seems unlikely; a more feasible approach is to substitute a limiting dilution in vitro culture system for the primary transplant recipient, utilizing either bone marrow stromal cells (PLOEMACHER et al. 1989; SUTHERLAND et al.1990) or growth factors derived from them (WITTE 1990). Ideally, a readout of in vitro colony formation as a substitute for the secondary CFU-S assay (DEXTER et al. 1977; PLOEMACHER et al. 1989; MCNIECE et al. 1990) would provide the basis for an assay system useful for the detection and quantitation of primitive hematopoietic stem cells in many species, including the human.

References

Barnes DWH, Evans EP, Ford CE, West BJ (1968) Spleen colonies in mice: karyotypic evidence of multiple colonies from single cells. Nature 219: 518–520

Bertoncello I, Hodgson GS, Bradley TR (1985) Multiparameter analysis of transplantable hemopoietic stem cells. I. The separation and enrichment of stem cells homing to marrow and spleen on the basis of rhodamine-123 flourescence. Exp Hematol 13: 999–1006

Bertoncello I, Hodgson GS, Bradley TR (1988) Multiparameter analysis of transplantable hemopoietic stem cells. II. Stem cells of long-term bone marrow-reconstituted recipients. Exp Hematol 16: 245–249

Brecher G, Neben S, Yee M, Bullis J, Cronkite EP (1988) Pluripotent stem cells with normal or reduced self renewal survive lethal irradiation. Exp Hematol 16: 627–630

Dexter TM, Allen TD, Lajtha LG (1977) Conditions controlling the proliferation of haemopoietic stem cells in vitro. J Cell Physiol 91: 335–344

Hellman S, Botnick LE, Hannon EC, Vigneulle RM (1978) Proliferative capacity of murine hematopoietic stem cells. Proc Natl Acad Sci USA 75: 490–494

Hodgson GS, Bradley TR (1979) Properties of haematopoietic stem cells surviving 5-fluorouracil treatment: evidence for a pre-CFU-S cell? Nature 281: 381–382

Hodgson GS, Bradley TR (1984) In vivo kinetic status of hematopoietic stem and progenitor cells as inferred from labeling with bromodeoxyuridine. Exp Hematol 12: 683–687

Hodgson GS, Bradley TR, Radley JM (1982) The organization of hemopoietic tissue as inferred from the effects of 5-fluorouracil. Exp Hematol 10: 26–35

Iscove NN, Yan X-Q (1990) Precursors (pre-CFC_{multi}) of multilineage hematopoietic colony-forming cells quantitated in vitro. Uniqueness of IL-1 requirement, partial separation from pluripotential colony-forming cells, and correlation with long term reconstituting cells in vivo. J Immunol 145: 190–195

Jones RJ, Celano P, Sharkis SJ, Sensenbrenner LL (1989) Two phases of engraftment established by serial bone marrow transplantation in mice. Blood 73: 397–401

Jones RJ, Wagner JE, Celano P, Zicha MS, Sharkis SJ (1990) Separation of pluripotent haematopoietic stem cells from spleen colony-forming cells. Nature 347: 188–189

Lewis JP, Trobaugh FE Jr (1964) Haematopoietic stem cells. Nature 204: 589–590

Lord BI, Molineux G, Schofield R, Humphreys ER, Stones VA (1989) On the late seeding of CFU-S to the spleen: 8- vs 12-day CFU-S. Exp Hematol 17: 836–842

Magli MC, Iscove NN, Odartchenko N (1982) Transient nature of early haematopoietic spleen colonies. Nature 295: 527–529

McCune JM, Namikawa R, Kaneshima H, Schultz LD, Lieberman M, Weissman IL (1988) The SCID-hu mouse: murine model for the analysis of human hematolymphoid differentiation and function. Science 241: 1632–1639

McNiece IK, Bertoncello I, Kriegler AB, Quesenberry PJ (1990) Colony-forming cells with high proliferative potential (HPP-CFC). Int J Cell Cloning 8: 146–160

Ploemacher RE, Brons NHC (1988) In vivo proliferative and differential properties of murine bone marrow cells separated on the basis of rhodamine-123 retention. Exp Hematol 16: 903–907

Ploemacher RE, Brons RHC (1989) Separation of CFU-S from primitive cells responsible for reconstitution of the bone marrow hemopoietic stem cell compartment following irradiation: evidence for a pre-CFU-S cell. Exp Hematol 17: 263–266

Ploemacher RE, van der Sluijs JP, Voerman JSA, Brons NHC (1989) An in vitro limiting-dilution assay of long-term repopulating hematopoietic stem cells in the mouse. Blood 74: 2755–2763

Priestley GV, Wolf NS (1985) A technique for the daily examination of spleen colonies in mice. Exp Hematol 13: 733–735

Rosendaal M, Hodgson GS, Bradley TR (1979) Organization of haemopoietic stem cells: the generation-age hypothesis. Cell Tissue Kinet 12: 17–29

Siminovitch L, McCulloch EA, Till JE (1963) The distribution of colony-forming cells among spleen colonies. J Cell Comp Physiol 62: 327–336

Siminovitch L, Till JE, McCulloch EA (1964) Decline in colony-forming ability of marrow cells subjected to serial transplantation into irradiated mice. J Cell Comp Physiol 64: 23–32

Spangrude GJ, Johnson GR (1990) Resting and activated subsets of mouse multipotent hematopoietic stem cells. Proc Natl Acad Sci USA 87: 7433–7437

Spangrude GJ, Scollay R (1990) A simplified method for enrichment of mouse hematopoietic stem cells. Exp Hematol 18: 920–926

Spangrude GJ, Heimfeld S, Weissman IL (1988) Purification and characterization of mouse hematopoietic stem cells. Science 241: 58–62

Sutherland HJ, Eaves CJ, Eaves AC, Dragowska W, Lansdorp PM (1989) Characterization and partial purification of human marrow cells capable of initiating long-term hematopoiesis in vitro. Blood 74: 1563–1570

Sutherland HJ, Lansdorp PM, Henkelman DH, Eaves AC, Eaves CJ (1990) Functional characterization of individual human hematopoietic stem cells cultured at limiting dilution on supportive marrow stromal layers. Proc Natl Acad Sci USA 87: 3584–3588

Till JE, McCulloch EA (1961) A direct measurement of the radiation sensitivity of normal mouse bone marrow cells. Radiat Res 14: 213–222

Van der Sluijs JP, de Jong JP, Brons NHC, Ploemacher RE (1990) Marrow repopulating cells, but not CFU-S, establish long-term in vitro hemopoiesis on a marrow-derived stromal layer. Exp Hematol 18: 893–896

Van Zant G (1984) Studies of hematopoietic stem cells spared by 5-fluorouracil. J Exp Med 159: 679–690

Witte ON (1990) Steel locus defines new multipotent growth factor. Cell 63: 5–6

Wolf NS, Priestley GV (1986) Kinetics of early and late spleen colony development. Exp Hematol 14: 676–682

Clonal Analysis of Hematopoietic Stem Cell Development In Vivo

G. KELLER

1 The Hematopoietic System

The hallmark of hematopoietic stem cells is their ability to maintain a functional hematopoietic system over extended periods of time in vivo (ABRAMSON et al. 1977; HARRISON et al. 1988; KELLER and SNODGRASS 1990; JORDAN and LEMISCHKA 1990; CAPEL et al. 1990). This capacity to sustain long-term hematopoiesis is directly related to the ability of the stem cell population to regenerate itself, a characteristic commonly referred to as self-renewal. In addition to possessing this self-renewal potential, hematopoietic stem cells are also pluripotential and thus able to generate progeny of all the myeloid (nonlymphoid) and lymphoid blood cell lineages (WU et al. 1968; ABRAMSON et al. 1977; KELLER et al. 1985; DICK et al. 1985; LEMISCHKA et al. 1986). It is their ability to function for long periods of time in vivo which distinguishes the stem cell population from virtually all other cells within the hematopoietic system and in fact provides a basis for their definition, namely cells which can maintain long-term hematopoiesis either in an intact animal or in one into which they have been transplanted. The term "stem cell" has been and still is frequently used for cells at various development at stages within the hematopoietic system, and as a consequence its use is at times confusing. In this review the term "stem cell" will be used only in reference to cells that display long-term repopulating capacity and will be used interchangeably

National Jewish Center, 1400 Jackson Street, Denver, CO 80206, USA

Current Topics in Microbiology and Immunology, Vol. 177

with the term "long-term repopulating stem cells" or simply "long-term repopulating cells".

Cells representing more advanced stages of development between the stem cells and the mature blood cells will be collectively referred to as "precursor cells". These differ from the more primitive stem cell population in that they are unable to self-renew and are restricted in their differentiation potential (Fig. 1). Many of the precursors within the myeloid lineages have been identified by their capacity to generate a clone of differentiated hematopoietic cells either in the spleen of an irradiated mouse (spleen colony-forming cell, CFU-S) (TILL and McCULLOCH 1961) or in semisolid cultures (in vitro colony-forming cell, CFC) in response to specific growth factors (reviewed in METCALF 1977, 1984). Most evidence would indicate that these precursor cells are organized in a hierarchy, ranging from the early ones which are able to generate progeny of multiple lineages (early precursors) to the most mature which are restricted to a single developmental pathway (late or committed). Associated with this increase in maturation is an increase in size of the respective precursor populations. Thus, those cells committed to a single lineage are found at a higher frequency than those representing the intermediate stages of development which, in turn, are more frequent than the most primitive precursors. The long-term repopulating

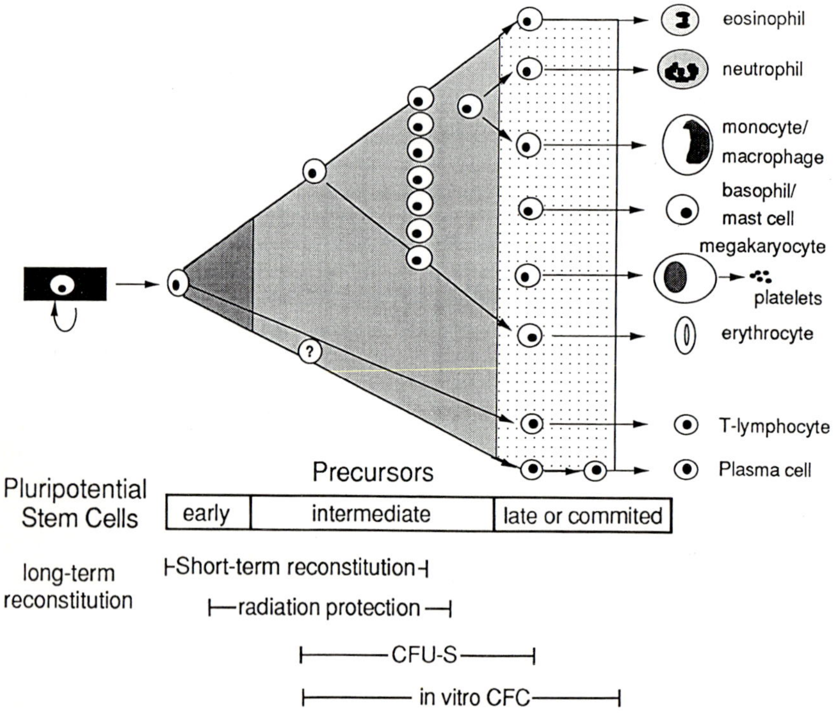

Fig. 1. Schematic representation of the mouse hematopoietic system

stem cells, the most primitive cells within the system, are found at a frequency that is considerably lower than any of the precursor populations.

Our understanding of the developmental potential of the various precursor populations and the factors which regulate their growth and differentiation has been greatly enhanced by availability of the above-mentioned clonal assays; in particular in the vitro colony assay. In contrast, our knowledge of many aspects of the stem cell population, including their physical characteristics as well as the machanisms which control their growth and differentiation remains limited. This lack of knowledge results from the fact that well-defined in vitro stem cell assays, comparable to those for the precursor populations, do not exist. In vitro assays do exist which measure cells that represent stages of development earlier than most clonable precursors (NAKAHATA and OGAWA 1982; PRAGNELL et al.1988; ISCOVE and YAN 1990; SUTHERLAND et al. 1990). However, conclusive evidence demonstrating that the cells detected in these assays are long-term repopulating cells does not yet exist. Consequently, the only means of assaying these cells is by their capacity to generate differentiated progeny over extended periods of time in vivo.

Initially it was thought that CFU-S were the cells that provided long-term reconstitution. However, it has become increasingly clear from a number of studies that CFU-S are unable to sustain a functional hematopoietic clone over long periods of time in vivo and most likely represent a more mature precursor population that is restricted to the myeloid lineages of development (PAIGE et al. 1979; KITAMURA et al. 1981; MAGLI et al. 1982; PLOEMACHER and BRONS 1989). The clearest demonstration that CFU-S and long-term repopulating stem cells are different has been provided by a recent study of JONES et al. (1990) in which these two populations were physically separated on the basis of size (see chapter by JONES, this volume).

An assay that is frequently used to measure stem cells is the radiation protection assay (VISSER et al. 1984; MÜLLER-SIEBURG et al. 1988; SPANGRUDE et al. 1988). The major drawback of this assay is that, by definition, it measures a cell population that can protect mice from radiation death. Consequently, the population being assayed must be able to generate sufficient numbers of mature cells within 2 weeks of transplantation to keep the recipients alive. It is likely that a population which contains significant numbers of precursors would function better in this assay than a pure population of long-term repopulating stem cells as the more advanced precursors are able to generate substantial numbers of mature cells in a shorter time span than are the primitive stem cells. Experimental evidence supporting this notion has been provided by JONES et al. (1990), in the same study as described above. In addition to separating CFU-S from long-term repopulating cells, they were also able to separate the population which provides radiation protection from these more primitive stero stem cells. The fraction which contained long-term repopulating cells but few clonable precursors (CFU-S and CFC) was unable to protect mice from radiation death. Radioprotective cells were found in the same fraction as the CFU-S, suggesting that they could represent cells at comparable stages of development. These findings provide the first clear demonstration that the cells which provide long-term repopulation do not protect mice from the acute effects of radiation and strongly suggest that this function is mediated by a more mature precursor population (Fig. 1).

The only assay which measures long-term repopulating stem cells is the long-term repopulation of the hematopoietic system of a recipient animal. To be useful in studying various aspects of stem cell biology, the assay must be quantitative, and the survival of the recipient should not depend upon the population being assayed. At least two such quantitative reconstitution assays exist. The first, the competitive repopulation assay (HARRISON 1980), is described in detail by HARRISON, this volume. This assay is based on the ability of an experimental population of hematopoietic cells to compete with a standard number of unmanipulated cells in the reconstitution of an irradiated recipient. The donor populations carry markers that enable one to distinguish them from one another as well as from the cells of the recipient. Administration of normal marrow to all recipients ensures that none will die from radiation death. Stem cell content of the experimental population is calculated relative to the normal competitor population. The second assay that is used for quantitating long-term repopulating cells is the reconstitution of genetically anemic W/W^v mice with limiting numbers of cells (BOGGS et al. 1982, 1984). As wild-type bone marrow cells have a growth advantage over those from the mutant animals, prior irradiation of the recipients is not required and thus relatively small numbers of cells can be assayed (reviewed in RUSSEL 1979). Using this approach, the frequency of long-term reconstituting stem cells (>6 months) in normal marrow has been calculated to be in the order of one in 3×10^4 cells (BOGGS et al. 1982, 1984, M. Kennedy and G. Keller, unpublished).

Although these long-term reconstitution assays provide a means for analyzing whole populations, they do not allow one to easily study individual cells. It is only through the reconstitution of large numbers of W/W^v recipients with limiting numbers of marrow that one can analyze the function of a single reconstituting stem cell. A more direct approach to studying individual cells is to uniquely mark them prior to transplantation. By identifying and tracing this maker in the recipient, it is possible to follow the fate of the uniquely marked cell. The first unique markers to be used were radiation-induced chromosomal translocations (WU et al. 1968; ABRAMSON et al. 1977). Studies using these markers directly demonstrated, for the first time, the existence of pluripotential stem cells capable of long-term reconstitution. More recently, the proviral integration site of recombinant retroviruses has been used as a unique marker for hematopoietic cells (KELLER et al. 1985; DICK et al. 1985; LEMISCHKA et al. 1986). This method of marking cells offers several advantages over the earlier chromosomal translocation method in that it is relatively easy to mark large numbers of cell with a retrovirus, and the genetic alteration to the marked cell is less detrimental.

The development of replication-defective recombinant retroviruses can be considered as a significant step forward in the field of stem cell biology as it provided a method for routinely marking these cells and following their fate when transplanted into recipient mice. Findings from studies using this approach have provided a clearer understanding of the populations representing the early stages of the hematopoietic hierarchy and in addition have further defined the potential of the long-term repopulating cells in vivo. The method of marking hematopoietic cells with retroviruses and a summary of the findings from these marking experiments is outlined in the following sections.

2 Marking Hematopoietic Cells with Retroviruses

Following construction of the retroviral vectors and generation of infectious virus by an appropriate packaging cell line, infection of hematopoietic cells is relatively straight forward. Although hematopoietic cells from a variety of sources can be used as targets for infection, the discussion in this chapter will focus on those studies involving the use of bone marrow cells. The best results are obtained with marrow from mice that have been pretreated with 5-fluorouracil (5-FU). 5-FU destroys the late- and intermediate-stage cycling precursors, but not the long-term repopulating cells and thus serves as a method for enriching for these primitive cells (HODGSON and BRADLEY 1979; LERNER and HARRISON 1990). In addition, it is assumed, but not yet experimentally demonstrated, that depletion of these precursor populations will induce the more primitive cells within the system to cycle. Inducing the stem cells to proliferate is critical as most evidence would suggest that a provirus will only integrate into cells which are cycling (VARMUS et al. 1977).

The cells are infected by exposing them to high concentrations of the recombinant virus either by co-culturing them with the virus-producing cells or by culturing them with supernatant from the virus-producing cells. Co-culture was the first method used to successfully infect long-term repopulating stem cells (KELLER et al. 1985; DICK et al. 1985) and it is still widely used for this purpose. It is most likely the most efficient method of infection as the cells are maintained continuously in the presence of freshly produced virus. However, co-culture does have several drawbacks which the supernatant infection protocol overcomes. First, any adherent population will be lost following co-culture as it is almost impossible to separate cells which adhere to the virus-producing cells from the virus-producing cell themselves. Following supernatant infection, adherent cells can be easily recovered from the culture flask by a short trypsinization or collagenase treatment. This is an important consideration as findings from a number of studies indicate that at least a subpopulation of long-term reconstituting cells are adherent (KERK et al. 1985; BEARPARK and GORDON 1989; Kiefer, Wagner, and Keller, unpublished). Second, fresh virus-producing cells have to be prepared for each experiment for co-culture. With supernatant infection, a large stock of virus can be produced, titered, and then frozen for use in a number of different experiments. This eliminates the variable introduced by continuously preparing fresh virus-producing cells for each experiment, and it allows one to use a constant, predetermined multiplicity of infection. We have found that a multiplicity of infection of fewer than three results is a low frquency of infection of marrow cells.

Provided that the virus used for infection contains a selectable gene, the infected population can be cultured in the appropriate selection conditions prior to transplantation. This preselection step significantly enriches the population for those cells wich are infected and expressing the transduced genes. For example, immediately following infection with a high-titer virus such as N2 which contains the *neo* gene (KELLER et al. 1985), 20%–50% of all precursors are G418 resistant (G418R). Following 72 h of preselection, the proportion G418R precursors in the population increases to 60%–95%.

The most difficult problem encountered when attempting to infect long-term repopulating cells with recombinant retroviruses is the loss of these cells throughout the infection and preselection steps. Although there is no evidence that interleukin 3 (IL-3) and IL-1 enhance the survival of these cells, they are routinely included in both the infection and preselection cultures as it has been shown that they do enhance the survival of precursor cells (ISCOVE and YAN 1990). One factor which may be beneficial for the survival of these primitive cells and is currently under investigation is the recently identified ligand to the *c-kit*-encoded receptor(HUANG et al. 1990; WILLIAMS et al. 1990; ZSEBO et al. 1990). In spite of the fact that conditions for survival of stem cells have not yet been optimized, a proportion of the population does survive the infection and preselection steps, and these cells are able to function when transplanted into recipient animals.

3 Analysis of the Development of Uniquely Marked Stem Cells In Vivo

At appropriate times following reconstitution, high molecular weight DNA is prepared from tissues and cell populations harvested from the reconstituted mice. This DNA is first digested with an enzyme that does not cleave within the proviral sequences, then analyzed by standard Southern blotting techniques, and the resulting blot hybridized with a virus-specific probe. In most cases each of the hybridizing bands represents a unique viral integration site, and, provided that only one virus has infected a cell, each band can be regarded as a single clone.

3.1 Infection of Pluripotential Stem Cells

The first significant observation to be made from experiments involving the reconstitution of mice with retovirus-marked cells was that the different mature hematopoietic populations of these recipients contained the same viral integration site, indicatiing that they originated from a common pluripotential cell (KELLER et al. 1985; DICK et al. 1985; LEMISCHKA et al. 1986). These findings confirmed earlier chromosomal translocation experiments (WU et al. 1968; ABRAMSON et al. 1977) in demonstrating the existence of pluripotential stem cells capable of long-term reconstitution. These studies also indicated that relatively few of these repopulating cells are sufficient for maintaining the entire hematopoietic system of an irradiated mouse, thereby highlighting their extensive proliferative potential. One potential problem with this interpretation is that only the largest infected clones are detected by this method. It is also possible that uninfected clones as well as smaller infected ones that are below the limits of detection of this method are contributing to the hematopoietic system of these recipients. This is unlikely, however, as a comparison of the intensity of the bands on the Southern blots (representing the large marked clones) to the intensity of the control band (representing a single copy of the transduced gene) indicate that a significant proportion (50%–90%) of the hematopoietic cells in these animals are indeed derived from relatively few cells.

The presence of small numbers of stem cells in these recipients was somewhat unexpected as in many instances large numbers of infected marrow cells were transplanted. The limited number of functional stem cells in these animals is likely the result of a number of factors. There include death of these cells during infection in culture, their adherence to the virus-producing cells and their susceptibility to infection. Initially, the demonstration of small numbers of long-term repopulating stem cells in these mice was used as evidence in support of a clonal succession model of hematopoiesis (LEMISCHKA et al. 1986). The model proposed that stem cells had a limited life span, and therefore, at any given time, only a subset of the total population would be active (KAY 1965). With time or under conditions of hematologic stress, new cells would be drawn from the inactive or dormant pool to replace those which are dying, and, as a consequence, hematopoiesis would be maintained by a succession of short-lived clones. This constant turnover in the active stem cell pool would lead to a change, with time, in the clonal make-up of the different hematopoietic lineages. This model gained support from a number of different studies which demonstrated some fluctuation within the clonal make-up of the hematopoietic system of reconstitued mice (MINTZ et al. 1984; LEMISCHKA et al. 1986). The question of clonal change within the hematopoietic system has been addressed in detail by retrovirus marking experiments which employed protocols that allowed the analysis of the clonal make-up of the hematopoietic lineages of an individual recipient at different time points following reconstitution (SNODGRASS and KELLER 1987; KELLER and SNODGRASS 1990; JORDAN and LEMISCHKA 1990). The outcome of these experiments is outlined below.

3.2 Temporal Analysis of the Clonal Make-up of the Hematopoietic System

3.2.1 Clonal Fluctuation

Findings from these studies demonstrated that early following reconstitution, the hematopoietic system undergoes clonal fluctuation resulting from the disappearance of some of the clones from some or all lineages and the expansion of others clones into the various lineages (SNODGRASS and KELLER 1987; KELLER and SNODGRASS 1990; JORDAN and LEMISCHKA 1990). With time, however, the number of clones contributing to the hematopoietic system decreases to a relatively small number which persist for the lifetime of the animal (KELLER and SNODGRASS 1990; JORDAN and LEMISCHKA 1990). One interpretation of these observations is that both mature precursor cells and primitive long-term repopulating stem cells were infected and are active early after transplantation. The clones derived from the precursors disappear due to the limited life span of these cells while the clones derived from stem cells increase in size and dominate the hematopoietic system.

3.2.2 Contribution from Precursors

All precursors will contribute, to some extent, to the hematopietic populations of a transplanted recipient. The intermediate and late-stage precursors will generate small clones that are below the limits of detection of this method and will likely persist for only a few weeks. Those precursors which contribute significantly to the various lineages would represent a relatively early stage of development as they must be able to generate a clone that is sufficiently large to be detected. This interpretation raises several key questions concerning the developmental state of these precursors. First, what is their differentiation potential? Second, how long can they function in a transplanted recipient?

It is often assumed that all pluripotential, cells are long-term repopulating stem cells. However, it is quite possible that at least some precursors are also pluripotential, but unable to self-renew. A clone originating from such a cell would be detected in the myeloid and lymphoid populations, but with time would disappear from some or all of the lineages. Several different patterns of clonal change observed in mice reconstituted with marked bone marrow cells would be consistent with the existence of this type of precursor. One interesting example of clonal change was observed in an animal that was analyzed at both 7 and 20 weeks following reconstitution (Fig. 2). At 7 weeks, three integration sites (indicated as S1, S2 and S4) were present in the myeloid lineages,represented by mast cells (MAST 1), but were almost undetectable in the lymphoid lineages. These three bands likely represented a single clone as their intensities were the same in the different populations. By week 20, this clone was no longer present in the myeloid lineages (MAST 2 and BM 2), but was now easily detectable in cells of lymphoid origin (e.g., T2, SPL 2, and LN2). The cell which generated this clone was obviously pluripotential. The loss of the clone from the myeloid lineages

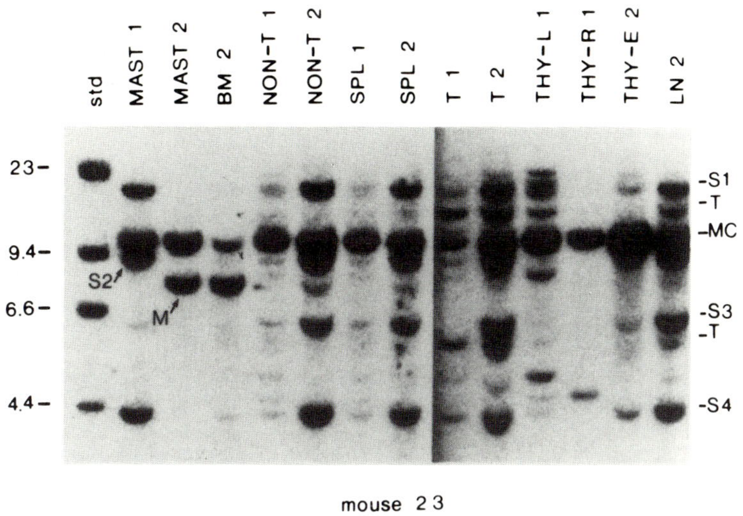

Fig. 2. Clonal analysis of the different hematopoietic populations of a repopulated animal at 7 weeks (1) and 5 months (2) following reconstitution. Southern blot analysis was performed with DNA from the various populations, and the resulting blot was hybridized with a *neo* specific probe. The clone designations are as follows: *MC*, major clone found in all lineages at both time points; *S*, clones originating from pluripotential cells, but not present in all lineages or tissues at both points in time; *M*, clone found predominantly in the myeloid lineages; *T*, clones found predominantly in the lymphoid lineages. The cell populations or tissues from which the DNA was isolated are indicated above each *track*. *MAST*, cells, *BM*, bone marrow, *NON-T*, splenocytes depleted of T cells; *SPL*, unfractioned splenocytes; *T*, spleenderived CON A blasts; *THY-L1*, *THY-L2*, thymic grafts under the left and right kidney capsule; *THY-E*, endogenous thymus; *LN*, lymph node. All tissues and cell populations from the 7-week time point are of splenic origin. (from SNODGRASS and KELLER 1987)

between the first and second analyses would be consistent with the interpretation that this clone was declining in size. It is possible that this clone would have also disappeared from the lymphoid lineages. A similar pattern was observed in another animal in which a clone was present in all lineages early following reconstitution and then could be found only in the lymphoid lineages at the later time point (SNODGRASS and KELLER 1987). In a more recent study JORDAN and LEMISCHKA (1990) followed clones in reconstituted mice for a longer period of time and found that those which disappeared initially from the myeloid population eventually also disappeared from the lymphoid populations. The cells which give rise to these clones can be considered as pluripotential precursors as they are able to generate progeny of the myeloid and lymphoid lineages, but are unable to sustain long-term hematopoiesis.

The loss of a clone from the myeloid lineages earlier than from the lymphoid populations is likely a reflection of the life span of cells from the respective lineages; most myeloid cells have a short life span, whereas a subpopulation of lymphocytes are long-lived (CRONKITE and VINCENT 1969; FREITAS et al. 1986). This interpretation could explain the finding that many of the resconstituted animals have integration sites restricted to the lymphoid lineages, in addition to those common to all of the hematopoietic populations (e.g., Fig. 2, clones marked "T"; Fig. 3, low molecular weight bands in tissues of mouse 34). Many of these clones which persist in the lymphoid lineages may have had a myeloid component at an earlier point in time. Myeloid restricted patterns are also found, however, at a lower frequency (Fig. 2, clone marked "m"). The finding that a clone can contribute to different lineages at different times clearly demonstrates that the presence of an integration site in a limited number of lineages is not necessarily an indication of a precursor or stem cell with restricted development potential. To demonstrate the existence of "resricted" stem cells, one would have to analyze the progeny of a particular cell at a number of different points in time to ensure that other lineages are not being produced.

Furhter evidence supporting the existence of primitive pluripotential precursor cells has been provided by a recent study of SPANGRUDE and JOHNSON (1990) in which they fractionated the $Thy^{lo}Lin^{-}Sca\text{-}1^{+}$ marrow population on the basis of the capacity of the cells to bind the fluorescent dye rhodamine-123(RH-123). The subpopulation which bound significant amounts of RH-123 ($RH\text{-}123^{hi}$) contained some multipotential cells that were able to repopulate the hematopoietic tissues of a primary recipient for short periods of time, but were unable to generate significant numbers of progeny to repopulate a secondary host. These $RH\text{-}123^{hi}$ cells can be considered as pluripotential precursors, able to generate progeny of myeloid and lymphoid lineages, but unable to sustain a functional hematopoietic clone for long periods of time. The population which bound very low or intermediate amounts of RH-123($RH\text{-}123^{lo+med}$) was able to reconstitute both the primary and secondary recipient and thus was likely to contain at least some long-term repopulating cells. The fact that these cells can be separated by physical parameters strongly supports the view that they represent distinct populations at different stages of development within the hematopoietic system.

The feature which distinguishes these multipotential precursors from the stem cells which provide long-term repopulation is their inability to maintain a functional hematopoietic clone for extended periods of time in vivo. The question of how long a clone derived from these precursor cells can persist will depend, to some extent, upon which lineages are being analyzed. From the analysis of a large number of clones over time, it appears as if the majority of them show some decline, either in some or in all lineages, within the first 2 months of reconstitution. However, as indicated earlier, they may persist in a limited number of lineages for considerably longer periods of time.

3.2.3 Establishment of Persistent Stem Cell-Derived Clones

As clones derived from the precursor population decline, those originating from the long-term repopulating stem cells increase in size to become the predominant clones of the hematopoietic system. The increase in size leading to the emergence of these clones in the various lineages is a factor which also contributes to the fluctuation observed early following reconstitution. The time required for the clones originating from these primitive stem cells to expand and dominate the hematopoietic system of a reconstituted animal is of key importance in defining a reconstitution assay that detects them rather than a precursor population. As with the precursor-derived clones, however, this timing will likely depend upon which lineages are being analyzed. To demonstrate the establishment and persistence of these clones, the hematopoietic system of the recipients must be analyzed over extended periods of time.

Figure 3 shows the results of the analysis from one such study. A primary recipient of retrovirus-marked bone marrow cells was analyzed at 2 and 10 months following reconstitution. At the 10-month time point, marrow was transplanted to a secondary recipient for a further 7 months. Clones originating from three different cells (or two cells, one with two integration sites) were present in the myeloid lineages of the primary recipient at two months following reconstitution. One of these was also present in the lymphoid populations of this animals at this time. The other two clones contributed very little to the lymphoid lineages at this stage of reconstitution. By 10 months following reconstitution, all three clones were predominant in all lineages and persisted in a secondary recipient for a further 7 months. The persistence of these clones over this extended period of time demonstrates that they were derived from long-term repopulating stem cells. The findings from this analysis indicate that clones from some of these stem cells are fully developed within 2 months of transplantation, whereas others are present in some but not all lineages at this time. In another animal, a developing clone was detected in the mast cells as well as in grafted thymus tissue by 7 weeks following reconstitution (SNODGRASS and

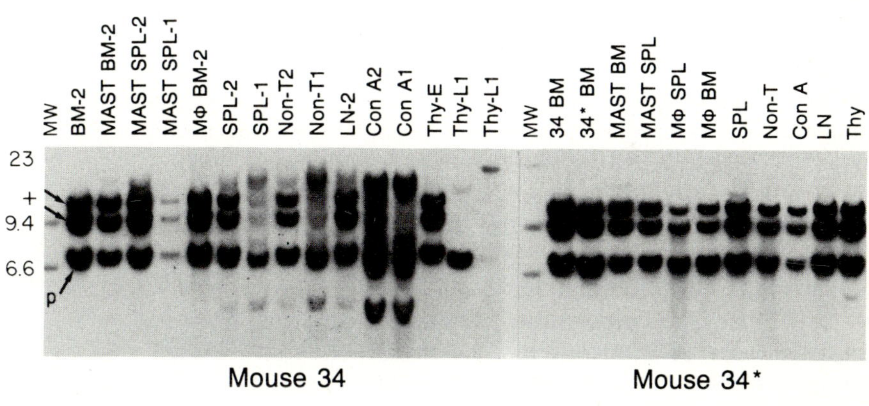

Fig. 3. Lineage analysis of a primary recipient (*mouse 34*) at 2 (1) and 10 (2) months following reconstitution and of a secondary recipient (*mouse* 34*) following 7 months of reconstitution. *Plus sign*, two clones that expanded in size between the first and second times of analysis, *p*, clone that is present in all tissues at both time points. The tissue and cell population designations are the same as for fig. 2 with the following expections: *MAST-BM*, bone marrow-derived mast cells; *MAST-SPL*, spleen-derived mast cells ; *MO BM*, bone marrow-derived macrophages; *MO SPL*, spleen-derived macrophages; *ConA* ConA-stimulated splenic T cells. (from KELLER and SNODGRASS 1990)

KELLER 1987). It was not present to any significant extent in the peripheral lymphoid lineages (splenic T and B cells) at this time. By 20 weeks following reconstitution, this clone had expanded into the peripheral lymphoid populations to become the predominant clone of the hematopoietic system. These findings as well as those from the study of JORDAN and LEMISCHKA (1990) indicate that clones originating from the stem cells can appear in different lineages at different points in time. The most common pattern of development is the establishment of a clone in the myeloid lineages prior to its presence in the peripheral lymphoid populations. This difference in the time of appearance might reflect a difference in the time required for the cells within these lineages to reach maturity, the lymphoid populations requiring a longer time due to the selection and deletion of appropriate clones.

Although the studies described above demonstrate that clones originating from long-term repopulating stem cells can appear in different lineages at different times, findings from the analysis of a large number of clones indicate that most become established in the various populations between 3 and 5 months following reconstitution (SNODGRASS and KELLER 1987; JORDAN and LEMISCHKA 1990). Beyond this point, they persist in a stable fashion for many months and in many instances, for the lifetime of the animal. It is possible that a hierarchy exist within this stem cell population, with more primitive ones being able to function for longer periods of time than the more mature ones. However, given the fact that most of these clones, once established, persist for many months, demonstration of this hierarcy would require analysis which extends over a period of several years.

The demonstration of stem cell with long-term repopulating capacity in these retrovirus marking studies is consistent with data from a number of other studies which also indicate that such cells exist (ABRAMSON et al. 1977, BOGGS et al. 1982, 1984, HARRISON et al. 1988). Taken together, these findings do not support a clonal succession model of hematopoiesis, but rather indicate that, at least in irradiated reconstituted mice, the hematopoietic system is maintained by a cohort of long-lived stem cells.

3.3 Expansion of the Stem Cell Pool

The final, and perhaps most intriguing observation from these unique marker studies, comes from experiments in which bone marrow cells from one primary recipient were used to reconstitute multiple secondary animals (KELLER and SNODGRASS 1990, CAPEL et al. 1990). In each instance, the clones present in the primary animals were found in the hematopoietic tissues of more than one secondary host, suggesting that the stem cell population which mediates long-term repopulation can actually expand during the course of hematopoietic reconstitution. The cells transplanted into the secondary animals appear to have true stem cell characteristics as they are pluripotential and able to function in these recipients for extended periods of time, in some instances for year (KELLER and SNODGRASS 1990; CAPEL et al. 1990). The extent to which this population can expand is presently unknown. In our initial study we observed the same clones in two different secondary recipients (KELLER and SNODGRASS 1990), while in another study, CAPEL et al. (1990) found the same clones in five secondary animals, indicating that the initial infected cell could generate a minimum of at least five new stem cells. This same approach has been used to demonstrate that the population which mediates short-term repopulation and possibly even the long-term repopulating cells undergoes some expansion in long-term bone marrow cultures (FRASER et al. 1990). These findings are the first indication that this stem cell population can actually expand in number. Future studies need to address questions concerning the extent to which this population can expand, the kinetics of expansion, and the capacity to which one can influence this expansion with various growth factors.

4 Conclusions

The findings from these retrovirus marker experiments together with those from a number of other studies (HARRISON et al. 1988; JONES et al. 1990; SPANGRUDE and JOHNSON 1990) have given us a better understanding of the structure of the early stages of development within the hematopoietic system. It is now clear that a population of long-term repopulating stem cells exists which can function for at least the lifetime of a mouse and perhaps longer. These cells can be considered as the true hematopoietic stem cells and are likely to represent the most primitive cells within the system. Owing to their early developmental stage, these primitive stem cells do not generate significant numbers of differentiated progeny immediately following transplantation to an irradiated recipient. As a consequence, these cells are unable to provide short-term reconstitution or protection from radiation death. Clones of differentiated cells originating from these long-term repopulating cells begin to emerge in the various lineages sometime between 1 and 2 months following transplantation. Their expansion to all lineages can take as long as 4–5 months. Thus, to assay the progeny of this primitive stem cell population, transplanted recipients must be analyzed at several time points beyond 2 months of reconstitution.

These studies have also provided evidence for the existence of a pluripotential precursor population that functions to provide short-term reconstitution, contributing progeny to all lineages (myeloid and lymphoid) early after transplantation. These precursors differ from the stem cells in that they are unable to sustain a functional hematopoietic clone for long periods of time in these recipients. The findings would indicate that the contribution from these precursors begins to wane by 2 months following reconstitution, although, as indicated earlier, they can persist in a limited number of lineages (lymphoid) for significantly longer periods of time. The existence of these pluripotential precursors with short-term repopulating potential indicates that pluripotentiality alone is not sufficient to define a stem cell population with long-term repopulating capacity. To date, the only unique characteristic of these stem cells is their ability to sustain hematopoiesis over extended periods of time, and the only means of assaying them is the long-term reconstitution of the hematopoietic system of a recipient animal.

Finally, the marking studies have provided evidence that the stem cell population can actually expand during the course of reconstitution, indicating that these cells can proliferate without differentiating. If this potential is substantial, it is possible that these cells can be maintained in culture in an undifferentiated state, given the appropriate conditions.

5 Future Directions

Although the unique marker studies were instrumental in defining the developmental potential of these early hematopoietic cells, the approach does have certain limitations which make it less useful for more detailed analysis of the stem cell population. The major disadvantage of the retrovirus marking approach is that only those cells which can be infected with the retrovirus can be analyzed, and thus it is not suitable for quantitative studies on the entire stem cell population. In addition, the progeny of the infected stem cell is analyzed following long periods of time in a recipient animal, making it impossible to address issues concerning lineage commitment at the level of the stem cell. To progress further in our understanding of the mechanisms underlying the control of proliferation and development of this primitive population of cells, other approachs must be followed. These include the development of protocols for further enrichment and purification of long-term reconstituting cells, the identification of conditions that will promote their growth in culture, and finally the development of quantitative in vitro assays for these cells.

One new approach which offers the unique opportunity to study developmental aspects of the hematopoietic system is the generation of hematopoietic cells from embryonic stem (ES) cells in culture. ES cells are totipotent stem cells that can be maintained as undifferentiated populations in culture (Evans and Kaufman 1981; Martin 1981). When introduced into mouse blastocysts these cultured ES cells are able to contribute to all of the lineages of the developing mouse (Bradley et al. 1984). In addition to their developmental potential in vivo, ES cells are also able to give rise to differentiated progeny in culture, and among these progeny one can identify cells of hematopoietic origin (Evans and Kaufman 1981; Doetschman et al. 1985).

In the initial studies, ES cells were induced to differentiate by maintaining them in liquid culture in the absence of feeder cells. Under these conditions they formed large fluid-filled sacs known as embryoid bodies (EB). Blood islands, consisting predominantly of embryonic erythroid cells, developed within some of these EB (DOETSCHMAN et al. 1985). Efficient development of these blood islands was dependent upon the addition of preselected batches of human chord serum to the cultures. More recent studies using a similar protocol for differentiation have demonstrated that, in addition to erythroid cells, precursors from most of the myeloid lineages develop within the EB (SCHMITT et al. 1991). The development of these precursors is however, also dependent upon the addition of human chord serum.

In an attempt to better define the conditions which can influence the development of hematopoietic cells from ES cells in culture, we have recently established a new protocol which promotes their differentiation (WILES and KELLER 1991). Rather than maintaining ES cells in liquid culture, they are seeded directly into methylcellulose-containing medium. In these semisolid cultures, the ES cells grow and differentiate to form a colony which consists of a variety of cell types including those of the hematopoietic system. The advantages of this semisolid culture system over the previous method of ES differentiatioin are that the culture conditions are relatively well defined, the developing colony is immobilized allowing quantitative analysis, a much higher proportion of the ES population will generate hematopoietic cells, and the kinetics of hematopoietic development is more reproducible.

Between 30% and 60% of the cells from the CCE ES line (ROBERTSON et al. 1986) seeded into methylcelluose will develop into a colony of differentiated cells, and of these 70%–90% will contain visible hematopoietic cells. Precursors of the macrophage (M-CFC) and erythroid (E-CFC) lineages can be detected with in these colonies by day 6 of differentiation. By day 9 of differentiation, in addition to the macrophage and erythroid precursors, pluripotential precursors able to generate multiple myeloid lineages (MIX-CFC), bipotential precursors commited to neutrophil and macrophage development (GM-CFC), as well as neutrophil committed precursors (G-CFC), are found within these ES-derived colonies. Mast cell precursors are the latest to develop, being present in colonies following 14 days of differentiation. This sequence of precursor development is highly reproducible and appears to be independent of exogenous growth factors, including those present in human chord serum. Their development could depend upon factors produced within these colonies. Current studies are now aimed at determining when the long-term repopulating stem cells generated within the ES-derived colonies.

Although these studies are in their early stages, the efficient development of hematopoietic cells from ES cells in the semisolid cultures provides a unique system with which to analyze the early events involved in the establishment of the hematopoietic system. In addition, it offers a novel method for introducing genes into hematopoietic cells and for addressing questions concerning the effect of the expression of various genes on the developing hematopoietic lineages.

Acknowledgements. I wish to thank Marion Kenndy, Kyunghee Choi, and Norman Iscove for helpful discussions and for critical reading of the manuscript.

References

Abramson S, Miller RG, Phillips RA (1977) The identification in adult bone marrow of pluripotent and restricted stem cells of the myeloid and lymphoid and systems. J Exp Med 145: 1567–1579

Bearpark AD, Gordon MY (1989) Adhesive properties distinguish sub-populations of haemopoietic stem cells with different spleen colony-forming and marrow repopulating capacities. Bone Marrow Transplant 4: 625–628

Boggs D, Boggs SS, Saxe DF, Gress LA, Canfield DR (1982) Hematopoietic stem cells with high proliferative potential. J Clin Invest 70: 242–253

Boggs DR, Saxe DF, Boggs SS (1984) Aging and hematopoiesis. II. The ability of bone marrow cells from young and aged mice to cure and maintain cure in W/W^v. Transplantation 37: 300

Bradley A, Evans M, Kaufman MH, Robertson E (1984) Formation of germ-line chimeras from embroy-derived teratocarcinoma cell lines. Nature 309: 255–256

Capel B, Hawley RG, Mintz B (1990) Long- and short-lived murine hematopoietic stem cell clones individually identified with retroviral integration markers. Blood 75: 2267–2270

Cronkite EP, Vincent PC (1969) Granulocytopoiesis. Ser Haematol 2: 3–43

Dick JE, Magli MC, Huszar D, Phillips RA, Bernstein A (1985) Introduction of a selectable gene into primitive stem cells capable of long-term reconstitution of the hemopoietic system of W/W^vv mice. Cell 42: 71–79

Doetschman TC, Eistetter H, Katz M, Schmidt W, Kemler R (1985) The in vitro development of blastocyst-derived embryonic stem cell lines: formation of visceral yolk sac, blood islands and myocardium. J Embryol Exp Morphol 87: 27–45

Evans MJ, Kaufman MH (1981) Establishment in culture of pluripotential cells from mouse embryos. Nature 292: 154–156

Fraser CC, Eaves CJ, Szilvassy SJ, Humphries RK (1990) Expansion in vitro of retrovirally marked totipotent hematopoietic stem cells. Blood 76: 1071–1076

Freitas AA, Rochaa B, Coutinho AA (1986) Lymphocyte population kinetics in the mouse. Immunol Rev 91: 5–37

Harrison DE (1980) Competitive repopulation: a new assay for long-term stem cell functional capacity. Blood 55: 77–81

Harrison DE, Astle CS, Lerner C (1988) Number and continuous proliferative pattern of transplanted primitive immunohematopoietic stem cells. Proc Natl Acad Sci USA 85: 822–826

Hodgson GS, Bradley RS (1979) Properties of hematopoietic stem cells surviving 5-fluorouracil treatment: evidence for a pre-CFU-S cell? 281: 381–382

Huang E, Nocka K, Beier DR, Chu T-Y, Buck J, Lahm HW, Wellner D, Leder P, Besmer P (1990) The hematopoietic growth factor KL is encoded by the Sl locus and is the ligand of the c-kit receptor, the gene product of the W locus. Cell 63: 225–233

Iscove NN, Yan X (1990) Precursors (pre-CFC_{multi} of multilineage hemtopoietic colony-forming cells quantitated in vitro. J Immunol 145: 190–195

Jones RJ, Wagner JE, Celano P, Zicha MS, Sharkis SJ (1990) Separation of pluripotent haematopoietic stem cells from spleen colony-forming cells. Nature 347: 188–189

Jorden CT, Lemischka IR (1990) Clonal and systemic analysis of long-term hematopoiesis in the mouse. Genes Dev 4: 220–232

Kay HEM (1965) How many cell-generations? Lancet ii: 418–419

Keller G, Snodgrass R (1990) Life span of multipotential hematopoietic stem cells in vivo. J Exp Med 171: 1407–1418

Keller G, Paige C, Gilboa E, Wagner EF (1985) Expression of a foreign gene in myeloid and lymphoid cells derived from multipotent haematopoietic precursors. Nature 318: 149–154

Kerk DK, Henry EA, Eaves AC, Eaves CJ (1985) Two classes of primitive pluripotent hemtopoietic progenitor cells: seperation by adherence. J Cell Physiol 125: 127–134

Kitamura Y, Yokoyama M, Matsuda H, Okno T, Mori JK (1981) Spleen-colony forming cells as common precursors for tissue mast cells and granulocytes. Nature 291: 159–160

Lemischka IR, Raulet DH, Mulligan RC (1986) Developmental potential and dynamic behavior of hematopoietic stem cells. Cell 45: 917–927

Lerner C, Harrison DE (1990) 5-Fluoruracil spares hematopoietic stem cells responsible for long-term repopulation. Exp Hematol 8: 114–118

Magli MC, Iscove NN, Odartchenko (1982) Transient nature of early haematopoietic spleen colonies. Nature 295: 527–529

Martin GR (1981) Isolation of a pluripotent cell line from early mouse embryos cultured in medium conditioned by teratocarcinoma stem cells. Proc Natl Acad Sci USA 78: 7634–7638

Metcalf D (1977) Hemtopoietic colonies. Invitrocoling of normal and leukemic cells. Springer, Berlin Heidelberg New York (Recent results in cancer research, vol 61)

Metcalf D (1984) The hematopoietic colony stimulating factors. Elsevier, Amsterdam

Mintz B, Anthony K, Litwin S (1984) Monoclonal derivation of mouse myeloid and lymphoid lineages from totipotent hematopoietic stem cells experimentally engrafted in fetal hosts. Proc Natl Acad Sci USA 81: 7835–7839

Müller-Sieburg CE, Townsend K, Weissman IL, Rennick D (1988) Proliferation and differentiation of highly enriched mouse hematopoietic stem cells and progenitor cells in response to defined growth factors. J Exp Med 167: 1825–1840

Nakahata T, Ogawa M (1982) Identification in culture of a class of hemopoietic colony-forming units with extensive capability to self-renew and generate multipotential hemopoietic colonies. Proc Natl Acad Sci USA 79: 3843–3847

Paige CJ, Kincade PW, Moore MAS, Lee G (1979) The fate of fetal and adult B-cell progenitors grafted into immunodeficient CBA/N mice. J Exp Med 150: 548–563

Ploemacher RE, Brons RHC (1989) Separation of CFU-S from primitive cells responsible for reconstitution of the bone marrow hematopoietic stem cell compartment following irradiation: evidence for a pre-CFU-S cell. Exp Hematol 17: 263–266

Pragnell IB, Wright EG, Lorimore SA, Adam J, Rosendaal M, Delamarter JF, Freshney M, Eckmann L, Sproul A, Wilkie N (1988) The effect of stem cell proliferation regulators demonstrated with an in vitro assay. Blood 72: 196–201

Robertson E, Bradley A, Kuehn M, Evans M (1986) Germ line transmission of genes introduced into cultured pluripotential cells by retoviral vector. Nature 323: 445–448

Russel ES (1979) Hereditary anemias of the mouse: a review for the geneticist. Adv Genet 20: 357–459

Schmitt RM, Bruyns E, Snodgrass HR (1991) Hematopoietic development of embryonic stem cells in vitro: cytokine and receptor gene expression. Genes Development 5: 728–740

Snodgrass R, Keller G (1987) Clonal fluctuation within the hematopoietic system of mice reconstituted with retrovirus-infected stem cells. EMBO J 6: 3955–3960

Spangrude GJ, Johnson GR (1990) Resting and activated subsets of mouse multipotent hematopoietic stem cells. Proc Natl Acad Sci USA 87: 7433–7437

Spangrude GJ, Heimfeld S, Weissman IL (1988) Purification and characterization of mouse hematopoietic stem cells. Science 241: 58–62

Sutherland HJ, Lansdrop PM, Henkelman DH, Eaves AC, Eaves CJ (1990) Functional characterization of individual human hematopoietic stem cells cultured at limiting dilution on supportive marrow stromal layers. Proc Natl Acad Sci 87: 3584–3588

Till JE, McCulloch EA (1961) A direct measurement of the radiation sensitivity of normal mouse bone marrow cells. Readiat Res 14: 213–222

Varmus HE, Padgett T, Heasley S, Simon G, Bishop JM (1977) Cellular functions are required for the synthesis and integration of avain sarcoma virus-specfic DNA. Cell 11: 307–319

Visser JWM, Bauman JGJ, Mulder AH, Eliason JF, Leeuw AM (1984) Isolation of murine pluripotent hemopoietic stem cells. J exp Med 59: 1576–1590

Wiles MVN, Keller G (1991) Multiple hematopoietic lineages develop from embryonic stem (ES) cells in culture. Development 11: 259–267

Williams DE, Eisenman J, Baird A, Rauch C, Van Ness K, March CJ, Park LS, Martin U, Mochizuki DY, Boswell HS, Burgess GS, Cosman D, Lyman SD (1990) Identification of a ligand for the c-*kit* proto-oncogene. Cell 63: 167–174

Wu AM, Till JE, Siminovitch L, McCulloch EA (1968) Cytological evidence for a relationship between normal hematopoietic colony-forming cells and cells of the lymphoid system. J Exp Med 127: 455–464

Zsebo KM, Wypych J, MCNiece IK, Lu HS, Smith KA, Karkare SB, Sachdev RK, Yuschenkoff VN, Birkett NC, Williams LR, Satyagal VN, Tung W, Bosselman RA, Mendiaz EA, Langley KE (1990) Identification, purification, and biological characterization of hematopoietic stem cell factor from buffalo rat liver-conditioned medium. Cell 63: 195–201

What We Have Learned From Retroviral Marking of Hematopoietic Stem Cells

I. R. LEMISCHKA

1 Introduction

In contrast to most vertebrate somatic cell populations the hematopoietic system requires that commitment and differentiation processes occur continuously throughout adult life. Although similar in this regard to the skin, intestinal epithelium, and spermatogenic systems, hematopoiesis is more complex in that at least eight cell lineages must be produced. The most primitive cell responsible for hematopoiesis is pluripotent in its ability to contribute to the various blood cell types (DEXTER and SPOONCER 1987; METCALF 1988). In addition to this remarkable ability, such a stem cell must also possess the capacity to self-renew in order to guarantee lifelong hematopoiesis. The central problem in hematopoiesis is, therefore, not only to understand mechanisms mediating differentiation, but also those that facilitate undifferentiated proliferation.

In most models of hematopoietic development, partitioning of developmental potential is linked to a decrease in self-renewal capacity (OGAWA et al. 1983). Stochastic (KORN et al. 1973; NAKAHATA et al. 1982) as well as inductive mechanisms (TRENTIN 1970) have been proposed to govern the establishment of the hematopoietic stem/progenitor cell hierarchy. Virtually all models of primitive stem cell behavior rely on extrapolation from in vitro studies. These experiments generally involve progenitors considered to be closely linked to fully

Department of Molecular Biology, Princeton University, Princeton, NJ 08544, USA

Current Topics in Microbiology and Immunology, Vol. 177

differentiated cells and therefore describe stem cells only indirectly (METCALF 1984). Critical to an understanding of hematopoietic development is, therefore, a *direct* in vivo description of primitive pluripotent stem cell behavior at *early* times in its clonal expansion.

2 In Vivo Stem Cell Assays

Measurement of stem cell properties relies on functional assays. Any valid assay system must assess at least two properties: (a) developmental potential; (b) self-renewal capacity. In practice, these assays involve transplantation of appropriately "marked" donor tissue into radiation or genetically ablated host hematopoietic systems. All properties of stem cells are retroactively inferred from analysis of mature myeloid and lymphoid progeny of transplanted stem cells, that is, the end products of a complex program of cellular differentiation.

Critical to any stem cell assay system are markers. Two categories of markers have been traditionally employed: (a) host vs. donor markers; and (b) markers that are clonotypic, that is, they are unique to each marked cell. The ability to measure clonal properties of individual stem cells using host vs. donor markers requires strict limiting-dilution engraftment. Such studies would be greatly facilitated by a pure population of stem cells. Numerous studies have deomonstrated a 10–1000-fold enrichment of stem cells (VISSER et al. 1984; JOHNSON and NICOLA 1984; MÜLLER-SIEBURG et al. 1988; SPANGRUDE et al. 1988; PLOEMACHER and BRONS 1989; SZILVASSY et al. 1989a; JORDAN et al. 1990; SMITH et al. 1991). However, convincing evidence for a homogeneous stem cell population is not available. An additional limitation of host vs. donor markers is that these do not allow an analysis of the possible interplay of individual stem cells in a given in vivo context. Thus, the dynamics of the stem cell population as a whole cannot be elucidated.

An alternate approach relies on the introduction of clonotypic markers into a population known to contain stem cells (ABRAMSON et al. 1977). These markers *uniquely* "label" each stem cell in a given population. The in vivo function of a stem cell is inferred *solely* by distribution of the unique marker in mature tissue. While indirect, this method is clonally precise and permits analyses of multiple stem cells in a given hematopoietic system.

In recent years, recombinant retroviruses have been employed as an efficient means of marking stem cells (DICK et al. 1985; KELLER et al. 1985; LEMISCHKA et al. 1986). Viral integration is largely random, therefore each stem cell carries a distinct proviral integrant. DNA analysis can be performed, in a quantitative fashion, on bulk populations (not limited to those in mitosis as in karyotype marker studies). Furthermore, retroviruses can be engineered to be immobile, thus insuring that all marking takes place only at the initiation of an experiment. Given the ease with which proviral markers were initially introduced into stem cells, it was anticipated that this technique would yield a detailed fate-map describing commitment decisions occurring early in hematopoietic development. This article will assess what has been learned in these studies, what we can expect to learn, and how this strategy may be extended.

3 Defining a Stem Cell in Terms of Proviral Markers

How are the properties of stem cells elucidated by employing proviral markers? Following engraftment of retrovirally marked cells into irradiated animals, mature tissues are analyzed for DNA segments containing the provirus. A detectable provirus is diagnostic of at least two events: (a) commitment to given pathway(s) of differentiation (because only mature tissues are sampled); (b) substantial proliferation, most likely following such a commitment(s) (because each unique provirus was initially present in a single cell and the ability to detect a marker relies on the ability of that cell to give rise to a sizeable clone). A stem cell is therefore defined by these criteria as any cell capable of giving rise to a clone of differentiated progeny large enough to result in a detectable proviral integrant [at least 5% of a sampled population (LEMISCHKA et al. 1986)]. The molarity of individual integrants quantitates the absolute fraction of a cell population derived from a single clone (LEMISCHKA et al. 1986). Given the limited degree to which most mature hematopoietic cells can proliferate, this also implies a degree of self-renewal activity following engraftment (see below).

The developmental potential of a stem cell is measured by the *distribution* of an integrant in a variety of cell lineages. The presence of the same proviral integrant in all sampled cell lineages (e.g., T cells, B cells, and macrophages) proves that the initially marked cell had the developmental potential to differentiate at least along these pathways. Similarly, an integrant observed only in certain lineages may be diagnostic of a lineage-restricted stem cell.

Given that a detectable provirus defines commitment and differentiation, how can self-renewal be measured in this system? Self-renewal, by definition, is the "regeneration" of stem cell potential which will contribute to hematopoiesis at a future time. Thus, it follows that self-renewal can only be measured as a function of time. A proviral integrant which persists over time in mature tissues is indicative of self-renewal activity. This is especially true if short-lived myeloid tissues are analyzed. Maintenance in other lineages, although suggestive of self-renewal at the most primitive stem cell level, must be cautiously interpreted, given the persistence of long-lived mature lymphocytes (FREITAS et al. 1986).

Strictly defined, the outcome of a self-renewal process is a clone of uncommitted stem cells. Such a clone can be demonstrated by retransplantation of marked marrow, obtained from a primary reconstituted host, into *multiple* secondary recipients. Detection of a proviral integrant present in the primary host in *more than one* secondary host serves as a direct demonstration of self-renewal.

In summary, the *presence*, *molarity*, and *distribution* of a proviral integrant in mature tissues, analyzed at a single point in time, define a stem cell in terms of its minimal defferentiation and proliferation potential. Analysis of a reconstituted system at multiple points in time and by retransplantation into secondary hosts can shed light on stem cell self-renewal.

In vivo assay systems which analyze only mature hematopoietic cell populations only measure the "readout" of a cell differentiation program. Is it also possible to gain insight into the nature of the actual mechanisms that govern this differentiation program? Detection of *fluctuations* (either qualitative or quantitative) in the behavior of a proviral integrant over time can provide mechanistic insights. As stated above, a proviral integrant can only be detected if the initially marked cell has undergone substantial clonal expansion. It follows, therefore, that integrant fluctuations occurring over time are likely to reflect mechanisms acting on highly clonogenic stem cells.

4 Early Retroviral Marking Studies

Initial viral marking experiments confirmed as well as extended classical studies. It was not a surprise that proviral markers confirmed the existence and prevalent role of pluripotent stem cells. More unexpected was the frequent activity of few (sometimes one) stem cell clones in contributing to the majority of cells of all lineages. Thus, single stem cells were sufficient to generate a hematopoietic system which was monoclonal at the time of analysis. This remarkable capability had been suggested previously by limiting-dilution (BOGGS et al. 1982; HARRISON et al. 1988) and karyotype marker studies (MICKLEM et al. 1983, 1987). The retroviral marking studies provided a minimal and direct quantitative estimate of the clonal proliferative capacity of the stem cell and its progeny.

The initial studies also described a wide variety of stem cell behaviors. In brief, proviral integrants were observed in single lineages, subsets of lineages, or members of a cell lineage obtained from different anatomic locations (LEMISCHKA et al. 1986). These restricted marker distributions could reflect gene transfer into restricted stem cells and thus define a wide spectrum of intrinsically distinct stem cell entities. The existence of stem cells restricted in developmental potential had been proposed previously (ABRAMSON et al. 1977). In addition, restricted cells constitute an obligate intermediate in most models of hematopoietic development (DEXTER and SPOONCER 1987). The hope of the early studies was to derive a fate-map of hematopoiesis by assuming the existence and retroviral infectibility of all products of commitment decisions in the donor tissue.

A particularly informative class of stem cell contributory behavior may be termed "quantitatively restricted " (LEMISCHKA et al. 1986). A given retroviral marker was observed to be much more prevalent in some lineages than in others. The unequal representation of a given proviral marker in different myeloid and lymphoid lineages suggested that a pluripotent stem cell may be skewed in its contribution to mature cell types, favoring some over others. Because more sensitive and quantitative analytical techniques were used in these studies than in older experiments, such stem cell clones might have been previously classed as absolutely restricted. An extension of this reasoning would argue that stem cell clones whose progeny are not visible in all myeloid and lymphoid lineages are not intrinsically restricted in potential, but, rather, may represent extreme examples of skewed contribution by an unrestricted pluripotent stem cell. Thus, it could be argued that hematopoietic activity in a reconstituted animal may represent, in large part, if not exclusively, a series of pluripotent stem cell clones arranged in a quantitative continuum with respect to manifesting their complete intrinsic potential. Therefore, models based on reconstitution studies, postulating the existence of distinct lymphoid stem cells and myeloid stem cells, must be viewed with caution.

The wide quantitative spectrum of stem cell activity, ranging from complete clonal reconstitution of all lineages to apparent reconstitution of single lineages, was most consistent with stochastic models of committment. Previously, such models were supported exclusively by short-term in vivo spleen colony-forming unit (CFU-S) studies (KORN et al. 1973) or by in vitro experiments (NAKAHATA et al. 1982). These models predict that, at any point in time, a pluripotent stem cell may "choose" to commit to single or subsets of lineages and that the exact "choices" may vary with time. Therefore, an accurate assessment of stem cell potential to proliferate and differentiate can only be obtained as a function of time.

Several other observations in these early studies suggested that a temporal mode of analysis was necessary. For example, it was found that immature and mature members of a given lineage, obtained from single hosts, contained different spectra of integrants and were thus repopulated by progeny of different stem cell clones (LEMISCHKA et al. 1986). These observations predicted that not all genetically marked stem cells assume hematopoietic function concurrently upon engraftment. This, in turn, suggested that dormant (undetectable by analysis of proviral integrants in mature cells) stem cells may exist in a given reconstituted host.

Earlier, classical studies had proposed that the hematopoietic system is organized in a clonally successive manner (KAY 1965; MICKLEM et al. 1983, 1987; MINTZ et al. 1984). The basic feature of all clonal succession models is that a small subset of the available stem cell compartment is hematopoietically actice at any given time. One prediction of such models may be that, while collectively stem cells in an organism have far greater than lifelong hematopoietic ability, individual stem cells have a much more limited lifespan. Alternatively, stem cells that have ceased to contribute may simply have returned to a quiescent state.

Early attempts to "uncover" dormant stem cells relied on retransplantation of bone marrow from primary hosts into secondary recipients (LEMISCHKA et al. 1986). Novel proviral integrants in the secondary host did indeed demonstrate the presence of relatively noncontributing stem cells in the primary recipients. In these same studies the maintenance of other integrants in both primary and secondary hosts argued that some stem cells had undergone a degree of self-renewal in the primary animal.

Two early studies, involving either partial removal of myeloid-lymphoid tissues (SNODGRASS and KELLER 1987) or sequential sampling of small volumes of peripheral blood (LEMISCHKA et al. 1986), examined the fate of marked stem cell clones in single hosts at different points in time. In both studies, dramatic clonal fluctuations over time were observed. These fluctuations could be categorized as increases, decreases, or lineage shifts in the contributory activity of individual stem cell clones. Broadly interpreted, these data provided support for clonal succession models of stem cell utilization. Cautiously interpreted, the observed fluctuations represented a hematopoietic system not yet functioning at equilibrium. In any case, in vivo dynamics among a population of individual marked stem cell clones or their progeny were directly established.

Taken together, the early studies had underscored the necessity of long-term, sequential, and lineage-specific analyses in deriving an accurate picture of stem cell behavior and in gaining insight into the overall systemic properties of reconstituted hematopoiesis.

5 Recent Retroviral Marking Studies

The first of the more recent studies (CAPEL et al. 1989) substantially extended the period of analysis to 9–11 months post-reconstitution. Virally marked bone narrow of fetal liver was engrafted into anemic (W/W^v) recipients. Animals were first sampled at 3 months, and later at 9–11 months. Identical proviral markers (observed at these two time points) suggested that individual stem cell clones can persist for extremely long time periods. Continued presence of an integrant in myeloid tissue further supported this interpretation. The data suggest that individual stem cell clones have the capability (self-renewal potential) to contribute to substantial portions of mature cells *continuously* over time. However, because these studies consisted of only two time points, rigorous conclusions regarding *continuous* stem cell activity were not possible. In addition, these experiments did not demonstrate that fluctuations can occur over the entire time period and therefore did not demonstrate clonal succession as a hallmark of steady-state, long-term hematopoiesis.

A subsequent study introduced multi-timepoint long-term analysis (JORDAN and LEMISCHKA 1990). A key feature of the experimental design was the ability to sample the four major peripheral blood cell lineages at 6–8 week intervals for periods of up to 16 months. A large number of marked clones (142 proviral integrants) were analyzed sequentially in the context of 63 reconstituted animals. Several important conclusions can be drawn regarding individual stem cell function, as well as the overall dynamics of the hematopoietic system. When viewed over long time intervals, hematopoiesis is often oligo-monoclonal in stem cell origin. That is, a small number of totipotent stem cell clones function *continuously* and *stably* to repopulate mature cell populations. Because this behavior is largely independent of stem cell source (bone marrow or fetal liver) and inoculum size, it may be a general feature of reconstituted hematopoiesis. All stable clones contribute equally to all lineages, although the absolute degree of such contribution can vary. This suggests that other stem cell clones exist whose hematopoietic activity is below detectable levels. Stem cell activity is therefore likely to be distributed in a quantitative continuum and not segregated into absolutely "on" or "off" states.

A key conclusion from these studies is that *individual* stem cells present at the time of engraftment are both necessary and sufficient for permanent hematopoiesis. Thus, at least in theory, it should be possible to engraft with single, physically isolated stem cells. The existence of minimally active stem cells further suggests that the entire complement of transplanted stem cells is not utilized either at single points in time or during near lifelong intervals, thus calling into question clonal succession models.

In addition to stable pluripotent stem cell activity, the current studies also described a variety of disparate contributions and clonal fluctuations. The observed increases or decreases in clonal contribution displayed a range of kinetic and lineage-specific differences. Indeed, all the phenomena observed in the earlier studies were also currently observed. Interestingly, nearly all of these occur in the first 4–6 months post-engraftment. Therefore, they most likely reflect a system that is not yet at equilibrium rather than intrinsically distinct classes of stem cells. As the time frame of the early experiments was also only 4–6 months, these results must be reinterpreted from this point of view.

Based on the recent studies, post-transplantation hematopoiesis can be divided into two very different phases: a 4–6 months interval of clonal fluctuation and a subsequent, much longer, interval characterized by stable activity of few pluripotent stem cell clones. Analysis of individual recipients during both time intervals reveals that long-term stability often results from the selection of a few initially engrafted stem cells for eventual dominant and permanent function. Such stem cells fall into three catagories: (a) those which remain stable during the entire time period; (b) those that display initial fluctuations in hematopoietic activity; (c) those which are initially minor or invisible contributors.

6 A Model for Stem Cell Behavior in Reconstituted Hematopoiesis

Is it possible to synthesize the temporal description of stem cell behavior into an overall model of reconstituted hematopoiesis? As stated previously, discrete proviral integrants define highly clonogenic cells. Therefore it is likely that the fluctuation or stability of proviral integrants over time is a consequence of mechanisms which act directly on these cells. A priori, such clonogenic cells may be committed to particular lineages or they may be pluripotent. However, the vast majority of analyzed clones are pluripotent. Thus, it is likely that these clonogenic cells are uncommitted and can self-renew. The observed coordinate nature of proviral fluctuation in clonally related myeloid and lymphoid lineages further argues that the reponsible mechanisms act on the most primitive cells. A valid model must therefore explain developmentally restricted clonal contribution, as well as clonal fluctuation, in terms of mechanisms acting on a population of intrinsically unrestricted stem cells. A simple model (shown in Fig. 1) accounts for most, if not all, temporally distributed developmental behavior. The temporal segregation of unstable and stable clonal behavior is consistent with stochastic commitment vs. self-renewal mechanisms operating on an expanding pool (clone) of target stem cells.

A stem cell which predominantly self-renews following engraftment would eventually give rise to a substantially large clone of pluripotent stem cells (Fig.

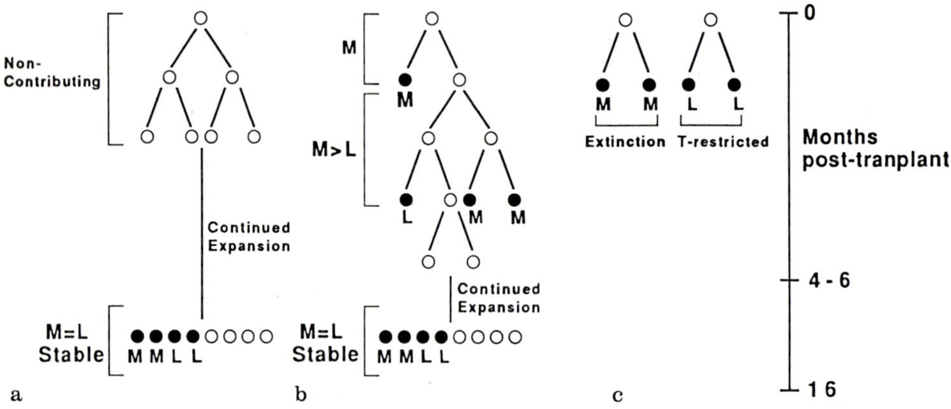

Fig. 1 a–c. A model of engrafted totipotent stem cell behavior. Categories of clonal behavior: **a** stable contribution as a consequence of self-renewal; **b** disparate/fluctuating contribution as a result of self-renewal and commitment; and **c** decreasing/restricted contribution due to commitment. *Open circles*, uncommitted stem cells; *closed circles*, committed progenitors; (*M*, myeloid committed; *L*, lymphoid committed). On the *left* or *underneath* each category, in *brackets*, are the classe of observed behavior and on the *right* is the time post-engraftment. **a** Continuous stable contribution is a consequence of an initial period of self-renewal (designated by an expansion of *open circles*) shortly after engraftment. Such clones would appear to be noncontributing during this time (prior to our first sampling point). Subsequent stochastic commitment of stem cells from an expanded pool of uncommitted cells gives rise to equal and stable lineage contribution. **b** Unequal or fluctuating contribution of a stem cell clone is observed when commitment decisions occur in parallel with significant self-renewal. Such clones would initially display restricted (*M* only) or disparate ($M > L$) contribution. Gradual expansion of the uncommitted pool of stem cells results in an equilibration of contribution during the first 4–6 months post-engraftment. **c** Decreasing or restricted contribution occurs in situations where either stochastic mechanisms or the demands of a radiation-ablated system result in stem cell commitment, without significant self-renewal. Stem cell commitment to myeloid lineages yields a decrease/extinction of a given clone due to short myeloid cell half-lives. Commitment along both myeloid and lymphoid lineages, or the lymphoid lineage alone, yields a persistence of lymphoid cell types due to the long-lived nature of some T lymphocytes. Reprinted with permission from Jordan and Lemischka (1990)

1a). Because clonal expansion is assumed to be exponential, a numerically large clone of stem cells could result from a subtle proliferative or self-renewal advantage following engraftment. The hematopoietic repopulating activity of such cells would appear minimal during this period of self-renewal. In time, because of numerical clone size, these cells would dominate in hematopoietic contribution. Clonal stability would also result from a large clone of uncommitted stem cells. Stochastic commitment mechanisms, functioning on a numerically large stem cell clone, would yield equal representation of that clone in all lineages. A small stem cell clone could result in extinction or lineage-restricted contribution (Fig. 1c). That is, all, or most, members of such a clone could make a commitment to differentiation as a simple consequence of stochastic mechanisms acting on a small number of stem cells. A balance of commitment and self-renewal expansion is depicted in Fig. 1b. Initially, stochastic mechanisms acting on a small clone of uncommitted cells could result in large disparity or fluctuation in lineage contribution. Gradually, as the stem cell clone expands (during the initial 4–6 months), contribution to all lineages would become homogeneous and stable.

Interestingly, these studies document that decreases in clonal contribution outnumber increases by a factor of two (JORDAN and LEMISCHKA 1990). This is in accord with early, quantitative CFU-S studies (VOGEL et al. 1968). These CFU-S experiments demonstrated a stochastic component in self-renewal vs. commitment decisions of this stem/progenitor cell class.

Stochastic models predict that pluripotent cells may contribute to single lineages or subsets of lineages. The model in Fig. 1 suggests that at long post-transplantation time points such occurrences should be extremely rare. In agreement with this, the studies report only a single example of such a phenomenon (JORDAN and LEMISCHKA 1990).

According to the model, self-renewal vs. commitment decisions in pluripotent stem cells are governed by stochastic mechanisms. However, the readout of such mechanisms may depend on lineage-specific demands, a nonrandom progression of exact commitment decisions and/or the microenvironment. In fact, the studies demonstrated a temporal pattern in the contribution of a given clone to specific lineages (JORDAN and LEMISCHKA 1990). Such a pattern is illustrated in Fig. 1b. A pluripotent stem cell clone initially contributes to myeloid tissue and only subsequently to lymphoid cell populations. The opposite situation (lymphoid contribution followed by myeloid contribution) was not observed. The most simple interpretation of these observations is that short-lived host myeloid cells or precursors disappear relatively quickly following irradiation, thus creating an immediate need for these cell populations.

An additional indicator of nonrandom lineage contribution is the observation of totipotent stem cell clones which cease to contribute specifically to the monocyte-macrophage lineage (JORDAN and LEMISCHKA 1990). These clones continue to repopulate the granulocyte and lymphoid cell lineages. Other lineage-specific decreases are not observed. Thus, it is possible that in vivo commitment and differentiation decisions occur in a relatively ordered fashion in

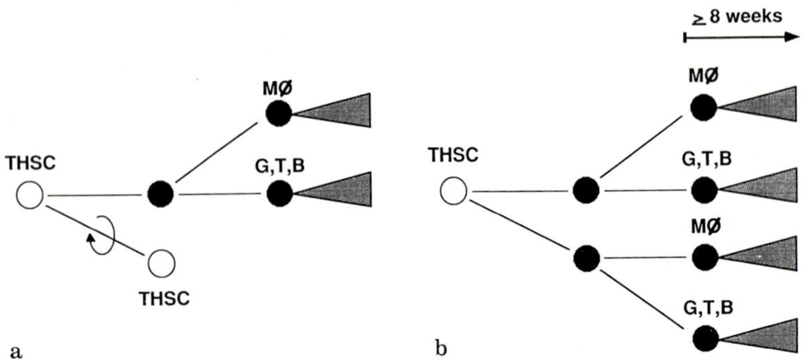

Fig. 2 a, b. Monocyte-macrophage-specific extinction. **a** A possible ordered segregation of developmental potential following a commitment decision by a pluripotent hematopoietic stem cell (*THSC*). The THSC has also self-renewed to yield an uncommitted cell. The segregation of macrophage (*M*) potential is suggested by in vivo experiments (JORDAN et al. 1990). The segregation of other developmental potentials [granulocyte (*G*), T-lymphoid (*T*) and B-lymphoid (*B*)] into a single cellular entity is speculative. **b** Commitment decisions in both daughter cells. Monocyte-macrophage contribution can extinguish after as few as 8 weeks. The extinction of the other lineages has not been analyzed. The clone represented in the panel may be destined to extinguish in all lineages after a finite time interval. *Open circles*, uncommitted cells; *closed circles*, committed cells; *shaded areas*, expansion and differentiation

spite of a stochastic mechanism mediating the decision to commit *or* self-renew. It is intriguing to compare these results with in vitro data. The existence of the non-self-renewing, granulocyte-macrophage (GM) precursor, detected in vitro as the CFU-GM (METCALF 1984), suggests that these two lineages are linked up to a relatively late developmental stage. The in vivo data, together with the likelihood that fluctuations (in this case, a decrease in contribution to monocytes) are the result of mechanisms acting on highly clonogenic cells, argue that in vivo macrophage and granulocyte repopulation programs are regulated individually.

The monocyte-macrophage specific decrease seen in our studies may also shed light on the self-renewal capacity of a lineage-restricted precursor or stem cell. It is possible that this phenomenon is the earliest indicator of a pluripotent stem cell clone destined for decrease in contribution to all hematopoietic lineages. Given that the contribution of a pluripotent stem cell to discrete lineages proceeds through(an) intermediate class(es) of committed cells, the rapid disappearance of clonal progeny in a given lineage directly argues that such restricted cells are not capable of extensive self-renewal (Fig. 2).

7 Conclusions and Future Directions

The initial expectations of retroviral marking studies have, only in part, been fulfilled. Clearly, as yet it has not been possible to derive a fate-map of hematopoietic development, nor has it been possible to evaluate clonal succession models. However, more importantly, the experiments have defined the temporal behavior of individual stem cells and have recast reconstituted hematopoiesis in a dynamic and systemic fashion. As such, the studies suggest a novel way to think about this system; that is, one where developmental behavior is a function of time and not necessarily a reflection of distinct stem cell classes. Indeed, the models presented above rely on a small number of assumptions and are considerably more simple than classical models. Importantly, several aspects of the models could be readily tested. For example, it should be possible to address the ordered progression of commitment decisions by retransplantation of marrow from primary recipients containing a clone which has ceased in macrophage contribution. Similarly, engraftment of animals in which lineage-specific deficiencies have been created should be informative. In addition, use of multiple retroviral markers in conjunction with retransplantation should yield valuable insights. The model (Fig. 1) predicts that hematopoietic equilibrium could be accelerated by infusion of factors which promote the self-renewal of primitive stem cells. One such candidate factor may be the product of the Steel locus, shown to have dramatic in vitro effects (reviewed by WITTE 1990). No doubt other molecules will be isolated in the future and could similarly be assessed in this system.

In a sense, retroviral marking studies have pointed out the limitations of the reconstitution assay. If an equilibrium situation is attained by a selective expansion of a few initially engrafted stem cells, then, in effect, a proviral marker becomes equivalent to a donor vs. host marker in long-term studies. Obviously, detection of fluctuation or clonal succession during this long-term interval is not possible. A means of introducing random markers during the time of initial clonal expansion would be invaluable, as would a technology for marking stem cells as they develop in situ in nonablated reconstituted animals.

In conclusion, it is fair to say that the first wave of retroviral marking studies has provided as much insight as it is likely to provide. In the future, the power of this system will reside in its effective combination with other approaches. In particluar, the combination of retroviral marking and stem cell isolation (SZILVASSY et al. 1989b; JORDAN et al. 1990) or long-term in vitro culture should facilitate direct elucidation of pluripotent hematopoietic stem cell properties.

Acknowledgements. The author wishes to thank Drs. Jean E. Schwarzbauer and Arnold J. Levine for critical reading of the manuscript. In addition, many hours of discussion over the last 4 years with Dr. Craig T. Jordan were instrumental in formulating the concepts in this article. The work was supported by NIH Grant CA45339, a Dupont Young Faculty Grant, and a Preceptorship from the American Cyanamid Company.

References

Abramson S, Miller RG, Phillips RA (1977) The identification in adult marrow of pluripotent and restricted stem cells of the myeloid and lymphoid systems. J Exp Med 145: 1567–1579

Boggs DR, Boggs SS, Saxe DF, Gress LA, Canfield DR (1982) Hematopoietic stem cells with high proliferative potential: assay of their concentration in marrow by the frequency and duration of cure of W/W^v mice. J Clin Invest 70: 242–252

Capel B, Hawley R, Covarrubias L, Hawley T, Mintz B (1989) Clonal contributions of small numbers of retrovirally marked hematopoietic stem cells engrafted in unirradiated W/W^v mice. Proc Natl Acad Sci USA 86: 4564–4568

Dexter TM, Spooncer E (1987) Growth and differentiation in the hemopoietic system. Ann Rev Cell Biol 3: 423–441

Dick JE, Magli MC, Huszar D, Phillip RA, Bernstein A (1985) Introduction of a selectable gene into primitive stem cells capable of long-term reconstitution of the hemopoietic system of W/W^v mice. Cell 42: 71–79

Freitas AA, Rocha B, Coutinho AA (1986) Lymphocyte population kinetics in the mouse. Immunol Rev 91: 5–37

Harrison DE, Astle CM, Lerner C (1988) Number and continuous proliferative pattern of transplanted primitive immunohematopoietic stem cells. Proc Natl Acad Sci USA 85: 822–826

Johnson GR, Nicola NA (1984) Characterization of two populations of CFU-S fractionated from fetal liver by flourescence activated cell sorting. J Cell Physiol 118: 45–52

Jordan CT, Lemischka IR (1990) Clonal and systemic analysis of long-term hematopoiesis in the mouse. Genes and Development 4: 220–232

Jordan, CT, McKearn JP, Lemischka IR (1990) Cellular and developmental properties of fetal hematopoietic stem cells. Cell 61: 953–963

Kay HEM (1965) How many cell-generations? Lancet ii: 418–419

Keller GC, Paige C, Gilboa E, Wagner EF (1985) Expression of a foreign gene in myeloid and lymphoid cells derived from multipotent hematopoietic precursors. Nature 318: 149–154

Korn AP, Henkelman RM, Ottensmayer FP, Till JE (1973) Investigations of a stochastic model of haemotopoiesis. Exp Hematol 1: 362–375

Lemischka IR, Raulet RH, Mulligan RC (1986) Developmental potential and dynamic bahavior of hematopoietic stem cells. Cell 45: 917–927

Metcalf D (1984) The hemtopoietic stimulating factors. Elsevier, New York

Metcalf D (1988) The molecular control of blood cells. Harvard University Press, Cambridge

Micklem HS, Ansell JD, Wayman JE, Forrester L (1983) The clonal organization of hematopoiesis in the mouse. Prog Immunol 5: 633–644

Micklem HS, Lennon JE, Ansell JD, Gray RA (1987) Numbers and dispersion of repopulating hematopoietic cell clones in radiation chimeras as functions of injected cell dose. Exp Hematol 15: 251–257

Mintz B, Anthony K, Litwin S (1984) Monoclonal derivation of mouse myeloid and lymphoid lineages from totipotent hematopoietic stem cells experimentally engrafted in fetal hosts. Proc Natl Acad Sci USA 81: 7835–7839

Müller-Sieburg CE, Townsend K, Weissman IL, Rennick D (1988) Proliferation and differentiation of highly enriched mouse hemopoietic stem cells and progenitor cells in response to defined growth factors. J Exp Med 167: 1825–1835

Nakahata T, Gross AF, Ogawa M (1982) A stochastic model of the primitive hemopoietic stem cells in culture. J Cell Physiol 113: 455–458

Ogawa M, Porter PN, Nakahata T (1983) Renewal and commitment to differentiation of hemopoietic stem cells. Blood 61: 823–829

Ploemacher RE, Brons RHC (1989) Separation of CFU-S from primitive cells responsible for reconstitution of the bone marrow hemopoietic stem cell compartment following irradiation: evidence for a pre-CFU-S cell. Exp Hematol 17: 263–266

Smith LG, Weissman, IL, Heimfeld S (1991) Clonal analysis of hematopoietic stem-cell differentiation in vivo. Proc Natl Acad Sci USA 88: 2788–2792

Snodgrass R, Keller G (1987) Clonal fluctuation within the haematopoietic system of mice reconstituted with retrovirus-infected stem cells. EMBO J 6: 3955–3960

Spangrude GJ, Heimfeld S, Weissman IL (1988) Purification and characteristics of mouse hematopoietic stem cells. Science 241: 58–62

Szilvassy SJ, Lansdorp PM, Humphries RK, Eaves AC, Eaves CJ (1989a) Isolation in a single step of a highly enriched murine hematopoietic stem cell population with competitive long-term repopulating ability. Blood 74: 930–939

Szilvassy SJ, Fraser CC, Eaves CJ, Lansdorp PM, Eaves AC, Humphries RK (1989b) Retrovirus-mediated gene transfer to purified hemotopoietic stem cells with long-term lympho-myelopoietic repopulating ability. Proc Natl Acad Sci USA 86: 8798–8802

Trentin JJ (1970) Influence of hematopoietic organ stroma (hematopoietic inductive microenvironments) on stem cell differentiation. In: Regulation of hematopoiesis (Gordon AS, ed.) Appleton-Century-Crofts, New York, p. 161

Visser JWM, Bauman JGJ, Mulder AH, Eliason JF, deLeeuw AM (1984) Isolation of murine pluripotent hemopoietic stem cells. J Exp Med 59: 1576–1590

Vogel H, Neiwisch H, Matioli G (1968) The self-renewal probability of hematopoietic stem cells. J Cell Physiol 72: 221–228

Witte O (1990) Steel locus defines new multi-potent growth factor. Cell 63: 5–6

Differentiation of Stem Cells into Lineage Restricted Precursors

Spleen Colony-Forming Unit: A Myeloid Stem Cell

R. J. JONES

1 Introduction

For years it has been suspected that the differentiated progeny of all the lymphohematopoietic cell lineages arise from a common cell. The initial evidence for pluripotent hematopoietic stem cells (PHSC) was derived from studies in patients with chronic myelogenous leukemia. The Philadelphia chromosome could be detected in all lymphohematopoietic cell lineages (WHANG et al. 1963), suggesting that these cells descended from a single leukemic stem cell. Definite proof for the existence of normal PHSC was provided by finding the same unique radiation-induced chromosome abnormality in all lymphohematopoietic cell lineages following murine bone marrow transplantation (BMT) (WU et al. 1965; NOWELL et al. 1970; ABRAMSON et al. 1977), indicating that all lineages were derived from a common stem cell. Recently, this finding was confirmed using retroviral gene markers (LEMISCHKA et al. 1986; CAPEL et al. 1990; SZILVASSY et al. 1989b). However, identification and isolation of the PHSC have been difficult because of the rarity of the cells, which account for only about 1 in 10^4 nucleated bone marrow cells (JONES et al. 1987; BOGGS et al. 1982) and because of the lack of a simple, definitive assay for the cells. Although spleen colony-forming units (CFU-S) (TILL and MCCULLOCH 1961) have long been considered to be a marker

---

The Johns Hopkins Oncology Center, Room 2-127, 600 North Wolfe Street, Baltimore, MD 21205, USA

Current Topics in Microbiology and Immunology, Vol. 177

for the mouse PHSC (SIMINOVITCH et al. 1963; SPANGRUDE et al. 1988), recent evidence suggests that neither day 8 CFU-S, nor the possibly more primitive day 12 CFU-S represent the PHSC (BERTONCELLO et al. 1988; PLOEMACHER and BRONS 1989; JONES et al. 1989, 1990).

2 Spleen Colony-Forming Units

The CFU-S assay was first described 30 years ago (TILL and MCCULLOCH 1961). CFU-S are cells capable of lodging in the spleen and forming macroscopic nodules 8–14 days following BMT. Ablation of host bone marrow is required to allow the transplanted bone marrow to form spleen colonies. Classically, marrow ablation has been achieved by total body irradiation, although marrow ablation by high-dose alkylating agents will also enable spleen colonies to form (SANTOS and HAGHSHENASS 1968). High doses of cytotoxic agents that produce bone marrow aplasia but not ablation will result in the generation of endogenous spleen colonies derived from residual host CFU-S that survived the treatment (MARSH et al. 1967; SANTOS and HAGHSHENASS 1968). Using chromosomal markers, it was shown that the spleen colonies are clonal: all the cells in an individual colony originate from a single CFU-S (BECKER et al. 1963; WU et al. 1967). The individual CFU-S have extensive proliferative capacity as demonstrated by the fact that the macroscopic spleen colonies will frequently contain over one million cells (SIMINOVITCH et al. 1963). Only a fraction of the CFU-S transplanted actually lodge in the spleen and form colonies; the rest of the CFU-S settle and proliferate elsewhere, particularly in the marrow. The fraction of the injected CFU-S that form macroscopic splenic nodules (the "f factor") is about 15%–20% (SIMINOVITCH et al. 1963).

2.1 Heterogeneity of CFU-S

It is now clear that CFU-S are very heterogeneous in respect to their proliferative capacity, self-renewal ability, and cell cycle status. It was recognized over 25 years ago that spleen colonies at day 14 after injection contain many more secondary CFU-S than spleen colonies obtained earlier (SIMINOVITCH et al. 1963). In addition, most "early-appearing" spleen colonies (those present at 7–9 days) contain cells of only one hematopoietic lineage (usually erythroid), while "late-appearing" spleen colonies (those present at 12–14 days) contain cells of more than one lineage (CURRY and TRENTIN 1967). The presumption was that the late spleen colonies represented further evolution of the same colonies that were present earlier.

However, spleen colonies continuously appear and disappear throughout the period from day 7 to day 14 after injection (MAGLI et al. 1982; WOLF and

PRIESTLEY 1986). MAGLI et al. (1982) showed that early-appearing spleen colonies are unipotential, have little self-renewal potential, and are destined to disappear within 3 days. This led to the hypothesis that the late-appearing spleen colonies represent multipotential, self-maintaining PHSC, whereas early-appearing spleen colonies represent committed hematopoietic progenitors more closely related to erythroid progenitors (BFU-E) or granulocyte-macrophage progenitors (CFU-GM). BAINES and VISSER (1983) provided further evidence that CFU-S that form spleen colonies on day 12 are separate and more primitive than day 8 CFU-S. They showed that day 12 CFU-S could be partially separated from day 8 CFU-S by fluorescence-activated cell sorting on the basis of DNA content, and that most of the day 12 CFU-S are in G_0 while day 8 CFU-S are primarily in the G_1 phase of the cell cycle. Furthermore, cell cycle-specific chemotherapeutic agents such as hydroxyurea and 5-fluorouracil, that preferentially eliminate proliferating cells while sparing noncycling cells, spare late-appearing spleen colonies at doses that eliminate early spleen colonies (ROSENDAAL et al. 1976; HODGSON and BRADLEY 1979). The distinction between day 8 CFU-S and day 12 CFU-S is not absolute, however. Some day 8 spleen colonies persist untill day 12, and these colonies contain as many secondary or daughter CFU-S as late-appearing CFU-S (WOLF and PRIESTLEY 1986).

2.2 Multipotential Nature of CFU-S

CFU-S, especially those producing late-appearing spleen colonies, exhibit many characteristics of PHSC: extensive proliferative capacity, ability for self-renewal and the capability of generating spleen colonies containing cells of multiple hematopoietic lineages. Moreover, most approaches that enrich for PHSC also enrich for day 12 CFU-S, usually to a much greater degree than other assayable hematopoietic progenitors (VISSER et al. 1984; SPANGRUDE et al. 1988). By isolating mouse bone marrow cells that bound wheat germ agglutinin and expressed class I histocompatibility antigens, VISSER et al. (1984) were able to radioprotect mice with just 170 ± 30 cells, about a 100-fold enrichment over the radioprotective ability of unseparated marrow (JONES et al. 1987; BOGGS et al. 1982). This isolation technique yielded a similar enrichment of day 12 CFU-S. SPANGRAUDE et al. (1988) found that <100 marrow cells, isolated by their absence of lineage-specific cell markers and their expression of the Thy-1 and Sca-1 antigens, would produce long-term repopulation of all lymphohematopoietic lineages in mice. This cell population was also markedly enriched for day 12 CFU-S, while much less enriched for day 8 CFU-S.

3 Separation of PHSC from CFU-S

Although day 12 CFU-S appear to be primitive progenitors with extensive multipotential proliferative capability, several observations have cast doubt on whether CFU-S actually represent PHSC. No study has shown that CFU-S are functional progenitors of the lymphoid system (ABRAMSON et al. 1977). Although clonal erythroid, granulocytic, monocytic, and megakaryocytic cells are found in a spleen colony (SIMINOVITCH et al. 1963; WU et al. 1967), clonal lymphoid cells have not been identified. Furthermore, CFU-S have not been shown to be capable of long-term lymphoid reconstitution in vivo. The frequency of PSHC in mouse bone marrow also appears to be much less than the frequency of day 12 CFU-S (JONES et al. 1987; BOGGS et al. 1982). Moreover, a number of studies have recently shown that PHSC can be separated from day 12 CFU-S (BERTONCELLO et al. 1988; JONES et al. 1989, 1990; PLOEMACHER and BRONS 1989; SPANGRUDE and JOHNSON 1990).

3.1 Serial BMT Studies

Serial BMT is a means to study the self-renewal capacity of hematopoietic progenitors. For these studies, bone marrow is transplanted into lethally irradiated mice, and, after engraftment, bone marrow is removed from the reconstituted mice and serially transplanted into another group of lethally irradiated mice. Serial passage of the bone marrow is continued as long as possible. It has been known for over 30 years that serially transplanted bone marrow will eventually lose its ability to reconstitute the lymphohematopoietic system (BARNES et al. 1962). The reason for the eventual loss of reconstituting ability with serial BMT was unknown. SIMINOVITCH et al. (1964) showed that the CFU-S content of mouse bone marrow also decreased with serial BMT. However, the CFU-S content of the serially transplanted marrow did not decrease to the same extent as the reconstituting ability of the marrow; substantial numbers of CFU-S were still present in the serially transplanted marrow even on the serial transfers that would not produce reconstitution (SIMINOVITCH et al. 1964; LAJTHA and SCHOFIELD 1971; HELLMAN et al. 1978).

Although the CFU-S content of marrow decreased with serial BMT, we also found that the numbers of CFU-S (both day 8 and day 12) and CFU-GM in serially transferred marrow grafts that could not produce donor engraftment remained at levels that should be associated with full hematologic reconstitution (JONES et al. 1989). Furthermore, on the serial transfers that would not allow donor reconstitution, there was initial engraftment from donor cells at 10–14 days after BMT which disappeared by 17–21 days (JONES et al. 1989). It appeared that serial BMT dissociated two waves of engraftment after BMT: an initial but unsustained wave from committed progenitors (CFU-S and CFU-GM) that was maintained with serial BMT, and a late, sustained wave from the PHSC that was eventually lost with serial BMT. Therefore, CFU-S (both day 8 and day 12) that seemed to be responsible for early engraftment could be separated from the PHSC which were responsible for late engraftment. We also found that prolonging the time interval between transfers from 1 month to 6 months overcame the loss of repopulating ability of serially transferred bone marrow (JONES et al. 1989). This suggested that the limited reconstitutive capacity of serially transplanted marrow results from dilution of the slowly proliferating PHSC by the more rapidly dividing committed progenitors whose numbers are relatively maintained with serial BMT. With repeated serial transfer, the concentration of PHSC in the bone marrow would decrease to levels that could no longer produce long-term engraftment. Increasing the time interval between transfers would allow the slowly proliferating PHSC time to regenerate their numbers and reestablish the ratio of PHSC to committed progenitors.

3.2 Studies with Rhodamine-123

A number of studies have used rhodamine-123 (Rh) to separate PHSC from day 12 CFU-S (PLOEMACHER and BRONS 1989; SPANGRUDE and JOHNSON 1990). Rh is a fluorescent dye that has relatively high affinity for mitochondrial membranes and therefore appears to be much better retained in cycling cells than in quiescent cells. PLOEMACHER and BRONS (1989) showed that marrow fractions displaying high Rh fluorescence intensity were most enriched for day 12 CFU-S, while marrow fractions with low Rh fluorescence were most enriched for pre-CFU-S cells as defined by their ability to generate secondary day 12 CFU-S. Although they did not directly demonstrate that PHSC could be at least partially separated from day 12 CFU-S, they did demonstrate that CFU-S precursors could be separated from day 12 CFU-S. SPANGRUDE and JOHNSON (1990) confirmed that PHSC could be separated from day 12 CFU-S on the basis of their Rh fluorescence. They analyzed a fraction of marrow cells, isolated by their lack of lineage-specific markers and their expression of Thy-1 and Sca-1, and previously shown to be enriched for both PHSC and day 12 CFU-S. The high Rh-fluorescent cells were most enriched for day 13 CFU-S. However, despite being relatively depleted of day 13 CFU-S, only the low Rh-fluorescent cells could transfer long-term hematopoietic reconstitution to mice upon serial BMT.

3.3 Counterflow Centrifugal Elutriation

Direct evidence that day 12 CFU-S are not PHSC was furnished by studies using counterflow centrifugal elutriation (CCE) (JONES et al. 1990), a procedure that sorts cells on the basis of size and density. With the centrifuge rotor speed held constant at 1260 g, marrow cells were sorted by changing the rate of medium flow. The fraction containing the largest cells was most enriched for CFU-GM (fourfold) and twofold enriched for both day 8 and day 12 CFU-S. The intermediate fraction was threefold enriched for both day 8 and day 12 CFU-S but depleted of CFU-GM. The fraction of smallest cells contained virtually all morphologic lymphocytes and was essentially completely depleted of CFU-GM and CFU-S (both day 8 and day 12).

In order to study the engraftment potential of the various CCE fractions, male marrow cells were transplanted into lethally irradiated female syngeneic mice, and the origin of the engrafted cells was determined by DNA hybridization with a Y-chromosome-specific probe (LAMAR and PALMER 1984). Mice transplanted with 2×10^4 cells from the large cell fraction showed only transient, early engraftment, despite the fact that 2×10^4 cells from this fraction contain about the same number of CFU-GM as 1×10^5 unfractionated marrow cells, which will produce complete, durable engraftment in 100% of lethally irradiated mice (JONES et al. 1987; MAUCH and HELLMAN 1989). Therefore, the large cell fraction was relatively depleted of PHSC, even though this fraction was enriched for committed progenitors and produced early engraftment. The transplantation 10^5 cells from the small cell fraction, containing all morphologic lymphocytes, would not rescue mice from lethal aplasia. Female mice were then transplanted with 2×10^4 male cells from the lymphoid fraction mixed with 2×10^4 female large cells to provide an early wave of transient engraftment. Although reconstituted female mice were completely derived from female cells 14 days after BMT, all hematologic lineages were primarily derived from male cells 2 months after BMT. Therefore, the lymphoid fraction, which was nearly completely depleted of committed progenitors (including day 12 CFU-S) and could not rescue mice from lethal aplasia, contained the PHSC.

These data demonstrate that neither day 8 CFU-S nor the possibly more primitive day 12 CFU-S represent the PHSC. Thus, durable long-term reconstitution of all lymphohematopoietic lineages in vivo following marrow-ablative therapy and BMT remains the only means to identify the PHSC. These data also confirm that there are two vital classes of cells necessary for normal engraftment after BMT: committed progenitors that provide initial but unsustained engraftment and PHSC—slowly proliferating cells that produce only delayed, but durable, engraftment. Furthermore, in order to demonstrate late engraftment from transplanted slowly proliferating PHSC, these cells must be combined with a distinguishable source of early engrafting cells (committed progenitors), which will allow the marrow-ablated host to survive initial aplasia.

4 Conclusions

CFU-S, particularly day 12 CFU-S, are primitive multipotential progenitors as illustrated by their extensive proliferative capacity, their ability for self- renewal, and their capability of forming spleen colonies containing cells of multiple hematopoietic lineages. Furthermore, CFU-S share many physical and cell surface characteristics with PHSC (VISSER et al. 1984; SPANGRUDE et al. 1988; SZILVASSY et al. 1989a). However, evidence now strongly suggests that CFU-S are at least one step more differentiated than PHSC. CFU-S cannot be shown to be progenitors for lymphocytes. More importantly, multiple studies have now demonstrated that CFU-S, both day 8 and day 12, can be physically separated from PHSC. The transplantation of fractions enriched for CFU-S, but depleted of PHSC, will rescue mice from lethal aplasia. However, these fractions produce only early hematopoietic engraftment that is unsustained (JONES et al. 1990). Thus, CFU-S appear to be primitive myeloid stem cells, most closely related to the in vitro assayable granulocyte-erythrocyte-macrophage-megakaryocyte CFU (CFU-GEMM) (OGAWA et al. 1983). Long-term reconstitution of all lymphohematopoietic lineages in vivo after marrow ablation and BMT remains the only definitive means of identifying the PHSC.

References

Abramson S, Miller RG, Phillips RA (1977) The identification in adult bone marrow of pluripotent and restricted stem cells of the myeloid and lymphoid systems. J Exp Med 145: 1567–1579

Baines P, Visser JWM (1983) Analysis and separation of murine bone marrow stem cells by H33342 fluorescence-activated cell sorting. Exp Hematol 11: 701–708

Barnes DWH, Loutit JF, Micklem HS (1962) "Secondary Disease" of radiation chimeras: a syndrome due to lymphoid aplasia. Ann NY Acad Sci 99: 374–385

Becker AJ, McCulloch EA, Till JE (1963) Cytological demonstration of the clonal nature of spleen colonies derived from transplanted mouse marrow cells. Nature 197: 452–454

Bertoncello I, Hodgson GS, Bradley TR (1988) Multiparameter analysis of transplantable hemopoietic stem cells. II. Stem cells of long-term bone marrow-reconstituted recipients. Exp Hematol 16: 245–249

Boggs DR, Boggs SS, Saxe, DF, Gress LA, Canfield DR (1982) Hematopoietic stem cells with high proliferative potential. J Clin Invest 70: 242–253

Capel B, Hawley RG, Mintz B (1990) Long- and short-lived murine hematopoietic stem cell clones individually identified with retroviral integration markers. Blood 75: 2267–2270

Curry JL, Trentin JJ (1967) Hematopoietic spleen colony studies. I. Growth and differentiation. Dev Biol 15: 395–400

Hellman S, Botnick LE, Hannon EC, Vigneulle RM (1978) Proliferative capacity of murine hematopoietic stem cells. Proc Natl Acad Sci USA 75: 490–494

Hodgson GS, Bradley TR (1979) Properties of haematopoietic stem cells surviving 5-fluorouracil treatment: evidence for a pre-CFU-S cell? Nature 281: 381–382

Jones RJ, Sharkis SJ, Celano P, Colvin OM, Rowley SD, Sensenbrenner LL (1987) Progenitor cell assays predict hematopoietic reconstitution after syngeneic transplantation in mice. Blood 70: 1186–1192

Jones RJ, Celano P, Sharkis SJ, Sensenbrenner LL (1989) Two phases of engraftment established by serial bone marrow transplantation in mice. Blood 73: 397–401

Jones RJ, Wagner JE, Celano P, Zicha MS, Sharkis SJ (1990) Separation of pluripotent hematopoietic stem cells from multipotent progenitors (CFU-S). Nature 347: 188–189

Lajtha LG, Schofield R (1971) Regulation of stem cell renewal and differentiation: possible significance in aging. Adv Gerontol Res 3: 131–146

Lamar EE, Palmer E (1984) Y-encoded, species-specific DNA in mice: evidence that the Y chromosome exists in two ploymorphic forms in inbred strains. Cell 37: 171–177

Lemischka IR, Raulet DH, Mulligan RC (1986) Developmental potential and dynamic behavior of hematopoietic stem cells. Cell 45: 917–927

Magli MC, Iscove NN, Ordartchenko N (1982) Transient nature of early haemopoietic spleen colonies. Nature 295: 527–529

Marsh JC, Boggs DR, Bishop CR, Chervamick PA, Cartwright GE, Wintrobe MM (1967) Factors influencing hematopoietic spleen colony formation in irradiated mice. I. The normal pattern of endogenous colony formation. J Exp Med 126: 833–850

Mauch P, Hellman S (1989) Loss of hematopoietic stem cell self-renewal after bone marrow transplantation. Blood 74: 872–875

Nowell PC, Hirsch BE, Fox DH, Wilson DB (1970) Evidence for the existence of multipotential lympho-hematopoietic stem cells in the adult rat. J Cell Physiol 75: 151–158

Ogawa M, Porter PN, Nakahata T (1983) Renewal and commitment to differentiation of hemopoietic stem cells (an interpretive review). Blood 61: 823–829

Ploemacher RE, Brons RHC (1989) Separation of CFU-S from primitive cells responsible for reconstitution of the bone marrow hemopoietic stem cell compartment following irradiation: evidence for a pre-CFU-S cell. Exp Hematol 17: 263–266

Rosendaal M, Hodgson GS, Bradley TR (1976) Haemopoietic stem cells are organised for use on the basis of their generation-age. Nature 264: 68–69

Santos GW, Haghshenass M (1968) Cloning of syngenic hematopoietic cells in the spleens of mice and rats pretreated with cytotoxic drugs. Blood 32: 629–637

Siminovitch L, McCulloch EA, Till JE (1963) The distribution of colony- forming cells among spleen colonies. J cell Comp Physiol 62: 327–336

Siminovitch L, Till JE, McCulloch EA (1964) Decline in colony-forming ability of marrow cells subjected to serial transplantation into irradiated mice. J Cell Comp Physiol 64: 23–31

Spangrude GJ, Johnson GR (1990) Resting and activated subsets of mouse multipotent hematopoietic stem cells. Proc Natl Acad Sci USA 87: 7433–7437

Spangrude GJ, Heimfeld S, Weissman IL (1988) Purification and characterization of mouse hematopoietic stem cells. Science 241: 58–62

Szilvassy SJ, Lansdrop PM, Humpries RK, Eaves Ac, Eaves CJ (1989a) Isolation in a single step of a highly enriched murine hematopoietic stem cell population with competitive long-term repopulating ability. Blood 74: 930–939

Szilvassy SJ, Fraser CC, Eaves CJ, Lansdrop PM, Eaves AC, Humphries RK (1989b) Retrovirus-mediated gene transfer to purified hemopoietic stem cells with long-term lympho-myelopoietic repopulating ability. Proc Natl Acad Sci USA 86: 8798–8802

Till JE, McCulloch EA (1961) A direct measurement of the radiation sensitivity of normal mouse bone marrow cells. Radiat Res 14: 213–222

Visser JWM, Bauman JGJ, Mulder AH, Eliason JF, de Leeuw AM (1984) Isolation of murine pluripotent hemopoietic stem cells. J Exp Med 59: 1576–1590

Whang J, Frei E, III Carbone PP, Brecher G (1963) The distribution of the Philadelphia chromosome in patients with chronic myelogenous leukemia. Blood 22: 664–673

Wolf NS, Priestley GV (1986) Kinetics of early and late spleen colony development. Exp Hematol 14: 676–682

Wu AM, Till JE, Siminovitch L, McCulloch EA (1965) Cytological evidence for a relationship between normal hematopoietic colony-forming cells and cells of the lymphoid system. J Exp Med 127: 455–464

Wu AM, Till JE, Siminovitch L, McCulloch EA (1967) A cytological study of the capacity for differentiation of normal hemopoietic colony-forming cells. J Cell Physiol 69: 177–184

Status of High Proliferative Potential Colony-Forming Cells in the Hematopoietic Stem Cell Hierarchy

I. BERTONCELLO

1 Introduction

The analysis of regenerating murine bone marrow following perturbation with 5-fluorouracil (5-FU) has contributed significantly to our understanding of the organization of the primitive hematopoietic stem cell compartment (HODGSON and BRADLEY 1979; BRADLEY et al. 1980; HODGSON et al. 1982; VAN ZANT 1984). Using this model, BRADLEY and HODGSON (1979) identified a unique subpopulation of hematopoietic cells in clonal agar culture which they designated high proliferative potential colony-forming cells (HPP-CFC). Operationally, HPP-CFC are distinguished from other hematopoietic progenitor cells by: (a) their relative resistance to the cytotoxic effects of 5-FU; (b) their obligatory requirement for multiple hematopoietic growth factors acting in synergy in order to clone in agar; and (c) their ability to form macroscopic colonies in agar culture greater than 0.5 mm in diameter containing at least 50000 cells.

HPP-CFC are among the most primitive hematopoietic cells yet identified in vitro, having many characteristics of stem cells with extensive hematopoiesis-reconstituting ability in vivo. HPP-CFC are rare, with a frequency of less than one in 400 normal murine bone marrow cells. They are highly quiescent (HODGSON and BRADLEY 1984) and multipotential and generate transplantable hematopoietic cells and committed progenitors of various hematopoietic lineages in vitro

Cell Biology Group, Peter MacCallum Cancer Institute, 481 Little Lonsdale Street, Melbourne, Victoria 3000, Australia

Current Topics in Microbiology and Immunology, Vol. 177

(McNiece et al. 1987). Their pattern of regeneration following 5-FU treatment is closely correlated with that of pre spleen colony-forming unit (pre-CFU-S) cells with marrow-repopulating ability (MRA) and platelet-repopulating ability (PRA) (Hodgson and Bradley, 1979; Bradley et al. 1980). Consequently, the HPP-CFC assay is a valuable tool for the analysis of the early events in hematopoiesis.

This chapter reviews recent investigations defining the growth factor requirements of HPP-CFC and examines their status in the hematopoietic stem cell hierarchy in relationship to primitive hematopoietic cells defined by other in vivo and in vitro assays. Tissue culture requirements for optimal HPP-CFC assay (Bradley et al. 1990) and the early HPP-CFC literature (McNiece et al. 1990) have been reviewed comprehensively elsewhere.

2 Identification and Functional Analysis of HPP-CFC Subpopulations

Two days after intravenous administration of 150–200 mg 5-FU/kg to normal mice, lineage-restricted progenitor cells of low proliferative potential (LPP-CFC), which clone in agar in the presence of colony-stimulating factor-1 (CSF-1) alone, and CFU-S assayed in lethally irradiated mice 10–13 days after transplantation are depleted to less than 0.1% of normal. In contrast, HPP-CFC and pre-CFU-S cells, assayed by their MRA and PRA, are depleted to only 40% of normal values (Bradley et al. 1980).

However HPP-CFC are not homogeneous. HPP-CFC surviving 5-FU treatment can only be detected in clonal agar culture in the presence of the combined stimulus of interleukin1α (IL-1α) plus IL-3 plus CSF-1 (Bartelmez et al. 1989). HPP-CFC with more restricted growth factor requirements are ablated by 5-FU and re-emerge sequentially in the bone marrow during hematopoietic regeneration following 5-FU treatment. These include HPP-CFC responsive to IL-3 + CSF-1 or IL-1α + CSF-1 (Bartelmez et al. 1989; Kriegler et al. 1990), GM-CSF + CSF-1 (McNiece et al. 1988; Kriegler et al. 1990), hematolymphopoietic growth factor 1 (HLGF-1) + CSF-1 (Quesenberry et al. 1987), and IL-4 + IL-9 (Rennick et al. 1989), as well as other growth factor combinations.

All HPP-CFC form macroscopic macrophage colonies of similar size and cellularity in agar despite differences in their hematopoietic growth factor requirements. However, functional analysis of HPP-CFC stimulated by different combinations of growth factors shows that they differ in their proliferative capacity and their differentiative potential:

1. When propagated in liquid culture, HPP-CFC responsive to IL-1α + IL-3 + CSF-1 are able to generate approximately 27×10^6 cells, compared to approximately 5×10^6 cells for IL-3 + CSF-1-responsive HPP-CFC; or 5000 cells for LPP-CFC.

2. Generation of HPP-CFC responsive to IL-3 + CSF-1 with concurrent depletion of HPP-CFC responsive to IL-1α + IL-3 + CSF-1 has been demonstrated by feedout in liquid culture (BRADLEY et al. 1985) and by recloning in agar (MCNIECE et al. 1986).
3. When developing colonies derived from IL-1α + IL-3 + CSF-1-responsive HPP-CFC, harvested when comprising 100–200 cells, are dispersed and assayed in vivo and in vitro they contain myeloid, erythroid, and megakaryocytic progenitors as well as transplantable hematopoietic stem cells. HPP-CFC responsive to IL-3 + CSF-1 contain predominantly lineage-restricted LPP-CFC responsive to CSF-1 alone (MCNIECE et al. 1987).

HPP-CFC share many of the markers which characterize primitive transplantable hematopoietic stem cells (Table 1). In normal bone marrow, HPP-CFC subpopulations previously identified in the regenerating bone marrow of 5-FU-treated mice can be enriched and resolved from lineage-restricted progenitor cells by negative immunomagnetic selection using cocktails of monoclonal antibodies directed against pre-B and B lymphocytes (B220), T cells (CD4, CD8), neutrophils, and activated macrophages (7/4) and erythroid cells (YW 25.12.7) (BERTONCELLO et al. 1989, 1991b). Positive flow cytometric cell sorting on the basis of the expression of various cell surface markers such as Qa-m7 (BERTONCELLO et al. 1986, 1987), Thy-1 (BOSWELL et al. 1984; WILLIAMS et al. 1985), or wheatgerm agglutinin (WGA) binding (BERTONCELLO et al. 1991b) yields highly enriched preparations of HPP-CFC. HPP-CFC have also been detected in high frequencies in lineage-negative (Lin^-), $Sca\text{-}1^+$ bone marrow cell preparations (Fig. 1, and LOWRY et al. 1991; WILLIAMS et al. 1992) similar to those described by SPANGRUDE et al. (1988) and SPANGRUDE and SCOLLAY (1990).

Table 1. Markers shared by HPP-CFC and transplantable stem cells with extensive hematopoiesis-reconstituting capacity in vivo

Marker	Phenotype	Reference
WGA	+	BERTONCELLO et al. 1991b
Rh123	Dull	BERTONCELLO et al. 1991a
Qa-m7	+	BERTONCELLO et al. 1986 BERTONCELLO et al. 1987
Thy-1	+/low	BOSWELL et al. 1984 WILLIAMS et al. 1985
Sca-1	+	LOWRY et al. 1991 WILLIAMS et al. 1992
Lyt.2 (CD4)	–	BERTONCELLO et al. 1991b
L_3T_4 (CD8)	–	BERTONCELLO et al. 1991b
B220	–	BERTONCELLO et al. 1989
7/4	–	BERTONCELLO et al. 1989
YW25.12.7	–	BERTONCELLO et al. 1991b
Gr-1	–	Unpublished data
Mac-1	–	Unpublished data

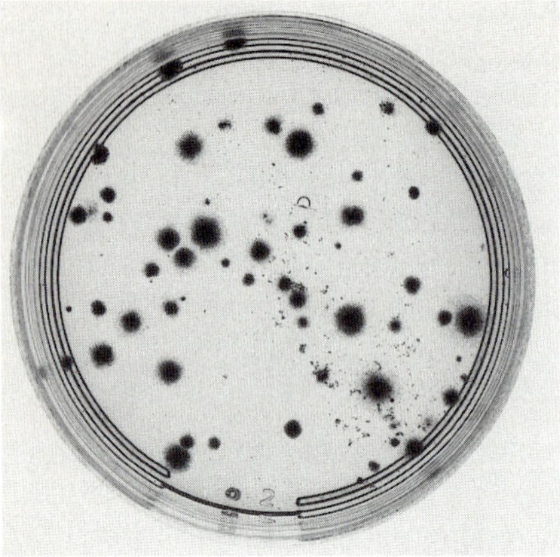

Fig. 1. Colonies derived from HPP-CFC in a lineage-negative, Sca-1-positive cell preparation assayed in clonal agar culture using IL-1α + IL-3 + CSF-1 as a stimulus. Colonies were stained in situ in 35-mm Petri dishes at day 14 of culture using the tetrazolium salt, INT (BOL et al. 1977)

Rhodamine-123 (Rh123) uptake as a parameter for the resolution and hierarchical ordering of transplantable hematopoietic stem cells was originally pioneered in this laboratory (BERTONCELLO et al. 1985). Rh123 has also proven to be one of the most useful probes for the resolution of HPP-CFC subpopulations with differing potentialities and growth factor requirements in normal bone marrow (BERTONCELLO et al. 1991a). Primitive transplantable hematopoietic stem cells with long-term hematopoiesis-reconstituting capacity in vivo are Rh-dull, whereas Rh-bright cells have rapid but limited hematopoiesis-reconstituting capacity (BERTONCELLO et al. 1985, 1988; MULDER and VISSER, 1987; VISSER and DE VRIES 1988; PLOEMACHER and BRONS 1988b, 1989; SPANGRUDE and JOHNSON 1990). HPP-CFC responsive to IL-1α + IL-3 + CSF-1 co-fractionate with transplantable cells with MRA and PRA and are Rh-dull. HPP-CFC responsive to IL-3 + CSF-1, and HPP-CFC of more restricted potentiality are Rh-bright (BERTONCELLO et al. 1991a).

3 HPP-CFC Nomenclature and Organization of the HPP-CFC Compartment

It is likely that the HPP-CFC compartment is a continuum in which gradual changes in growth factor responsiveness or requirements will occur as HPP-CFC progressively mature and differentiate. At least some of these HPP-CFC can be hierarchically ordered on the basis of their sequential regeneration following 5-FU treatment, their Rh123-staining characteristics, and their differing potentialities determined by liquid culture feedout and recloning. In this scheme, HPP-

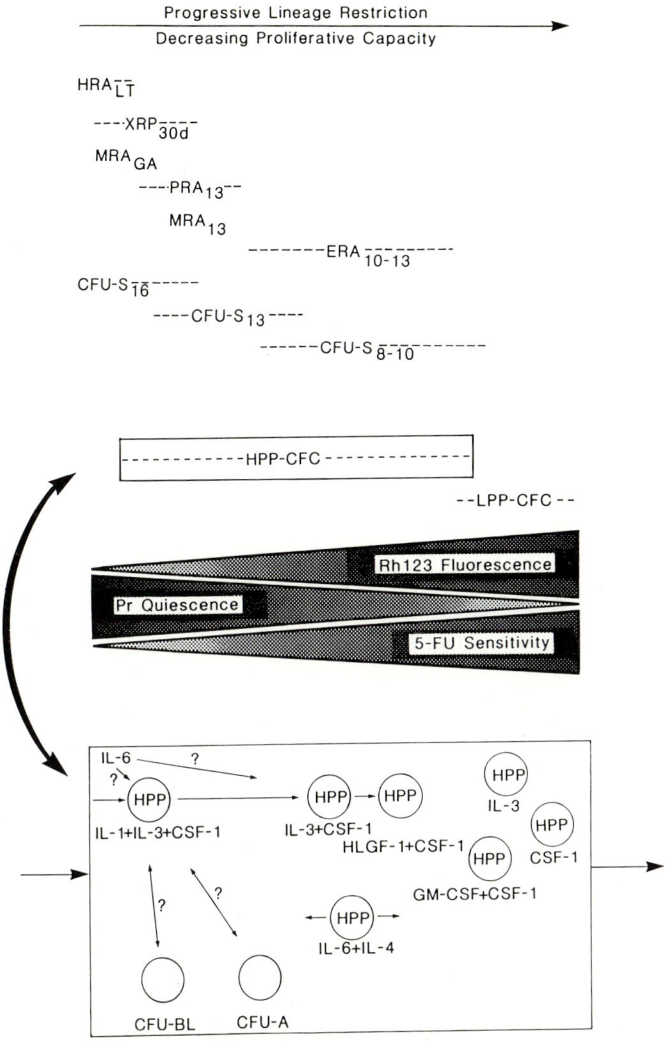

Fig. 2. A schematic representation of the interrelationships and hierarchical ordering of transplantable hematopoietic stem cells, HPP-CFC subpopulations, and related hematopoietic precursor cells detected in vitro as inferred from flow cytometric analysis and sorting, and their sensitivity to 5-FU in vivo. HRA_{LT}, transplantable cells with long-term hematopoiesis-reconstituting ability; XRP_{30d}, transplantable cells assayed by radioprotective capacity (survival 30 days after lethal irradiation and transplantation), MRA_{13d}, cells with marrow repopulating ability (assayed 13 days after transplantation, as the femoral content of CSF-1-responsive CFC per 10^5 cells injected); MRA_{GA}, cells with the potential to generate MRA cells in vivo (assayed by retransplantation followed by MRA assay in secondary recipients); *PRA*, *ERA*, platelet and erythroid repopulating ability, respectively (assayed by estimation of platelet or reticulocytes per milliliter of peripheral blood, at the specified times after transplantation)

CFC with an obligatory requirement for IL-1α + IL-3 + CSF-1 are the most primitive HPP-CFC subpopulation which can be detected (Fig 2). Accordingly, it has been proposed that HPP-CFC responsive to IL-1α + IL-3 + CSF-1, and HPP-CFC responsive to IL-3 + CSF-1 be designated HPP-CFC-1 and HPP-CFC-2, respectively (McNiece et al. 1986, 1990). It can also be inferred that other HPP-CFC subpopulations ablated by 5-FU and not detected in regenerating marrow until as many as 10 days after 5-FU treatment (Kriegler et al. 1990) are more closely related to committed LPP-CFC than they are to HPP-CFC responsive to IL-1α + IL-3 + CSF-1.

With the exception of HPP-CFC-1 and HPP-CFC-2, the exact interrelationship of HPP-CFC stimulated by other combinations of purified growth factors or conditioned media is unclear. Analysis of fractionated hematopoietic cell populations, using different combinations of known and novel growth factors and conditional media may demonstrate further heterogeneity in the HPP-CFC compartment and expose more primitive HPP-CFC than HPP-CFC-1. Therefore, it could be argued that HPP-CFC should be strictly designated by the growth factor stimulus used to elicit their growth in order to avoid confusion and ambiguity in interpreting HPP-CFC assays. For example, IL-6 acts by recruiting highly quiescent primitive hematopoietic cells into cycle (Ikebuchi et al. 1987; Okano et al. 1989). The addition of IL-6 to IL-1α + IL-3 + CSF-1 approximately doubles the incidence of detectable Rh-dull HPP-CFC (Bertoncello et al. 1990). Are these HPP-CFC with a requirement for IL-6 + IL-1α + IL-3 + CSF-1 more primitive than HPP-CFC-1, or does addition of IL-6 result in the recruitment or priming of highly quiescent HPP-CFC-1? Similarly, the novel growth factor, stem cell factor (SCF, or *c-kit* ligand) synergizes with other hematopoietic growth factors to stimulate HPP-CFC in the marrow of 5-FU-treated mice (Zsebo et al. 1990; Martin et al. 1990), and enriched normal bone marrow (Lowry et al. 1991; Williams et al. 1992). The status of these HPP-CFC has not been determined.

4 Relationship of HPP-CFC, CFU-A, and CFU-Bl

Two other assays which detect primitive multipotential hematopoietic cells of high proliferative potential the blast colony-forming cell (CFU-Bl) assay (Nakahata and Ogawa 1982) and the CFU-A assay (Pragnell et al. 1988; Eckmann et al. 1988) have also been described. CFU-Bl proliferate to form small multipotential blast cell colonies in methylcellulose culture which comprise up to 1000 cells at 16 days of culture when stimulated by factors such as IL-3 and IL-6 acting in synergy (Ikebuchi et al. 1987). CFU-A proliferate to form macroscopic macrophage colonies in agar greater than 2 mm in diameter when stimulated by conditioned medium from the AF1-19T cell line (transformed normal rat kidney cell line) acting in synergy with CSF-1.

When highly enriched WGA^+ bone marrow cells depleted of B and T lymphocytes, macrophages, granulocytes, and erythroid cells (BERTONCELLO et al. 1991b) were assayed for both HPP-CFC and CFU-A content, HPP-CFC responsive to IL-1α + IL-3 + CSF-1 could not be dissociated from CFU-A stimulated by AF1-19T-conditioned medium + CSF-1 (KRIEGLER et al. 1990). HPP-CFC responsive to IL-1α + IL-3 + CSF-1 and CFU-A were present in similar frequencies in post.5-FU bone marrow and exhibited indentical patterns of regeneration following 5-FU and identical proliferative capacities when propagated in liquid culture (KRIEGLER et al. 1990). A number of hematopoietic growth factors have been detected in AF1-19T-conditioned medium, including GM-CSF, IL-1, and IL-6 (KRIEGLER et al. 1990). Although PRAGNELL et al. (1988) have suggested that GM-CSF could replace AF1-19T-conditioned medium in the CFU-A assay, the two factors do not stimulate identical target cells. HPP-CFC stimulated by GM-CSF + CSF-1 are a relatively mature HPP-CFC subpopulation (MCNIECE et al. 1988; KRIEGLER et al. 1990).

CFU-Bl and HPP-CFC have a number of common functional characteristics, and there is considerable circumstantial evidence which suggests that HPP-CFC responsive to IL-1α + IL-3 + CSF-1 are closely related, if not identical to CFU-Bl. Both are highly quiescent and relatively resistant to 5-FU (SUDA et al. 1983; IKEBUCHI et al. 1987, 1988). Both require multiple factors, including IL-6 and IL-1 acting in synergy with IL-3 for optimal stimulation (IKEBUCHI et al. 1987, 1988; KOIKE et al. 1988; WONG et al. 1988). Like IL-1α + IL-3 + CSF-1-responsive HPP-CFC, CFU-Bl are multipotential, and on replating generate colonies of single and multiple lineages in vitro as well as transplantable hematopoietic cells (NAKAHATA and OGAWA 1982; SUDA et al. 1983). In order for HPP-CFC to proliferate to form macro scopic colonies the macrophage lineage-specific growth factor CSF-1 must be present. In the absence of CSF-1 colony growth is limited, and highly enriched HPP-CFC develop into CFU-Bl-like colonies of up to 100–200 cells. In delayed addition experiments performed in our laboratory, the combination of IL-6 + IL-1α + IL-3 will maintain highly purified HPP-CFC in clonal agar culture at low densities with minimal proliferation for at least 7 days prior to the addition of CSF-1 (unpublished observation).

5 HPP-CFC and Transplantable Hematopoietic Stem Cells: Are HPP-CFC an Index of Cells with Long-term Hematopoiesis-Reconstituting Capacity In Vivo?

Analysis of murine bone marrow following 5-FU treatment initially established the close relationship of HPP-CFC and transplantable stem cells with long-term hematopoietic reconstituting ability in vivo (HODGSON and BRADLEY 1979; BRADLEY et al. 1980). This relationship between HPP-CFC and transplantable hematopoietic cells was also supported by the demonstration that colonies developing from HPP-CFC generated CFU-S, and cells with MRA, PRA, and erythroid reconstituting ability (ERA) in vivo, as well as committed progenitor cells of various lineages (McNiece et al. 1987). In addition, HPP-CFC and candidate pre-CFU-S populations in the marrow share many phenotypic characteristics (Table 1), and co-fractionation of HPP-CFC and transplantable hematopoietic cells has been reported in purified bone marrow cell preparations obtained using various cell-separative protocols:

1. Cell sorting on the basis of Qa-m7 surface antigen expression yields fractions highly enriched for HPP-CFC and transplantable hematopoietic stem cells from normal (Bertoncello et al. 1986) and post-5-FU bone marrow (Bertoncello et al. 1987). Cell preparations containing up to 30% HPP-CFC, 50% CFU-S_{13}, and cells reconstituting the MRA, PRA, and ERA compartments of the marrow of lethally irradiated mice have been obtained.
2. Co-fractionation of transplantable stem cells with long-term hematopoiesis-reconstituting capacity and IL-1α+IL-3+CSF-1-responsive HPP-CFC on the basis of Rh123 uptake was alluded to in section 3 above.
3. HPP-CFC responsive to IL-1α + IL-3 + CSF-1 are detectable in high frequencies in enriched Lin^-, $Thy\text{-}1^{lo}$, $Sca\text{-}1^+$ bone marrow cell populations which contain cells capable of long-term hematopoietic reconstitution of lethally irradiated mice (unpublished observation).

Cumulatively, these data clearly show that HPP-CFC with a requirement for IL-1α + IL-3 + CSF-1 are among the most primitive hematopoietic cells yet detected in vitro, but have not conclusively established the equivalence of HPP-CFC and transplantable hematopoietic cells. On the other hand, the data did support the inference that the HPP-CFC assay may be a useful index of primitive transplantable hematopoietic stem cells, possibly predicting hematopoietic reconstitution following transplantation, or hematopoietic perturbation.

This hypothesis was tested in a series of experiments which compared the bone marrow of primary recipient mice following lethal irradiation and transplantation with that of littermate controls (Bertoncello et al. 1988). On retransplantation, 3 months after the primary transplant, the marrow of primary recipients had an impaired hematopoiesis-reconstituting capacity in comparison to littermate control marrow, although differences could not be detected in 5-FU sensitivity, Rh123-staining characteristics, or CFU-S_{13} and HPP-CFC content. Thus, the HPP-CFC assay appeared to be unable to predict the hematopoietic reconstituting capacity of transplanted bone marrow. Although the outcome of this study was not encouraging, the possibility that an as yet unresolved HPP-CFC subpopulation could predict hematopoiesis-reconstituting capacity should not be excluded. Such a HPP-CFC subpopulation may require additional growth factors for detection, or may be so highly quiescent that the 14 day culture period used for HPP-CFC assay may not be sufficient to enable primitive highly quiescent cells to express their true proliferative potential. There is some evidence for the latter hypothesis:

1. The ability of IL-6 to augment the detection of HPP-CFC in cultures of Rh-dull bone marrow cells containing IL-1α + IL-3 + CSF-1 (BERTONCELLO et al. 1990) has already been discussed.
2. In preliminary experiments in which Lin^-, $Sca\text{-}1^+$ cells were assayed for HPP-CFC, significantly more HPP-CFC were detected after 21 days in culture than were detected at day 14 (T.R. Bradley, unpublished observation).
3. In vivo, primitive stem cells with the ability to endow lasting long-term reconstitution of the hematopoietic system cannot be adequately assayed by short-term functional assays. They can only be measured by amplification assays such as the MRA assay which measures MRA by the potential of transplanted cells to replenish the progenitor cell compartment of primary hosts (HODGSON BRADLEY 1979; and BRADLEY et al. 1980; HODGSON et al. 1982), or by retransplantation of hematopoietic cells and the measurement of the ability of primary transplanted marrow to generate primitive progenitor cells in secondary host mice (PLOEMACHER and BRONS 1988a; SPANGRUDE and JOHNSON 1990).

6 Conclusion

The HPP-CFC compartment of murine bone marrow comprises a number of subpopulations which can be distinguished on the basis of their differing hematopoietic growth factor requirements, and their proliferative and differentiative capacity. The status of various HPP-CFC subpopulations and their inferred relationship to other primitive hematopoietic cells is illustrated schematically in Fig. 2. To date, HPP-CFC responsive to IL-1α + IL-3 + CSF-1 are among the most primitive hematopoietic cells yet detected in vitro. They share many of the functional characteristics and markers of primitive hematopoietic stem cells with long-term reconstituting capacity in vivo (HRA_{LT}) and may be closely related, if not identical, to CFU-Bl detected in vitro. However, despite their close relationship to primitive transplantable hematopoietic cells, the HPP-CFC assay does not appear to predict the hematopoietic reconstituting ability of transplanted cells. Some HPP-CFC subpopulations with more restricted growth factor requirements can be resolved from each other and hierarachically ordered on the basis of their functional properties, but the full extent of the heterogeneity within the HPP-CFC compartment is yet to be determined.

HPP-CFC have almost exclusively been defined in murine bone marrow using CSF-1 as the lineage-restricted hematopoietic growth factor. There is no reliable data available on the measurement or status of HPP-CFC detected using erythroid, or lymphoid lineage-restricted growth factors instead of CSF-1.

This review has been limited to a discussion of murine HPP-CFC. HPP colonies have also been described in human bone marrow (see review by MCNIECE et al. 1990). They exhibit some of the characteristics of HPP-CFC in the murine system, but it is too early to conclude that they are equivalent to murine HPP-CFC.

References

Bartelmez SH, Bradley TR, Bertoncello I, Mochizuki DY, Tushinski RJ, Stanley ER, Hapel AJ, Young IG, Kriegler AB, Hodgson GS (1989) Interleukin-1 plus interleukin-3 plus colony stimulating factor-1 are essential for clonal proliferation of primitive myeloid bone marrow cells. Exp Hematol 17: 240–245

Bertoncello I, Bartelmez SH, Bradley TR, Stanley ER, Harris RA, Sandrin MS, Kriegler AB, McNiece IK, Hunter SD, Hodgson GS (1986) Isolation and analysis of primitive hemopoietic progenitor cells on the basis of differential expression of Qa-m7 antigen. J Immunol 136: 3219–3224

Bertoncello I, Bartelmez SH, Bradley TR, Hodgson GS (1987) Increased Qa-m7 antigen expression is characteristic of primitive hemopoietic progenitors in regenerating marrow. J Immunol 139: 1096–1103

Bertoncello I, Bradley TR, Hodgson GS (1989) The concentration and resolution of primitive hemopoietic cells from normal mouse bone marrow by negative selection using monoclonal antibodies and dynabead monodisperse magnetic microspheres. Exp Hematol 17: 171–176

Bertoncello I, Bradley TR, Hodgson GS, Dunlop JM (1991a) The resolution, enrichment and organisation of normal bone marrow high proliferative potential colony-forming cell subsets on the basis of rhodamine 123 fluorescence. Exp Hematol 19: 174–178

Bertoncello I, Bradley TR, Watt SM (1991b) An improved negative immunomagnetic selection strategy for the purification of primitive hemopoietic cells from normal bone marrow. Exp Hematol 19: 95–100

Bertoncello I, Hodgson GS, Bradley TR (1985) Multiparameter analysis of transplantable hemopoietic stem cells: I. The separation and enrichment of stem cells homing to the marrow and spleen on the basis of rhodamine 123 fluorescence. Exp Hematol 13: 999–1006

Bertoncello I, Hodgson GS, Bradley TR (1988) Multiparameter analysis of transplantable hemopoietic stem cells: II. Stem cells of long-term bone marrow-reconstituted recipients. Exp Hematol 16: 245–249

Bertoncello I, Kriegler AB, Bradley TR, Dunlop JM, McNiece IK (1990) The effect of interleukin-6 on hemopoietic stem cell subsets resolved by rhodamine 123. (Abstract). Exp Hematol 18: 550

Bol S, Van Den Engh G, Visser J (1977) A technique for staining hemopoietic colonies in agar cultures. Exp Hematol 5: 551–553

Boswell HS, Wade PM, Quesenberry PJ (1984) Thy-1 antigen expression by murine high-proliferative capacity hematopoietic progenitor cells I. Relation between sensitivity to depletion by Thy-1 antibody and stem cell generation potential. J Immunol 133: 2940–2949

Bradley TR, Hodgson GS (1979) Detection of primitive macrophage progenitor cells in mouse bone marrow. Blood 54: 1446–1450

Bradley TR, Hodgson GS, Bertoncello I (1980) Characteristics of macrophage progenitor cells with high proliferative potential: relationship to cells with marrow repopulating ability in 5-fluorouracil treated mouse bone marrow. In: Experimental hematology today 1979, Baum SJ, Ledney GD, van Bekkum DW (eds), Karger, New York pp 285–297

Bradley TR, Hodgson GS, Kriegler AB, McNiece IK (1985) Generation of CFU- S13 in vitro. In Cronkite EP, Dainiak N, McCaffrey RP, Palek J, Quesenberry PJ (eds) Hematopoietic stem cell physiology. Alan Liss, New York pp 39–56

Bradley TR, Hodgson GS, Bertoncello I (1990) High proliferative potential colony forming cells. In: Pollard JW, Walker JM (eds) Animal cell culture. Humana, Clifton, pp 289–297 (Methods in molecular biology, vol 5)

Eckmann L, Freshney M, Wright EG, Sproul A, Wilkie N, Pragnell IB (1988) A novel in vitro assay for murine haematopoietic stem cells. Br J Cancer 58, [Suppl I] 9:36: 40

Hodgson GS, Bradley TR (1979) Properties of haemopoietic stem cells surviving 5-fluorouracil treatment: evidence for a pre-CFU-S cell? Nature 281: 381–382

Hodgson GS, Bradley TR (1984) In vivo kinetic status of hematopoietic stem and progenitor cells as inferred from labeling with bromodeoxyuridine. Exp Hematol 12: 683–687

Hodgson GS, Bradley TR, Radley JM (1982) The organisation of hemopoietic tissue as inferred from the effects of 5-fluorouracil. Exp Hematol 10: 26–35

Ikebuchi K, Wong GG, Clark SC, Ihle JN, Hirai Y, Ogawa M (1987) Interleukin 6 enhancement of interleukin 3-dependent proliferation of multipotential hemopoietic progenitors. Proc Natl Acad Sci USA 84: 9035–9039

Ikebuchi K, Ihle JN, Hirai Y, Wong GG, Clark SC, Ogawa M (1988) Synergistic factors for stem cell proliferation: further studies of the target stem cells and the mechanism of stimulation by interleukin-1, interleukin-6, and granulocyte colony-stimulating factor. Blood 72: 2007–2014

Koike K, Nakahata T, Takagi M, Kobayashi, Ishiguro A, Tsuji K, Naganuma K, Okano A, Akiyama Y, Akabane T (1988) Synergism of BSF-2/interleukin-6 and interleukin 3 on development of multipotential hemopoietic progenitors in serum-free culture. J Exp Med 168: 879–890

Kriegler AB, Bradley TR, Bertoncello I, Hamilton JA, Hart PH, Piccoli DS, Hodgson GS (1990) Progenitor cells in murine bone marrow stimulated by growth factors produced by the AF1-19T rat cell line. Exp Hematol 18: 372–378

Lowry PA, Zsebo KM, Deacon DH, Eichman CE, Quesenberry PJ (1991) Effects of rrSCF on multiple cytokine responsive HPP-CFC generated from SCA^+ Lin^- murine hematopoietic progenitors Exp Hematol: 19: 994–996

Martin FH, Suggs SV, Langley KE, Lu HS, Ting J, Okino KH, Morris CF, McNiece IK, Jacobsen FW, Mediaz EA, Birkett NC, Smith A, Johnson MJ, Parker VP, Flores JC, Patel AC, Fisher EF, Erjavec HO, Herrara CJ, Wypych J, Sachdev PK, Pope JA, Leslie I, Wen D, Lin C, Cupples R, Zsebo KM (1990) Primary structure and functional expression of rat and human stem cell factors DNAs. Cell 63: 203–211

McNiece IK, Bradley TR, Kriegler AB, Hodgson GS (1986) Subpopulations of mouse bone marrow high-proliferative-potential colony-forming cells. Exp Hematol 14: 856–860

McNiece IK, Williams NT, Johnson GR, Kriegler AB, Bradley TR, Hodgson GS (1987) Generation of murine hematopoietic precursor cells from macrophage high-proliferative-potential colony-forming cells. Exp Hematol 15: 972–977

McNiece IK, Robinson BE, Quesenberry PJ (1988) Stimulation of murine high proliferative potential colony forming cells by the combination of GM-CSF and CSF-1. Blood 72: 191–195

McNiece IK, Bertoncello I, Kriegler AB, Quesenberry PJ (1990) Colony-forming cells with high proliferative potential (HPP-CFC). Int J Cell Cloning 8: 146–160

Mulder AH, Visser JWM (1987) Separation and functional analysis of bone marrow cells separated by rhodamine 123 fluorescence. Exp Hematol 15: 99–104

Nakahata T, Ogawa M (1982) Identification in culture of a class of hemopoietic colony-forming units with extensive capability to self-renew and generate multipotential hemopoietic colonies. Proc Natl Acad Sci USA 79: 3843–3847

Okano A, Suzuki C, Takatsuki F, Akiyama Y, Koike K, Ozawa K, Kirano T, Kishimoto T, Nakahata T, Asano S (1989) In vitro expansion of the murine pluripotent hemopoietic stem cell population in response to interleukin 3 and interleukin 6. Application to bone marrow transplantation. Transplantation 48: 495–498

Ploemacher RE, Brons NHC (1988a) Isolation of hemopoietic stem cell subsets from murine bone marrow: II. Evidence for an early precursor of day-12 CFU-S and cells associated with radioprotective ability. Exp Hematol 16: 27–32

Ploemacher RE, Brons NHC (1988b) Cells with marrow and spleen repopulating ability and forming spleen colonies on day 16, 12, and 8 are sequentially ordered on the basis of increasing rhodamine 123 retention. J Cell Physiol 136: 531–536

Ploemacher RE, Brons NHC (1989) Separation of CFU-S from primitive cells responsible for reconstitution of the bone marrow hemopoietic stem cell compartment following irradiation: evidence for a pre-CFU cell. Exp Hematol 17: 263–266

Pragnell IB, Wright EG, Lorimore SA, Adam I, Rosendaal M, Delamarter JF, Freshney M, Eckmann L, Sproul A, Wilkie N (1988) The effect of stem cell regulators demonstrated with an in vitro assay. Blood 72: 196–201

Quesenberry PJ, Song Z, McGrath E, McNiece IK, Shadduck R, Waheed A, Baber G, Kleeman E, Kaiser D (1987) Multilineage synergistic activity produced by a murine adherent marrow cell line. Blood 69: 827–835

Rennick D, Jackson J, Yang G, Wideman J, Lee F, Hudak S (1989) Interleukin-6 interacts with interleukin-4 and other hematopoietic growth factors to selectively enhance the growth of megakaryocytic, erythroid, myeloid, and multipotential progenitor cells. Blood 73: 1828–1835

Spangrude GJ, Johnson GR (1990) Resting and activated subsets of mouse multipotent hematopoietic stem cells. Proc Natl Acad Sci USA 87: 7433–7437

Spangrude GJ, Scollay R (1990) A simplified method for enrichment of mouse hematopoietic stem cells. Exp Hematol 18: 920–926

Spangrude GJ, Heimfeld S, Weissman IL (1988) Purification and characterization of mouse hematopoietic stem cells. Science 241: 58–62

Suda T, Suda J, Ogawa M (1983) Proliferative kinetics and differentiation of murine blast cell colonies in culture: evidence for variable G_O periods and constant doubling rates of early pluripotent hemopoietic progenitors. J Cell Physiol 117: 308–318

Van Zant G (1984) Studies of hematopoietic stem cells spared by 5-fluorouracil. J Exp Med 159: 679–690

Visser JWM, de Vries P (1988) Isolation of spleen-colony forming cells (CFU-s) using wheat germ agglutinin and rhodamine 123 labeling. Blood Cells 14: 369–384

Williams DE, Boswell HS, Floyd AD, Broxmeyer HE (1985) Pluripotential hematopoietic stem cells in post-5-fluorouracil murine bone marrow express the Thy-1 antigen. J Immunol 135: 1004–1011

Williams N, Bertoncello I, Kavnoudias H, Zsebo K, McNiece IK (1992) Recombinant rat stem cell factor stimulates the amplification and differentiation of fractionated mouse stem cell populations. Blood 74: 1–7

Wong GG, Witek-Gianotti JS, Temple PA, Kriz R, Ferenz C, Hewick RM, Clark SC, Ikebuchi K, Ogawa M (1988) Stimulation of murine hemopoietic colony formation by human IL-6. J Immunol 140: 3040–3044

Zsebo KM, Wypych J, McNiece IK, Lu HS, Smith KA, Karkare SB, Sachdev RK, Yuschenkoff VN, Birkett NC, Williams LR, Satyagal VN, Tung W, Bosselman RA, Mendiaz EA, Langley KE (1990) Identification, purification, and biological characterization of hematopoietic stem cell factor from Buffalo rat liver-conditioned medium. Cell 63: 195–201

Characterization of Several Classes of Mouse Hematopoietic Progenitor Cells

S. HEIMFELD and I. L. WEISSMAN

1 Introduction

One of the most remarkable and mysterious programs of developmental biology is the elaboration of the vertebrate hematopoietic system. In adult animals the bone marrow serves as the primary site for blood cell formation. Contained within this tissue are the progenitor cells which ultimately give rise to the lymphoid, and erythroid lineages. Beginning with the now classic work of TILL and MCCULLOCH (1961), it was recognized that progenitor cells are very heterogeneous with regard to their proliferative and differentiative properties. The most primitive progenitors, termed pluripotent hematopoietic stem cells (PHSC), are defined operationally by their capacity for extensive self-renewal, their ability to rescue animals from lethal irradiation, and their potential to differentiate into all hematopoietic cell types. It is generally believed that PHSC give rise to more restricted types of precursor cells, which in turn generate unipotent progenitors through an unknown series differentiation steps (TILL and MCCULLOCH 1961; WU et al. 1968; ABRAMSON et al. 1977; TILL and MCCULLOCH 1980; MCCULLOCH 1983). Not surprisingly, given its ultimate role in the generation and reconstitution of the hematopoietic systems, the primary focus of most experimental study has been the PHSC. Much less attention has been paid to the more committed precursor cells. Thus, the early stages of differentiation that

Department of Pathology and the Howard Hughes Medical Institute, Stanford Medical Center, Stanford, CA 94305, USA

Current Topics in Microbiology and Immunology, Vol. 177

stem cells progress through on their way toward end cell maturation remain largely undefined. However, understanding this progression is critical to answering some basic developmental questions such as lineage restriction or the loss of self-renewal capacity, and ultimately in determining what assays can be used to uniquely define and quantitate PHSC. In this chapter we have detailed our findings on the identification and characterization of several classes of restricted progenitor cells, and the extent to which these non-stem cells overlap in their developmental properties with the PHSC in a variety of in vivo and in vitro experimental assays.

2 Developmental Potential of Bone Marrow Progenitors In Vivo

We have characterized several subsets of normal mouse bone marrow cells which are enriched for hematopoietic precursor activity. These subsets were obtained by combining monoclonal antibody staining for lineage-restricted antigens (see Fig. 1) with a two-stage immunomagnetic bead enrichment protocol and fluorescence-activated cell sorting (FACS) to yield highly purified subpopulations. Most progenitor cell activity fell within the 1% of mouse bone marrow which expressed a ten fold lower level of Thy-1 (REIF and ALLEN 1964) than that found on mature mouse T cells (Fig. 1A). These cells have been designated as the Thy^{lo} fraction (MÜLLER-SIEBURG et al. 1986). Using additional

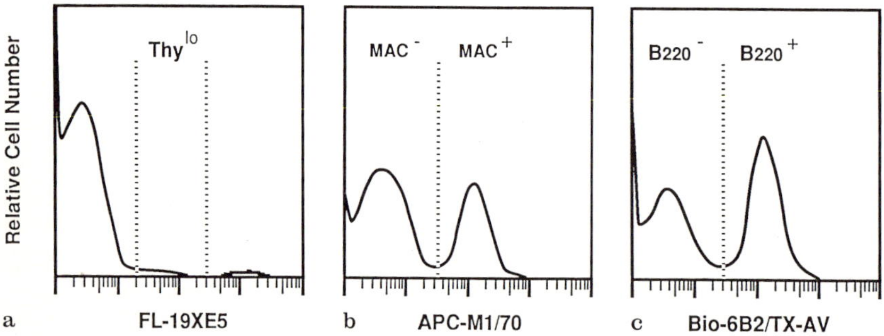

Fig. 1 a–c. The staining profiles of normal mouse bone marrow with Thy-1, B220, and Mac-1. Thy-1 is found on all mouse T cells (REIF and ALLEN 1964) and, as shown in (**a**) a small percentage (1%–4%) of bone marrow cells are Thy-1^+. This positive population can be split into two subsets. Cells with the brightest fluorescence are mature T cells, while those cells having approximately a tenfold lower level of expression have been designated as the Thy^{lo} fraction. B220 is the B cell- specific form of the T-200 family of molecules and is expressed on both the immature pre-B and mature B cells (COFFMAN and WEISSMAN 1983). A high proportion of the bone marrow cells are positive for this marker (**b**) Mac-1 primarily marks cells of the macrophage and granulocyte lineages (SPRINGER et al. 1979) and some 25%–35% of the bone marrow stains positive for this marker (**c**)

markers (see Fig. 1B, C) these Thy^{lo} cells have been further fractionated into five different subpopulations (see Table 1). Several different classes of precursors have been identified. These include progenitors restricted to the B cell lineage (Thy^{lo}, $B220^+$, Mac^-), precursors for the myelomonocytic pathway (Thy^{lo}, $B220^-$, Mac^+ and Thy^{lo}, Lin^-, Sca^-), a novel population which collectively appears multipotent for B, T, and myeloid differentiation, but has lost the capacity for extensive self-renewal (Thy^{lo}, $B220^+$, Mac^+), and a fraction which is highly enriched for PHSC activity (Thy^{lo}, Lin^-, Sca^+) (MÜLLER-SIEBURG et al. 1988; SPANGRUDE et al. 1988a; HEIMFELD et al. 1990, 1991a). Two additional bone marrow subsets have also been characterized: Thy^-, Lin^+ which includes the bulk of the mature lymphocytes, granulocytes, and monocytes; and Thy^-, Lin^- which contains the erythrocyte lineages. This Thy^--Lin^- population is also enriched for precursors that form single lineage colonies in response to colony-stimulating factors (MÜLLER-SIEBURG et al. 1988). With the exception of the mature T cells, these seven subpopulations encompass the entire bone marrow compartment. Each subset was then independently tested for its ability to reconstitute irradiated mice in several different progenitor cell assays.

2.1 Spleen and Thymus Colony-Forming Unit Activity

The frequencies of these subsets in normal bone marrow are listed in Table 1. All the Thy^{lo} populations are relatively rare, ranging from one in 150 for the Thy^{lo}, B^+, M^- cells to one in 2000 for Thy^{lo}, L^-, S^+ cells. Spleen colony-forming activity was greatly reduced among the two Thy^- fractions relative to whole bone marrow (Table 1). In contrast, each of the Thy^{lo} fractions shows enrichment for spleen colony-forming unit (CFU-S) activity, with the significant exception of the pre-B committed Thy^{lo}, B^+, M^- cells (TIDMARSH et al. 1989; HEIMFELD et al. 1991a). There were some differences among the populations in their ability to generate early versus late spleen colonies, with the more primitive Thy^{lo}, B^+, M^+ and Thy^{lo}, Lin^-, Sca^+ cells giving rise to more day 12 colonies (MÜLLER-SIEBURG et al. 1986; SPANGRUDE et al. 1988a; HEIMFELD et al. 1990). These results are consistent with the findings of other investigators that many types of progenitor cells can form spleen colonies, that there is some correlation between day 12 CFU-S and a more immature phenotype, and that at least some PHSC can contribute to spleen colony formation (SIMINOVITCH et al. 1963; WORTON et al. 1969; MAGLI et al. 1982; R. Jones, this volume).

Unlike the spleen, the thymus predominantly supports the outgrowth of mature T cells. Following irradiation, direct intrathymic injection of bone marrow progenitors leads to reconstitution of the thymocyte pool. Using a congenic marker system to distinguish donor from host cells, only two fractions of the bone marrow, the Thy^{lo}, B^+, M^+ and the Thy^{lo}, L^-, S^+ cells, showed any significant capacity to generate T cells (Table 1) (SPANGRUDE et al. 1988b; HEIMFELD et al.

Table 1. Differentiation of bone marrow subsets in vivo

Cell type	%	CFU-S activity		CFU-T	Short term reconstitution[a]			2° Transfer[b] ability
		Day 8	Day 12		T cells	B cells	M cells	
WBM	100	+	+	+	+	+	+	+
Thy^-, Lin^+	70	–	–	–	–	–	–	–
Thy^-, Lin^-	25	+/–	–	–	ND	ND	ND	ND
Thy^{lo}, B^+, M^-	0.5	–	–	–	–	++	–	–
Thy^{lo}, B^-, M^+	0.2	++	+/–	–	–	+/–	++	–
Thy^{lo}, B^+, M^+	0.1	++	++	++	++	++	++	–
Thy^{lo}, L^-, S^-	0.1	+++	+	–	–	+/–	++	–
Thy^{lo}, L^-, S^+	0.05	+	+++	+++	+++	+++	+++	+++

[a] Determined by injection of 10^2–10^5 cells of the indicated phenotype into irradiated Ly 5 congenic animals along with 10^5 syngeneic whole bone marrow cells to insure survival. Repopulation was measured by FACS analysis of peripheral blood at 4–8 weeks post-irradiation, staining for donor type versus lineage markers for B, T, and myeloid (M) cells

[b] Secondary reconstitution ability was assayed by transfer at 8 weeks of whole bone marrow from reconstituted animals described above. Repopulation was again measured by FACS analysis of peripheral blood. See text for further explanations and the specific references from which the data were compiled

WBM, whole bone marrow; Thy, Thy-1; Lin, L, mixture of lineage markers, B220, Mac-1, Gr-1, CD4, and CD8; B, B220; M, Mac1; S, Sca-1. Values are given as (+) and are meant to indicate the relative enrichment for activity in the specific assay listed as compared to unseparated bone marrow, (–) indicates no significant activity; ND, not done

1990). This implies that this assay for pre- T cell activity is fairly specific for the more primitive cell populations, although at least one other cell type besides the PHSC can give rise to T cells. Again, these results are consistent with the data of other investigators that progenitor cells other than the multipotent stem cell can reconstitute the thymus (MULDER et al. 1985; SPANGRUDE et al. 1988b).

2.2 Multilineage Reconstitution

A better measure of the developmental potential of the different bone marrow subsets was provided by assaying for multiple lineage reconstitution in irradiated animals. In this system, donor-derived cells were detected by an allelic difference at the T-200 leukocyte common antigen. T-200 is an antigen expressed on all hematopoietic lineages with the exception of erythroblasts and erythrocytes (SCHEID and TRIGLIA 1979). A minimal amount of syngeneic bone marrow was coinjected to insure survival of the host. Thus, this assay represents a modification of the competitive repopulation system (see D. E. Harrison, this volume). By FACS analysis of peripheral white blood cells one can detect whether the progenitors give rise to B cells, T cells, or myeloid cells. Several important points should be emphasized about these types of experiments:

1. During the course of recovery from lethal irradiation the three lineages appear in the peripheral blood at different times. Donor myeloid cells can be observed beginning at 2–3 weeks, while B cells arise at 4–6 weeks, and T cells at 6–8

weeks. Thus, several timepoints are required to fully assess the differentiation potential.

2. Frequently, reconstitutions were transient, i.e., donor cells could be detected at early timepoints, but were absent later.
3. The progenies detected are not necessarily clonally derived. Thus, finding donor cells in different lineages only indicates that the population being tested was multipotent. It does not mean any individual cell in that population was capable of multilineage differentiation.

A summary from a series of reconstitution experiments is shown in Table 1. The mature cells (Thy^-, Lin^+) did not contribute detectable levels of donor cells in these irradiated animals. Thy^-, Lin^- cells have not yet been tested in this system. A diverse set of responses was seen among the Thy^{lo} fractions, ranging from single lineage repopulations which were very transient (Thy^{lo}, B^-, M^+) to multilineage donor-derived progenies which were detected for the entire course of the experiment (Thy^{lo}, L^-, S^+) (MÜLLER-SIEBURG et al. 1986; SPANGRUDE et al. 1988a; HEIMFELD et al. 1990, 1991a). For the Thy^{lo}, L^-, S^+ cells, multilineage reconstitution has now been shown at the single cell level (SMITH et al. 1991). These results indicate that, as in the CFU-S and CFU-T assays, populations of bone marrow cells other than the PHSC can reconstitute irradiated animals.

A further assessment of self-renewal or maintenance capacity of these bone marrow fractions was done also. Injection of limiting numbers of fractionated cells only, without syngeneic bone marrow, failed to rescue mice from lethal irradiation. Furthermore, staining the bone marrow of competitively repopulated animals described above with Thy-1-specific antibodies indicated an absence of donor-type Thy^{lo} cells, suggesting a lack of capacity to maintain that cellular phenotype. The one exception noted was the Thy^{lo}, L^-, S^+ subset that was able to rescue lethally irradiated animals and generated donor-type $Thy\text{-}1^{lo}$ cells (SPANGRUDE et al. 1988a; HEIMFELD et al. 1990). Finally, secondary transfer of bone marrow from repopulated animals to new hosts also failed to yield subsequent donor reconstitution, except in some of those animals which had been repopulated by Thy^{lo}, L^-, S^+ cells (SPANGRUDE et al. 1988a; HEIMFELD et al. 1990; SMITH et al. 1991). Taken together, these results strongly indicate that none of the other bone marrow fractions except for Thy^{lo}, L^-, S^+ cells has an extensive capacity for self-renewal, which forms the basis for describing this subset as containing the PHSC compartment.

3 Developmental Potential of Bone Marrow Progenitors in Vitro

As summarized from the in vivo reconstitution experiments, the hematopoietic progenitor cell compartment is complex, consisting of cell types which can vary greatly in their self-renewal, proliferative, and differentiative capacities. Unfortunately, quantifying these differences among precursor cells in whole animal assays is not practical, given the vagaries of seeding efficiencies, the different microenvironments that the progenitor cells might migrate to, and the prohibitively large numbers of animals required to do true limit dilution clonal analysis. Also, these in vivo experiments are clearly not applicable to studies with cells from other species such as humans. Given these shortcomings, it becomes important to define alternative in vitro methods which could potentially substitute for these in vivo assays. Of particular relevance to the topic of this volume, in vitro systems are needed which can distinguish between the earliest multipotent stem cells and the more committed progenitors.

3.1 Long-Term Bone Marrow Cultures

Attempts to develop procedures which could mimic the normal bone marrow microenvironment resulted in the development of long-term marrow cultures (LTMC). DEXTER et al. (1976) have described culture conditions in which a bone marrow-derived stromal layer supports and maintains multipotent progenitors for over 6 months. These conditions allow for differentiation down the myeloid and erythroid pathways, but lymphoid differentiation does not occur. In contrast, Whitlock–Witte cultures, another type of LTMC, primarily support pre-B lymphopoiesis, and much less myeloid differentiation (WHITLOCK et al. 1984). It has been shown previously that LTMC cultures, once established, can be depleted of endogenous progenitor cell activity (DORSHKIND et al. 1986). The remaining adherent cells retain the capacity to support the proliferation of newly added progenitor cells for several weeks. A variation on this has been the establishment of cloned stromal cell lines which can substitute for the primary cultures in supporting hematopoietic development. We have tested the seven bone marrow fractions for colony-forming activity on these adherent layers. Limiting numbers of cells were seeded onto confluent LTMC cultures in 96 well plates in both Dexter and Whitlock–Witte conditions in order to determine response frequencies. At weekly intervals individual wells were screened for the presence of proliferating cells.

Within a few days after reconstitution, the formation of cobblestone areas was noted, in which clusters of cells were found to have integrated within or beneath the stromal adherent layer. Continued observation revealed that many of these cobblestone areas were sites of active proliferation, achieving up to several thousand cells per cluster over several weeks. Such zones were never observed in irradiated cultures which had not been recharged with fresh cells. For the Whitlock–Witte cultures, determinations as to the types of cells in the wells, pre-B and/or myeloid, were made by microscopic examination and by FACS analysis. The results are summarized in Table 2. Overall, the patterns of differentiation obtained corresponded very well with the lineage restrictions seen in vivo. Thus, the Thy^{lo}, B^{+}, M^{-} fraction generated short-lived B lineage cells, while Thy^{lo}, B^{-}, M^{+} and Thy^{lo}, L^{-}, S^{-} were restricted to myeloid differentiation,

Table 2. Differentiation of bone marrow subsets in vitro

Cell type	Long-term bone marrow culture					Methylcellulose culture colony types[c]		
	Whitlock–Witte[a]		Dexter[b]					
	B cells	Macrophages	Early	Middle	Late	Single	Mixed	HPP
WBM	+	+	+	+	+	+	+	+
Thy^{-}, Lin^{+}	−	−	−	−	−	−	−	−
Thy^{-}, Lin^{-}	+/−	+/−	+++	−	−	+	+/−	−
Thy^{lo}, B^{+}, M^{-}	++	−	−	−	−	−	−	−
Thy^{lo}, B^{-}, M^{+}	−	++	++	+	−	+/−	+/−	−
Thy^{lo}, B^{+}, M^{+}	+	++	+	++	−	+	−	−
Thy^{lo}, L^{-}, S^{-}	−	++	++	+	−	++	+++	+++
Thy^{lo}, L^{-}, S^{+}	+++	+++	++	+++	+++	++	+++	+++

[a] Growth in co-culture with cloned stromal lines under conditions which favor long-term lymphopoiesis. The relative ability of the indicated population to generate pre-B cells and/or macrophages is indicated

[b] Growth in co-culture with established, pre-irradiated mixed stromal cells under conditions which favor myelopoieis. The relative capacity of the sorted fractions to give rise to early- (1–2 weeks), middle- (2–3 weeks) and late- (3–4 weeks) appearing colonies are indicated

[c] Growth in methylcellulose cultures in the presence of various combinations of interleukins 1, 3, 6 and GM, G, and M colony-stimulating growth factors. "Single" indicates the generation of colonies containing only one cell lineage, "mixed" indicates at least three lineages per colony, "HPP" refers to high proliferative potential colonies of greater than 0.5 mm after 2 weeks of culture. See text for further explanations and the specific references from which the data were compiled

Abbreviations are given in the legend to Table 1

and Thy^{lo}, B^{+}, M^{+} and Thy^{lo}, L^{-}, S^{+} gave rise to both cell types. However, only the Thy^{lo}, L^{-}, S^{+} subset was capable of generating long-term pre-B cell cultures, indicating that perhaps only the PHSC is capable of sustained growth in vitro (Müller-Sieburg et al. 1986; Heimfeld et al. 1991a).

Similar differences were detected in the lifespan of colonies in the Dexter LTMC (Table 2). Most of the subsets contained progenitors which could proliferate on the depleted adherent layers. However, the vast majority of these reconstituted cultures had a limited lifespan in vitro. Only the Thy^{lo}, L^{-}, S^{+} fraction, which contains the PHSC, could generate longer-lived cultures. Upon further analysis, we also found that only the Thy^{lo}, L^{-}, S^{+}-reconstituted cultures contained progenitor cells which could grow in methycellulose culture (Weilbaecher et al. 1991). Ploemacher et al. (1989), using different separation protocols, have reported similar results. These experiments strongly suggest that it is predominantly, if not exclusively, the stem cell population which gives rise to long-lived colonies in both the Dexter and Whitlock–Witte culture systems (see C.E. Müller-Sieburg et al., this volume). Thus, reconstitution of LTMC may serve as a relatively quantitative in vitro assay for PHSC, and Sutherland et al. (1990) have recently reported on the use of such an assay in the human system.

3.2 Methylcellulose Cultures

Studies on progenitor cells cultured in semisolid media such as methylcellulose have suggested that hematopoiesis can be regulated by a group of glycoprotein molecules loosely termed hematopoietic growth factors. Several types of factors have now been purified, their genes cloned, and active recombinant molecules produced (for review see METCALF 1989). Growth of the more primitive precursors is enhanced by specific combinations of factors. In the presence of these factors the progenitor cells yield a complex pattern of growth in which individual colonies vary in their distribution of differentiated cell types, in their size and number of cells produced, and in the time of onset of proliferation. Attempts have been made to correlate these in vitro growth characteristics with the developmental properties of the progenitor cells in vivo. Thus, mixed colonies (containing cells in several different lineages), very large colonies (termed high proliferative potential, HPP), and blast cell colonies (derived from progenitors which do not initiate division for several days) have all been described as being specifically derived from PHSC proliferation (BRADLEY and HODGSON 1979; BARTELMETZ et al. 1989; IKEBUCHI et al. 1987; MCNIECE et al. 1988). However, whether such colony-forming attributes are truly unique to PHSC, and, conversely, whether all PHSC form only those types of colonies are two questions which have remained largely unanswered. The ability to purify specific classes of restricted and pluripotent progenitor cells allows one to address some of these points.

As in the LTMC cultures, most of the bone marrow subsets showed some colony formation in the presence of growth factors (Table 2) (MÜLLER-SIEBURG et al. 1988; HEIMFELD et al. 1991b). However, differences were found in regard to the types of colonies which formed. Of relevance to the questions raised above, the capacity to give rise to either mixed or HPP-type colonies was not restricted to the PHSC contained within the Thy^{lo}, L^-, S^+ fraction. Thy^{lo}, L^-, S^- cells, which are committed myeloid/erythroid progenitors, also generated significant numbers of these types of colonies (Table 2). In fact, quantitative calculations suggest that, from normal bone marrrow, non-stem cell progenitors contribute the majority of the mixed and HPP colonies (HEIMFELD et al. 1991b). Furthermore, a significant percentage of the colonies derived from the Thy^{lo}, L^-, S^+ cells were not mixed but contained only cells in a single lineage. Such restricted differentiation has generally not been observed for these cells in vivo. When individual Thy^{lo}, L^-, S^+ cells were injected into irradiated recipients, the predominant finding was multilineage reconstitution (SMITH et al. 1991). These data imply that, in vivo, each Thy^{lo}, L^-, S^+ cell is multipotent (SPANGRUDE et al. 1988a; SMITH et al. 1991). The formation of single lineage colonies from this population suggests some artificial limitations on the differentiation process in these culture conditions. This demonstrates the important point that all in vitro assays are selective in terms of what cell lineages can grow, and thus conclusions regarding the true differentiation potential of the responding cells must be treated with caution. Given the lack of stem cell-specific differences in colony formation and this restriction in differentiation, it seems unlikely that the methylcellulose culture assay by itself can serve as a quantitative measure of PHSC.

One colony attribute which may indicate proliferation from a stem cell is a delayed onset of growth. Ogawa and colleagues have described a blast cell colony assay in which certain progenitors remain quiescent for several days. These progenitors appear to have some self-renewal capacity (NAKAHATA and OGAWA 1982; IKEBUCHI et al.1987). While the blast cell assay has not yet been applied to all the bone marrow progenitors listed in Table 2, it appears that only the Thy^{lo}, L^-, S^+ fraction, and not the other subsets, contains such a quiescent population (HEIMFELD et al. 1991b). Only a subset of these cells showed delayed growth, which could indicate some heterogeneity in this population. The nature of the quiescent Thy^{lo}, L^-, S^+ subpopulation remains unresolved, although it does not appear to be related to resting versus activated stem cells as detected by rhodamine-123 uptake (HEIMFELD et al. 1991b).

4 Conclusion

To summarize, we have identified several classes of pluripotent and restricted progenitor cells in normal mouse bone marrow. Most of the precursor populations are characterized by low-level expression of Thy-1. The most primitive stem cells are negative for the cocktail of "lineage" differentiation antigens, but do have high levels of Sca-1 on their surface (MÜLLER-SIEBURG et al. 1986; SPANGRUDE et al. 1988a; HEIMFELD et al. 1990, 1991a). Using these progenitor cell subsets as tools, we have probed the stem cell specificity of various in vivo and in vitro assays. As shown by many others, neither CFU-S, pre-T cell, nor multilineage reconstitution activity is unique to the PHSC. However, it should be emphasized that some PHSC may contribute to this activity. To positively determine whether a stem cell has reconstituted an irradiated animal requires a clonal analysis and the transfer of donor progenitors to a new secondary host. Repopulation with a single isolated cell (SMITH et al. 1991) or the use of unique retroviral markers (eg. LEMISCHKA et al. 1986; I.R. Lemischka, this volume) are examples of such an analysis.

Given the inherent difficulties in quantitating in vivo assays, and the lack of such assays for human cells, alternative methods to define PHSC are required. Our results suggest that many of the properties used to differentiate colonies in the methycellulose culture system, such as HPP or mixed lineage colonies, can not distinguish PHSC from other progenitor cells. However, the ability to remain quiescent for a period of time may be unique property of PHSC. In LTMC only the Thy^{lo}, L^-, S^+ stem cell fraction was capable of long-term growth, and additionally of generating new progenitor cells. This suggests that reconstitution of LTMC can be used as an assay to distinguish PHSC from committed progenitors and may serve as a useful alternative to the reconstitution and secondary transfer of progenitors into irradiated animals. Furthermore, these in vitro assays should provide a better means of studying the earliest stages in hematopoietic development and how regulation of the processes of self-renewal and differentiation are achieved.

References

Abramson S, Miller RG, Phillips RA (1977) The identification in adult bone marrow of pluripotent and restricted stem cells of the myeloid and lymphoid systems. J Exp Med 145: 1567–1577

Bartelmez SH, Bradley TR, Bertoncello I, Mochizuki DY, Tushinski RJ, Stanley ER, Hapel AJ, Young IG, Kriegler AB, Hodgson GS (1989) Interleukin 1 plus interleukin 3 plus colony-stimulating factor 1 are essential for clonal proliferation of primitive myeloid bone marrow cells. Exp Hematol 17: 240–245

Bradley TR, Hodgson GS (1979) Detection of primitive macrophage progenitor cells in mouse bone marrow. Blood 54: 1446–1452

Coffman RL, Weissman IL (1983) Immunoglobulin gene rearrangement during pre-B cell differentiation. J Mol Cell Immunol 1: 31–37

Dexter TM, Testa NG, Lajtha LG (1976) Conditions controlling the proliferation of hematopoietic stem cells in vitro. J Cell Physiol 91: 335–342

Dorshkind K, Johnson A, Collins L, Keller G, Phillips R (1986) Generation of purified stromal cell cultures that support lymphoid amd myeloid precursors. J Immunol Methods 89: 37–47

Heimfeld S, Guidos CJ, Holzmann B, Siegelman MH, Weissman IL (1990). Developmental analysis of the mouse hematolymphoid system. Cold Spring Harbor Symp Quant Biol 54: 75–85

Heimfeld S, Spangrude GJ, Smith L, Weissman IL (1991a) Identification and characterization of several classes of mouse hematopoietic progenitor cells (submitted)

Heimfeld S, Hudak S, Weissman IL, Rennick D (1991b) The in vitro response of phenotypically defined stem cells and myeloerythroid progenitors to single or defined growth factors. Proc Natl Acad Sci USA (in press)

Ikebuchi K, Wong GC, Clark SC, Ihle JN, Hirai Y, Ogawa M (1987) Interleukin 6 enhancement of interleukin 3 dependent proliferation of multipotent progenitors. Proc Natl Acad Sci USA 84: 9035–9039

Lemischka IR, Raulet DH, Mulligan RC (1986) Developmental potential and dynamic behavior of hematopoietic stem cells. Cell 45: 917–921

Magli MC, Iscove NN, Odartchenko N (1982) Transient nature ofearly hematopoietic spleen colonies. Nature 295: 527–531

McCulloch EA (1983) Stem cells in normal and leukemic hemopoiesis. Blood 62: 1

McNiece IK, Stewart FM, Deacon DM, Quesenberry PJ (1988) Exp Hematol 16: 383–387

Metcalf D (1989) The molecular control of cell division, differentiation, commitment and maturation in haemopoietic cells. Nature 339: 27–30

Mulder AH, Visser JWM, van den Engh GJ (1985) Thymus regeneration by bone marrow cell suspensions differing in the potential to form early and late spleen colonies. Exp Hematol 13: 768–772

Müller-Sieburg CE, Whitlock CA, Weissman IL (1986) Isolation of two early B lymphocyte progenitors from mouse bone marrow: a committed Pre-pre-B cell and a clonogenic Thy-1^{lo} hematopoietic stem cell. Cell 44: 653–662

Müller-Sieburg CE, Townsend K, Weissman IL, Rennick D (1988) Proliferation and differentiation of highly enriched mouse hematopoietic stem cells and progenitor cells in response to defined growth factors. J Exp Med 167: 1825–1840

Nakahata T, Gros AJ, Ogawa M (1982) A stochastic model of self-renewal and commitment to differentiation of the primitive hemtopoietic stem cell in culture. J Cell Physiol 113: 455–458

Ploemaecher RE, van der Sluis JP, Voerman JS, Brons NH (1989) An in vitro limiting dilution assay of long-term repopulating hematopoietic stem cells in the mouse. Blood 74: 2755–2761

Reif AE, Allen JMV (1964) The AKR thymic antigen and its distribution in leukemias and nervous tissues. J Exp Med 120: 413

Scheid M, Triglia D (1979) Further descriptions of the Ly-5 system. Immunogenetics 9: 423–428
Siminovitch L, McCulloch EA, Till JE (1963) The distribution of colony-forming cells among spleen colonies. J Cell Comp Physiol 62: 327–336
Smith LG, Weissman IL, Heimfeld S (1991) Clonal analysis of hematopoietic stem cell differentiation in vivo. Proc Natl Acad Sci 88: 2788–2792
Spangrude GJ, Heimfeld S, Weissman IL (1988a) Purification and characterization of mouse hematopoietic stem cells. Science 241: 58–62
Spangrude GJ, Müller-Sieburg CE, Heimfeld S, Weissman IL (1988b) Two rare populations of mouse Thy-1^{lo} bone marrow cells repopulate the thymus. J Exp Med 167: 1671–1677
Springer T, Galfre G, Secher DS, Milstein C (1979) Mac-1: a macrophage differentiation antigen identified by monoclonal antibody. Eur J Immuol 9: 301
Sutherland HJ, Landsdorp PM, Henkelman DH, Eaves AC, Eaves CJ (1990) Functional characteristics of individual human hematopoietic stem cells cultured at limiting dilution on supportive marrow stromal layers. Proc Natl Acad Sci 87: 3584–3588
Tidmarsh GF, Heimfeld S, Whitlock CA, Weissman IL, Müller-Sieburg CE (1989) Identification of a novel bone marrow-derived B-cell progenitor population that coexpresses B220 and Thy-1 and is highly enriched for Abelson Leukemia virus targets. Mol Cell Biol 9: 2665–2671
Till JE, McCulloch EA (1961) A direct measurement of the radiation sensitivity of normal mouse bone marrow cells. Raidiat Res 14: 213–222
Till JE, McCulloch EA (1980) Hemopoietic stem cell differentiation. Biochim Biophys Acta 605: 431–459
Weilbaecher K, Weissman IL, Blume K, Heimfeld S (1991) Culture of phenotypically defined hematopoietic stem cells and other progenitors at limiting dilution of Dexter monolayers. Blood 78: 945–952
Whitlock CA, Robertson D, Witte ON (1984) Murine B cell lymphopoiesis in long-term culture. J Immunol Methods 67: 353–359
Worton RG, McCulloch EA, Till JE (1969) Physical separation of hematopoietic stem cells differing in their capacity for self-renewal. J Exp Med 130: 91–103
Wu AM, Till JE, Siminovitch L, McCulloch EA (1968) Cytological evidence for a relationship between normal and hematopoietic colony-forming cells and cells of the lymphoid system. J Exp Med 127: 455–464

Pluripotent Stem Cells and Early B Lymphocyte Precursors in Mice

C. E. MÜLLER-SIEBURG, N. MARTINA, and J. P. WINEMAN

1 Introduction

The hematopoietic system is developmentally active throughout the lifespan of an animal, thus providing an extraordinary model for the elucidation of events that regulate differentiation and lineage commitment. It is well established that a single pluripotent stem cell can generate cells of all hematopoietic lineages via series of complex differentiation steps. Stem cells not only give rise to mature cells, but also self-renew to maintain a pool of themselves. Great progress has been made in determining the lineage relationships in the hematopoietic differentiation cascade. This is particularly true for the myeloid and erythroid lineages, due to the availability of a variety of clonal in vivo and in vitro assays which define distinct differentiation stages in these lineages (see R.J. Jones, I. Bertoncello, I.B. Pragnell, this volume). However, a number of important issues remain to be resolved. For example, how the balance between differentiation and self-maintenance is regulated is poorly understood. Little is known about the process of commitment, how a stem cell decides to differentiate along a certain lineage. It is difficult to address these issues without separating the relevant cells from the complex mixture found in bone marrow. This is particularly important since some mature cells, such as T lymphocytes and stromal cells, can

Medical Biology Institute, 11077 North Torrey Pines Road, La Jolla, CA 92307, USA

Current Topics in Microbiology and Immunology, Vol. 177

profoundly influence the differentiation of cells at the early maturation stages. Furthermore, in studies aimed at defining regulatory signals for stem cells, the response of more mature cells might obscure stem cell activity. Additionally, mature cells and committed precursors might enhance or interfere with stem cell activity in bone marrow transplantation. Thus, a major goal in the study of hematopoiesis has been to isolate populations of pluripotent stem cells free of committed precursors or mature cells.

Of equal importance is the characterization of developmentally early precursors committed to differentiation along the hematopoietic lineages. Particularly, the study of lymphoid differentiation from stem cells has been hindered by the paucity of clonal assays and the lack of markers for early lymphoid precursors. Thus, many questions about early lymphocyte differentiation remain to be clarified. For example, it is unclear whether stem cells commit directly to the B and T cell lineages or via a common lymphoid precursor (FULOP and PHILLIPS 1989; SPANGRUDE et al. 1988a; SNODGRASS and KELLER 1987). Furthermore, information is scarce on the timing and regulation of the molecular events, such as rearrangement of the genes encoding the antigen receptors, that accompany lymphoid development.

The availability of markers that allow the isolation of stem cells or early lymphoid precursors, and assay systems that define these cells are prerequisites for the analysis of the events that regulate self-maintenance, commitment, and differentiation. In this chapter we will describe the currently available cell surface markers that allow the enrichment and characterization of pluripotent stem cells and early precursors restricted to the B lymphocyte lineage. Furthermore, we will discuss the features of some of the assay systems that detect and define these cells.

2 Stem Cell-Enriched Populations

2.1 Detection of Pluripotent Stem Cells

The characterization of pluripotent stem cells has received much attention due to its central role in hematopoiesis and its importance in clinical application. Unfortunately, the term "stem cell" has been used for many different cell types. In fact, this volume was conceived in the hope of generating a consensus about "what is a pluripotent stem cell, and how is it best measured?" For the sake of this chapter, we define pluripotent stem cells functionally, by their capacity to give rise to and to maintain all hematolymphoid lineages for at least the majority of the life span of an animal. Since in vitro assays that support the growth and differentiation of all hematopoietic cells are not available currently, stem cells are generally analyzed in vivo by their capacity to repopulate a compromised host. The host can be genetically deficient, like WW mice which have a stem cell defect, or like SCID mice which lack mature lymphocytes. The latter mice accept grafts from other species (MOSIER et al. 1988; MCCUNE et al. 1988). Alternatively, marrow ablation in normal recipients can be achieved by cytotoxic drugs or irradiation rendering the animals receptive to engraftment by exogenous stem cells.

A number of parameters can influence the outcome of in vivo repopulation assays. For example, the severity of marrow deficiency or the extent of marrow ablation clearly affects the capacity of stem cells to engraft (DOWN et al. 1991). Depending on the severity of the mutation, repopulation of some lineages is delayed or suppressed. For example, few donor-derived myeloid cells can be detected in bone marrow-reconstituted SCID mice unless the hosts are irradiated (FULOP and PHILLIPS 1989). In some WW strains, lymphoid development is slower than in irradiated mice (HARRISON and ASTLE 1991). In mice that have received marrow ablation regimens, primitive host stem cells can survive. It appears that these host stem cells have a proliferative advantage over transplanted stem cells. Donor stem cells presumably are compromised by manipulation, and the seeding efficiency to the appropriate host microenvironment may be low after intravenous injection (HARRISON and ASTLE 1982). Thus, host-derived cells can contribute a substantial fraction of the hematopoietic system post transplantation, particularly if low numbers of cells are injected. Furthermore, hosts that have been conditioned by lethal irradiation require at least two populations of cells for survival and full engraftment. Besides primitive stem cells, responsible for long-term repopulation, cells that rapidly provide mature hematopoietic cells are needed for the survival of the animal (JONES et al. 1990). It has been reported that the latter cells are pluripotent, but have limited proliferative capacity (SPANGRUDE and JOHNSON 1990). If a stem cell-depleted source of these cells is injected, lethally irradiated mice can survive with little or no long-term hematopoiesis derived from the transplanted cells (SPANGRUDE and JOHNSON 1990; JONES et al. 1990). On the other hand, irradiated mice that receive stem cells without "radioprotecting" cells may die. Therefore, when the stem cell content of marrow subpopulations is assessed, it is often benefical to co-inject untreated marrow, to ensure the survival of the hosts. This assay, referred to as the competitive repopulation assay, can be used to quantitate stem cells in limiting dilution. However, it has been reported that committed precursors can contribute substantially to post-transplantation hematopoiesis for several months (see D.E. Harrison, this volume, for a detailed description of the competitive assay). Thus, in all in vivo repopulation assays it is necessary to determine the extent of repopulation by donor cells to accurately assess the true stem cell potential in the graft.

A variety of systems have been used to discriminate donor and host hematopoietic cells. In allogeneic transplantation, histocompatibility markers can serve for this purpose (MINTZ et al. 1984). In syngeneic or congenic transplantation, chromosomal aberrations have been used (ROSS et al. 1982; WU et al. 1968). Allelic differences affecting the size or charge of proteins or genes, and most recently retroviral markers, have been employed to detect donor cells (HARRISON et al. 1989; MÜLLER-SIEBURG et al. 1986; WILLIAMS et al. 1984; KELLER et al. 1985; NAKANO et al. 1989). These methods tend to require large numbers of cells for detection, and donor cells can be quantitated only indirectly. In contrast, the transplantation system that utilizes Ly-5 congenic mice, allows the accurate measurement of the frequency of donor type cells (SPANGRUDE et al. 1988b). C57B1/6 mice express the Ly-5.2 allele of this cell surface antigen (using the revised nomenclature for Ly loci; MORSE et al. 1987). A congenic mouse strain that carries the Ly-5.1 allele on the C57BL/6 background is available. The allelic difference in the Ly-5 antigen does not affect transplantation. Monoclonal antibodies (mAb) specific for each of the two Ly-5 alleles have been derived (SHEN 1981; SPANGRUDE and SCOLLAY 1990a) which allow detection of donor and host cells by immunofluorescence. Since the Ly-5 antigen is expressed on all hematolymphoid cells with the exception of red blood cells (SCHEID and TRIGLIA 1979), the extent of repopulation in the B, T, and myeloid lineages can be assessed readily by two-color staining using antibodies specific for these lineages in addition to allele-specific anti-Ly-5 antibodies. Transplanted mice can be tested repeatedly as the immunofluorescence analysis can be performed on peripheral white blood cells. Thus, the Ly-5 congenic transplantation system is ideally suited for repeated, sensitive, and reliable quantification of the progeny of stem cells and precursors.

2.2 Cell Surface Markers for Stem Cells

Considerable progress has been made in obtaining populations highly enriched for hematopoietic stem cells from normal mouse bone marrow and fetal liver (VISSER et al. 1984; BERTONCELLO et al. 1985, 1986; MÜLLER-SIEBURG et al. 1988; SPANGRUDE et al. 1988a; JORDAN et al. 1990). Following reports that spleen colony-forming units (CFU-S) express the cell surface antigen Thy-1 (BASCH and BERMAN 1982), we showed that all pluripotent stem cells that repopulate lethally irradiated mice are contained in a population of bone marrow cells that expresses low levels of Thy-1, but lacks high levels of a variety of markers expressed on mature hematopoietic cells (Thy-1^{lo}Lin$^-$ cells). These markers include: the B lymphocyte-specific marker B220, the T lymphocyte-specific markers CD-4 and CD-8, the macrophage marker Mac-1, and the Gr-1 marker that is expressed on granulocytes and eosinophils. Using this combination of cell surface markers, stem cells that resuce and completely repopulate lethally irradiated hosts for $\geqq 3$ months can be enriched about 300-fold in a single-step cell sorting procedure (MÜLLER-SIEBURG et al. 1988). Recent data from our laboratory and from others (CEMS, unpublished; SPANGRUDE and SCOLLAY 1990a) show that stem cells also lack CD-5 and the transferrin receptor.

Similar enrichments for pluripotent stem cells have been obtained by selecting low-density cells that express high levels of class I antigen and the receptor for wheat-germ agglutinin (VISSER et al. 1984). Stem cells also express the Ly-6A/E antigen which has been named Sca-1 by SPANGRUDE et al. (1988a). Stem cells can be enriched roughly 100-fold in a population characterized as

Sac-1^+Lin$^-$ (SPANGRUDE and SCOLLAY 1990b). Furthermore, the Sca-1 marker can be used to dissect the Thy- 1^{lo}Lin$^-$ population into stem cell-rich and stem cell-depleted subpopulations, resulting additional enrichment for stem cells. Other markers on pluripotent stem cells include Qa-m7 (BERTONCELLOet al. 1986), the antigen defined by mAb AA4.1 (JORDAN et al. 1990), and CD-44 detected by mAb Pgp-1 (TOWNBRIDGE et al. 1982; SPANGRUDE and SCOLLAY 1990a). The dye rhodamine-123, which preferentially labels metabolically active cells, can be used to further define stem cell-enriched populations. Primitive stem cells with extensive repopulation capacity are small resting cells and do not label well with this dye (BERTONCELLO et al. 1985; SPANGRUDE and JOHNSON 1990). These markers clearly detect overlapping marrow populations as they all label stem cells. However, each new marker, or combinations thereof, have revealed new aspects about stem cells and their immediate progeny, the lineage restricted precursors. The separation of long-term repopulating stem cells and radioprotecting precursors (JONES et al. 1990) is a prime example of important advances in our understanding of stem cell biology reached through the introduction of new markers. Thus, it is reasonable to assume that additional new markers will provide further insights into hematopoietic differentiation stages.

We have recently described a novel stem cell antigen defined by mAb Fall-3 (MÜLLER-SIEBURG 1991). Fall-3 is one of a series of mAbs that we have generated against Lin$^-$ bone marrow cells. It stains about 15%–30% of nucleated bone marrow cells. Two-color immunofluorescence with mAb Fall-3 and lineage-specific markers suggests that Fall-3 is expressed predominantly on immature cells. The majority of Thy-1^+or Sca-1^+ cells in bone marrow lack the Fall-3 antigen, indicating that mAb Fall-3 defines a novel stem cell antigen. Fall-3^+ cells, isolated from bone marrow, contain most, if not all, pluripotent stem cells that rescue and completely repopulate all lineages in lethally irradiated hosts for $\geqq$5 months. Little, if any, repopulating capacity is found in the Fall-3^- subset as assessed in both the radioprotection and in the competitive repopulation assays (MÜLLER-SIEBURG 1991, and unpublished results). Interestingly, the Fall-3^+ populations can be subdivided into Fall-3^{hi} and Fall-3^{lo} cells. Preliminary results suggest that pluripotent stem cells are contained within both subsets. However, it appears that Fall-3^{hi} and Fall-3^{lo} stem cells have remarkably different kinetics of repopulation, suggesting that mAb Fall-3 could be used to separate functionally distinct subsets of pluripotent stem cells.

2.3 Hematopoietic Activities in Stem Cell Enriched Populations

All stem cell-enriched populations described to date contain cells that are active in a variety of in vitro and in vivo assays. For example, stem cell-enriched populations have been reported to be enriched for CFU-S (VISSER et al. 1984; MÜLLER-SIEBURG et al. 1986; SPANGRUDE et al. 1988a; BERTONCELLO et al. 1986; SZILVASSY et al. 1989), to give rise to osteoclasts in vitro (SCHEVEN et al. 1986), to generate T lymphocytes upon intrathymic injection (SPANGRUDE et al. 1988a, b), to respond to the stroma in both Dexter and Whitlock–Witte conditions (MÜLLER-SIEBURG et al. 1986; SUTHERLAND et al. 1989; S. Heimfeld and I.L. Weissman, this volume), and to give rise to mixed myeloid-erythroid colonies in vitro upon stimulation by IL-3 (MÜLLER-SIEBURG et al. 1988). With the help of new markers, it has been shown clearly that some of these assays do not detect stem cells, but define lineage-restricted precursors, particularly in the myeloid-erythroid lineages. This recent success in defining early lineage- restricted precursors in stem cell-enriched populations prompted us to examine the Thy-1^{lo}Lin^{-} populations for the presence of lymphocyte-restricted precursors. The Thy-1^{lo}Lin^{-} population contains cells that are active in most of the assay systems described above. Most notably, this population is highly enriched for cells that respond to the stroma in lymphoid bone marrow culture by differentiating along the B cell lineage (MÜLLER-SIEBURG et al. 1986). Both stem cells and stroma-responsive cells are completely contained within the Thy-1^{lo}Lin^{-} population and also within the Thy-1^{lo}Lin^{-}Sca-1^{+} subpopulation (S. Heimfeld and I.L. Weissman, this volume). Thus, it was tempting to speculate that the capacity to initiate these cultures was a reflection of stem cells themselves. However, recent data from our laboratory indicate that the stroma-responsive cell is an early precursor restricted to the B lymphocyte lineage that can be separated from pluripotent stem cells.

3 B Cell Lineage Differentiation

3.1 Stages of B Lineage Differentiation

Early B cell differentiation in the adult animal occurs predominantly in the bone marrow. Therefore, the bone marrow provides microenvironments not only for stem cell self-maintenance and differentiation, but also for the growth and differentiation of B lymphocytes. The cellular and molecular events in B lymphocyte maturation, beginning with the pro-B cell, are well understood (for review see PHILLIPS 1989; KINCADE 1984).

The functionally most important step in B cell differentiation is the re-arrangement of the immunoglobulin (Ig) gene segments that ultimately leads to the expression and secretion of antibodies (ALT et al. 1986). The first step in Ig gene rearrangement is believed to occur at the level of the pro-B cell when the D and J segments of the Ig heavy chain are joined. Subsequently, at the large pre-B cell stage, the Ig-H V segment is joined to this complex to generate a complete Ig-H chain gene which is expressed in the cytoplasm of pre-B cells. At the next step, small pre-B cells rearrange and express their Ig light chain genes to become B cells that bear complete Ig molecules on their cell surfaces. B cells then can mature to antibody- secreting plasma cells after receiving the appropriate stimuli by T cells and/or cytokines.

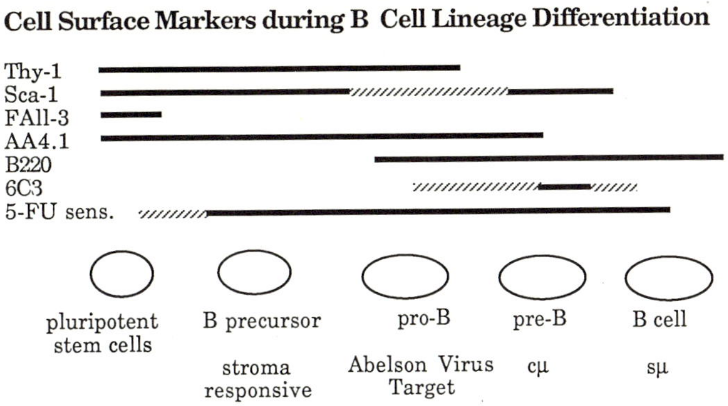

Fig. 1. Cell surface markers during B lineage differentiation. The classification of B lineage stages is based on PHILLIPS, (1989). For references and explanation of markers see text. *Broken lines*, expression of markers on these cell stages has not been formally examined, however appears likely. *5-FU sens.*, sensitive to 5-FU; *cμ*, cytoplasmic μ chain of immunoglobulin; *sμ* surface IgM

The analysis of B cell differentiation has been facilitated by the availability of cell surface markers; mAb specific for these antigens allow the discrimination of B lineage cells from other cells and the isolation of distinct stages of B lineage cells. For example, the B220 antigen is expressed on B lineage cells from the large pre-B cells to mature B cells (COFFMAN 1983). Large pre-B cells co-express B220 and low levels of Thy-1 (TIDMARSH et al. 1989). The 6C3/BP-1 antigen is found on a subset of bone marrow pre-B cells and on most cultured pre-B cells (TIDMARSH et al. 1985; COOPER et al. 1986). The antigens defined by mAb AA4.1 and Sca-1 are expressed on a subset of pre-B cells (SPANGRUDE and SCOLLAY 1990a; MCKEARN et al. 1984). It has been reported that pro-B cells, or perhaps even earlier B cell stages, also bear the AA4.1 antigen (PALACIOS et al. 1984). Thus, the AA4.1 antigen appears to be expressed on early B lineage cells beginning with the stem cells (JORDAN et al. 1990) through the pre-B cell. The distribution of Sca-1 on immature B lineage differentiation stages has not been examined. The occurrence of these markers during B lineage differentiation are depicted in Fig. 1.

A variety of in vitro systems has been described that permit analysis of the requirements for growth and differentiation of many of these B cell stages. Most of these assays detect the maturation of $B220^{+}$ pre-B or B cells to plasma cells (reviewed by PAIGE and WU 1989; KINCADE 1984). A short-term culture that allows the maturation of $B220^{-}$ bone marrow cells into $B220^{+}$ pre-B cells in the presence of a stromal cell-derived cytokine has been reported by LANDRETH and DORSHKIND (1988). Perhaps the most complete system that allows the study of B lineage differentiation, beginning with a very early B lineage precursor all the way through to mature B cells, has been described by WHITLOCK and WITTE (1982).

3.2 B Lymphopoiesis in Whitlock–Witte Cultures

These long-term lymphoid bone marrow cultures (WHITLOCK and WITTE 1982) consist of two major components: an adherent stromal cell layer and nonadherent small lymphocytes. If fresh bone marrow cells are seeded onto preestablished stromal cells, a transient burst of myeloid cells is followed by long-term (> 6 month) lymphopoiesis. About 10% of these cells are mature B cells as they bear surface IgM and, to a lesser extent, surface IgD. Roughly 30% of the cells express detectable levels of cytoplasmic Ig heavy chains, about 50% of the cells bear the B lineage-specific marker B220, and 60%–90% of the cells express 6C3/BP-1 (WHITLOCK et al. 1985; WITTE et al. 1987). Analysis of the status of the immunoglobulin genes indicate that most, if not all, cells have undergone D to J locus rearrangement. Furthermore, these cells can reconstitute the B cell lineage in immunodeficient SCID mice and mature in these hosts to plasma cells (KURLAND et al. 1984; NAGASAWA et al. 1985). Taken together, these data place the nonadherent cells found in Whitlock–Witte cultures firmly in the B cell lineage. Thus, these cultures provide an in vitro model of early B lymphopoiesis.

The isolation of monoclonal stromal cells from the adherent cells has expanded the utility of the culture system (WHITLOCK et al. 1987; HUNT et al. 1987; COLLINS and DORSHKIND 1987; RENNICK et al. 1987; ZIPORI et al. 1985). These stromal cell lines sustain B lymphopoiesis as well as the heterogeneous stroma layer, although most of the lines do not promote the acquisition of surface IgM by B lineage cells (WHITLOCK et al. 1987). Maturation from pre-B cell to B cell requires macrophages (GISLER et al. 1987) or a distinct class of stromal cells (HENDERSON et al. 1990). Stromal cell lines provide a constant microenvironment from culture to culture which is a prerequisite for limiting dilution analysis of hematopoietic precursors.

3.3 Limiting Dilution Analysis in Whitlock–Witte Cultures

The capacity to initiate Whitlock–Witte cultures in limiting dilution has been used as an assay system for early hematolymphoid precursors (MÜLLER-SIEBURG et al. 1986, 1991). When bone marrow cells are seeded onto monoclonal stromal cell lines in limiting dilution, about one in 500 to one in 1000 cells responds by forming a colony of B lineage cells. When the fraction of negative wells is plotted against the number of seeded cells, a straight line through the origin of the graph is obtained (MÜLLER-SIEBURG et al. 1986). This suggests that the system provides all factors necessary for B lineage differentiation, and that it is measuring one limiting cell, namely the B cell precursor that initiates the cultures. That the colonies formed are in fact B lineage cells has been shown by immunofluorescence following expansion of these cells. All colonies tested contain cells that express the B220 antigen (see Fig. 2b) (MÜLLER-SIEBURG et al. 1986).

One drawback of the limiting dilution system is that the cultures are unstable. The apparent frequency of cells that initiate cultures declines over time

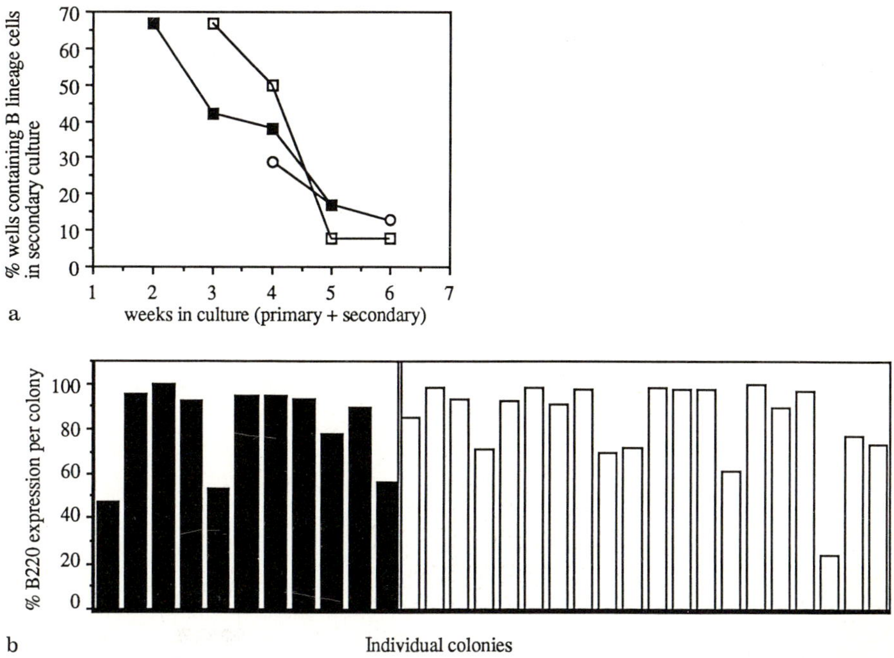

Fig. 2 a, b. Growth and phenotype of colonies in limiting dilution in Whitlock–Witte cultures. **a** Replating efficiency: Bone marrow (200 cells per well) was seeded on pre-established AC-3.GG, a monoclonal stromal cell line, in 96 well plates. In weekly intervals, individual colonies were harvested from the 96 well plates and transferred to 24 well plates containing fresh AC-3.GG. The number of secondary cultures that contained live, small, non-adherent cells (at least 10^4 cells/well) were monitored weekly. Depicted is the precentage of secondary cultures obtained from colonies picked at 1 week (■), 2 weeks (□), and 3 weeks (○) versus total culture time. **b** Level of B220 antigen expression on colonies picked from 96 well plates at 1 week (■) and 2 weeks (□) in limiting dilution. All cultures were grown for 4 weeks in secondary culture before the level of B220 antigen expression was determined by immunofluorescence

(WHITLOCK et al. 1987). Decline over time has been observed also in myeloid bone marrow cultures (PLOEMACHER et al. 1989; SUTHERLAND et al. 1989). In this system, the decline has been interpreted to reflect the gradual disappearance of precursors with restricted proliferative capacity until only colonies derived from highly proliferative primitive stem cells remain. To test whether the decline observed in Whitlock–Witte cultures reflects heterogeneity of the cells that initiate these cultures, we assessed the proliferative capacity of cells cultured for different times. Two hundred bone marrow cells per well were seeded onto the stromal cell line AC-3.GG. Colonies were picked after 1, 2 and 3 weeks of culture and transferred to new vessels containing AC-3.GG. If the colonies found at later times in limiting dilution were derived from precursors with high proliferative capacity, these cultures should have a high replating potency. However, we found that, regardless of the time spent in the primary culture, the capacity to initiate secondary cultures that lasted for at least 6 weeks was roughly the same (Fig. 2a). Furthermore, the level of B220 antigen expression in individual colonies was independent of the time in primary culture (Fig. 2b). Thus, the decline in colonies over time seen in Whitlock–Witte cultures may be due to the culture condition, rather than reflecting heterogeneity of the precursors. Long-term proliferation in bone marrow cultures may be a two-step event. The initial proliferation in the first few weeks most likely reflects the proliferative capacity of the bone marrow precursor, while proliferation after this time could be due to a rare event of "culture adaption". In fact, Whitlock–Witte cultures that are kept for several months become pauciclonal (WHITLOCK et al. 1985), and the cultured cells differ in phenotype and growth requirements from their normal counterparts in bone marrow (MÜLLER-SIEBURG et al. 1986).

Nevertheless, the capacity to form a colony in response to the stromal elements in short-term Whitlock–Witte cultures has proven to be a reliable and sensitive assay for early hematolymphoid cells. The capacity to initiate Whitlock–Witte cultures in limiting dilution resides exclusively in the stem cell-enriched Thy-1^{lo}Lin$^-$ bone marrow population (MULLER-SIEBURG et al. 1986). Purified bone marrow cells with the phenotype Thy-1^-Lin$^-$ or B220$^+$ pre-B cells cannot initiate limiting dilution cultures. However, when these cells were seeded at 10^4 cells per vessel they were maintained for several weeks in Whitlock–Witte culture. Although a subset of these cells proliferated for a limited time, little, if any, expansion of the cultures was observed. Probably, their proliferation was counterbalanced by cell death. This restricted capacity to proliferate in culture correlates well with the limited number of cell division that these cells undergo during maturation in vivo (for review see OSMOND and PARK 1987). In summary, the limiting dilution assay detects only cells that are able to expand significantly in response to the stromal elements.

While the cells that initiate Whitlock–Witte cultures in limiting dilution co-enrich with stem cells in the Thy-1^{lo}Lin$^-$ and the Thy-1^{lo}Lin$^-$Sca-1^+ populations, it became apparent that they were not the same cells. The first evidence that suggested that long-term repopulating stem cells and stroma-responsive cells were distinct came from studies that measured the kinetics of recovery after injection of 5-fluorouracil (5-FU). A single injection of the cytotoxic drug 5-FU affected the stroma-responsive cells much more severely than stem cells. Furthermore, the stroma-responsive cells recovered later than stem cells, suggesting a precursor-progeny relationship between stem cells and stroma-responsive cells. That these cells are distinct was substantiated when we showed that the antigen defined by mAb Fall-3 is expressed on most stem cells, but not on the majority of stroma-responsive cells (MÜLLER-SIEBURG 1991). Thus, the combination of a novel marker (Fall-3) with a sensitive assay system allowed us to define a novel precursor restricted to the B lineage. This cell is most likely the most primitive precursor restricted to the B cell lineage yet described. The absence of B lineage-specific markers, its extensive proliferative capacity. its failure to be transformed by Abelson virus in vitro (TIDMARSH et al. 1989), and its delayed differentiation into mature B cells place this cell very early in the B cell lineage (MÜLLER-SIEBURG et al. 1986). As we are able to separate this B cell precursor from primitive stem cells, it now will be possible to analyze the developmental potential of both of these cells more thoroughly in vivo and in vitro. Furthermore, it will be interesting to determine whether the Fall-3$^-$ cells can give rise to T lymphocytes.

4 Conclusion

The analysis of early hematolymphoid differentiation has been hampered by the heterogeneity of the hematopoietic tissues, the low frequency of stem cells and precursors, and the complex interactions between hematopoietic cells. Consequently, much effort has been and is dedicated to reducting this complexity to render the hematopoietic system amenable for analysis. This endeavor is dependent on the availability of *both* specific cell markers and well-defined assay systems. Markers, be they cell surface antigens, or differences in density or metabolic states of cells, are prerequisites for separating and isolating distinct cells in the hematopoietic differentiation cascade. It is intriguing to watch the progress in the attempts to purify pluripotent stem cells. Early on, stem cells where enriched by depleting mature cells by injection of cytotoxic drugs, cell density separation, or cell sorting (Lin$^-$). Recently, very early lineage- committed precursors in the myeloid and B lymphocyte lineages have been separated from pluripotent stem cells. Thus, while the currently available stem cell markers still detect a variety of committed cells in addition to stem cells, it can be assumed that the introduction of more markers will lead to the purification of stem cells to homogeneity.

Clearly, the purification of stem cells and precursors is equally dependent on reliable, well-defined assay systems. Most of the currently available assays that measure clonogenic cells had been described for the first time many years ago. In the last several years, much progress has been made in defining the components of many of the in vitro assays, particularly those that define myeloid and erythroid precursors. Most notably, the purification and molecular cloning of many cytokines are a direct result of this analysis.

Assays that define pluripotent stem cells and very early precursors in the lymphoid lineages have been more difficult to develop. Currently, the most comprehensive and stringent way to assess these cells is to measure their proliferation and differentiation capacity in vivo. Unfortunately, the in vivo assays are time consuming, require large numbers of animals, and most importantly, cannot be used in the human system. Long-term bone marrow cultures, which mimic aspects of the in vivo hematopoiesis probably are the most promising substitutes for in vivo systems. In fact, the capacity to initiate myeloid bone marrow cultures (Dexter conditions) has been shown to reside in cells that are phenotypically closely related to pluripotent stem cells in mouse and humans (PLOEMACHER et al. 1989; SUTHERLAND et al. 1989). The identification of an early B lineage-restricted precursor by its response to the stroma in Whitlock–Witte cultures (MÜLLER-SIEBURG 1991) is another example of the usefulness of bone marrow cultures to detect early hematolymphoid cells. Thus, it can be anticipated that variations or refinements of these cultures may provide definitive in vitro assays for stem cells and early precursors.

Acknowledgements. This work was supported by National Institutes of Health Grant DK41214 and by an American Leukemia Society Scholar Award to CEMS. I thank Drs. C. Cowing, K. Dorshkind, D. Katz, Sieburg, and H.B. Sieburg for critical review of this manuscript and numerous suggestions and discussions.

References

Alt FW, Blackwell TK, DePinho RA, Reth MG, Yancopoulos GD (1986) Regulation of genome rearrangement events during lymphocyte differentiation. Immunol Rev. 89: 5–30

Basch RS, Berman JW (1982) Thy-1 determinants are present on many murine hematopoietic cells other than T cells. Eur J Immunol 12: 359–364

Bertoncello I, Hodgson GS, Bradley TR (1985) Multiparameter analysis of transplantable hemopoietic stem cells: I. The separation and enrichment of stem cell homing to the marrow and spleen on the basis of rhodamine 123 fluorescence. Exp Hematol 13: 999–1006

Bertoncello I, Bartelmez SH, Bradley TR, Stanley ER, Harris RA, Sandrin MS, Kriegler AB, McNiece IK, Hunter SD, Hodgson GS (1986) Isolation and analysis of primitive progenitor cells on the basis of differential expression of Qa-m7 antigen. J Immunol 136: 3219–3224

Coffman RL (1983) Surface antigen expression and immunoglobulin gene rearrangement during mouse pre-B cell development. Immunol Rev 69: 5–23

Collins L, Dorshkind K (1987) A stromal cell line from myeloid long-term bone marrow cultures can support myelopoiesis and B lymphopoiesis. J Immunol 138: 1082–1087

Cooper MD, Mulvaney D, Coutinho A, Cazenave PA (1986) A novel cell surface molecule on early B-lineage cells. Nature 321: 616–618

Down JD, Tarbell NJ, Thames HD, Mauch PM (1991) Syngeneic and allogeneic bone engraftment after total body irradiation: dependence on dose, dose rate, and fractionation. Blood 77: 661–669

Fulop GM, Phillips RA (1989) Use of scid mice to identify and quantitate lymphoid-restricted stem cells in long-term bone marrow cultures. Blood 74: 1537–1544

Gisler RH, Soderburg A, Kamber M (1987) Functional maturation of murine B lymphocyte precursors. II. Analysis of cells required from bone marrow microenvironmental. J Immunol 138: 2433–2438

Harrison DE, Astle CM (1982) Loss of stem cell repopulating ability with transplantation. Effects of donor age, cell number and transplant procedure. J Exp Med 156: 1767–1779

Harrison DE, Astle CM (1991) Lymphoid repopulation delayed in B6 congenic W-anemic mice, a new unirradiated recipient. Exp Hematol 19: 374–377

Harrison DE, Astle CM, Stone M (1989) Number and function of transplantable primitive immuno hematopoietic stem cells. Effects of age. J Immunol 142: 3833–3840

Harrison DE, Stone M, Astle CA (1990) Effect of transplantation on primitive immunohematopoietic stem cells. J Exp Med 172: 431–437

Henderson AJ, Johnson A, Dorshkind K (1990) Functional characterization of two stromal cell lines that support B lymphopoiesis. J Immunol 145: 432–428

Hunt P, Robertson D, Weiss D, Rennick D, Lee F, Witte ON (1987) A single bone marrow-derived stromal cell type supports the in vitro growth of early lymphoid and myeloid cells. Cell 48: 997–1007

Jones RJ, Wagner JE, Celano P, Zicha MS, Sharkis SJ (1990) Separation of pluripotent haematopoietic stem from spleen colony-forming cells. Nature 347: 188–189

Jordan CT, McKearn JP, Lemischka IR (1990) Cellular and developmental properties of fetal hematopoietic stem cells. Cell 61: 953–963

Keller G, Paige C, Gilboa E, Wagner EF (1985) Expression of a foreign gene in myeloid and lymphoid cells derived from multipotent hematopoietic precursors. Nature 318: 149–154

Kincade PW (1984) Formation of B lymphocytes in fetal and adult life. Adv Immunol 31: 177–245

Kurland JI, Ziegler SF, Witte OW (1984) Long-term cultured B lymphocytes and their precursors reconstitute the B lymphocyte lineage in vivo. Proc Natl Acad Sci USA 81: 7554–7558

Landreth K, Dorshkind K (1988) Pre-B cell generation potentiated by soluble factors from a bone marrow derived stromal cell line. J Immunol 140: 845–852

McCune JM, Namikawa R, Kaneshima H, Shultz LD, Lieberman M, Weissman IL (1988) The SCID-hu mouse: murine model for the analysis of human hematolymphoid differentiation and function. Science 241: 1632–1639

McKearn JP, Baum C, Davie JM (1984) Cell surface antigen expression by subsets of pre-B cells and B cells. J Immunol 132: 332–338

Mintz B, Anthony K, Litwins S (1984) Monoclonal derivation of mouse myeloid and lymphoid lineages from totipotent hematopoietic stem cells experimentally engrafted in fetal hosts. Proc Natl Acad Sci USA 8: 7835–7839

Morse HC III, Shen F-W, Hämmerling U (1987) Genetic nomenclature for loci controlling mouse lymphocyte antigens. Immunogenetics 25: 71–78

Mosier DE, Gulizia RJ, Baird FM, Wilson DB (1988) Transfer of a functional human immune system to mice with severe combined immunodeficiency. Nature 335: 256–259

Müller-Sieburg CE (1991) Separartion of pluripotent stem cells and early B lymphocyte precursors with antibody Fall-3. J Exp Med 174: 161–168

Müller-Sieburg CE, Whitlock CA, Weissman IL (1986) Isolation of two early B lymphocyte progenitors from mouse bone marrow: a committed pre-pre-B cell and a clonogenic Thy-1^{lo} hematopoietic stem cell. Cell 44: 653–662

Müller-Sieburg CE, Townsend K, Weissman IL, Rennick D (1988) Proliferation and differentiation of highly enriched mouse hematopoietic stem cells and progenitor cells in response to defined growth factors. J Exp Med 167: 1825–1840

Nagasawa R, Kanagawa O, Title TV, Chiller JM (1985) In vivo maturation of pre-B cells derived from long-term cultured bone marrow. J Immunol 135: 965–972

Nakano T, Waki N, Asai H H, Kitamura Y (1989) Lymphoid differentiation of the hematopoietic stem cell that reconstitutes total erythropoieis of a genetically anemic W/W^{v} mouse. Blood 73: 1175–1179

Osmond DG, Park YH (1987) B lymphocyte progenitors in mouse bone marrow. Int Rev Immunol 12: 241–261

Paige CJ, Wu GE (1989) The B cell repertoire. FASEB J 3: 1818–1824

Palacios R, Henson G, Steimetz M, McKearn JP (1984) Interleukin-3 supports growth of mouse pre-B-cell clones in vitro. Nature 309: 126–131

Phillips RA (1989) Development and regulation of the B lymphocyte lineage: an interpretive overview. Prog Immunol 7: 305–315

Ploemacher RE, van der Sluijs JP, Voerman JSA, Brons NHC (1989) An in vitro limiting-dilution assay of long-term repopulating hematopoietic stem cells in the mouse. Blood 74: 2755–2763

Rennick D, Yang G, Gemmel L, Lee F (1987) Control of hemopoiesis by a bone marrow stromal cell clone: lipopolysaccharide- and interleukin-1-inducible production of colony-stimulating factors. Blood 69: 682–691

Ross EAM, Anderson N, Micklem HS (1982) Serial dilution and regeneration of the murine hematopoietic system. Implication for hematopoietic organization and the study of cellular aging. J Exp Med 155: 432–444

Scheid MP, Triglia D (1979) Further description of the Ly-5 system. Immunogenetics 9: 423–433

Scheven B AA, Visser JWM, Nijweide PJ (1986) In vitro osteoclast generation from different bone marrow fraction, including a highly enriched hematopoietic stem cell population. Nature 323: 166–169

Shen FW (1981) Monoclonal antibodies to mouse lymphocyte differentiation antigens. In: Hämmerling GJ, Hämmerling U, Kearney JF (eds) Monoclonal antibodies and T cell hybridomas: perspectives and technical advances. Elsevier, New York, pp 25–31

Snodgrass R, Keller G (1987) Clonal fluctuation within the heamatopoietic system of mice reconstituted with retrovirus-infected stem cells. EMBO J 6: 3955–3960

Spangrude GJ, Johnson GR (1990) Resting and activated subsets of mouse multipotent hematopoietic stem cells. Proc Natl Acad Sci USA 87: 7433–7437

Spangrude GJ, Scollay R (1990a) Differentiation of hematopoietic stem cells in irradiated mouse thymic lobes. Kinetics and phenotype of progeny. J Immunol 145: 3661–3668

Spangrude GJ, Scollay R (1990b) A simplified method for enrichment of mouse hematopoietic stem cells. Exp Hematol 18:920

Spangrude GJ, Heimfeld S, Weissman IL (1988a) Purification and characterization of mouse hematopoietic stem cells. Science 241:58–62

Spangrude GJ, Müller-Sieburg CE, Heimfeld S, Weissman IL (1988b) Two rare populations of mouse Thy-1^{lo} bone marrow bone cells repopulate the thymus. J Exp Med 167:1671–1683

Sutherland HJ, Eaves CJ, Eaves AC, Dragowska W, Lansdorp PM (1989) Characterisation and partial purification of human marrow cells capable of initiating long-term hematopoiesis in vitro. Blood 74:1563–1570

Szilvassy SJ, Lansdorp PM, Humphries RK, Eaves AC, Eaves CJ (1989) Isolation in a single step of a highly enriched murine hematopoietic stem cell population with competitive long-term repopulation ability. Blood 74:930–939

Tidmarsh GF, Dailey MO, Whitlock CA, Weissman IL (1985) Transformed lymphocytes from Abelson diseased mice express levels of a B-lineage transformation-associated antigen elevated from that found on normal lymphocytes. J Exp Med 162:1421–1434

Tidmarsh GF, Heimfeld S, Whitlock CA, Weissman IL, Müller-Sieburg CE (1989) Identification of a novel bone marrow-derived B-cell progenitor population that coexpresses B220 and Thy-1 and is highly enriched for Abelson Leukemia virus targets. Mol Cell Biol 9:2665–2671

Townbridge IS, Lesley J, Schulte R, Hyman R, Trotter J (1982) Biochemical characterization and cellular distribution of a polymorphic, murine cell- surface glycoprotein expressed on lymphoid tissues. Immunogenetics 15:299–312

Visser JWM, Bauman JGJ, Mulder AH, Eliason JF, de Leeuw AM (1984) Isolation of murine pluripotent hemopoietic stem cells. J Exp Med 59:1576–1590

Whitlock CA, Witte ON (1982) Long-term culture of B lymphocytes and their precursors from murine bone marrow. Proc Natl Acad Sci USA 79:3608–3612

Whitlock CA, Denis K, Robertson D, Witte O (1985) In vitro analysis of murine B cell development. Annu Rev Immunol 3:213–235

Whitlock CA, Tidmarsh GF, Müller-Sieburg CE, Weissman IL (1987) Bone marrow stromal cells with lymphopoietic activity express high levels of a pre-B neoplasia associated molecule. Cell 48:1009–1021

Williams DA, Lemischka IR, Nathan DG, Mulligan RC (1984) Introduction of new genetic material into pluripotent heamatopoietic stem cells of the mouse. Nature 310:476–480

Witte PL, Robonson M, Henley A, Low MG, Stiers DL, Perkins S, Fleischman RA, Kincade PW (1987) Relationships between B-lineage lymphocytes and stromal cells in long-term bone marrow cultures. Eur J Immunol 17:1473–1484

Wu AM, Till JE, Simovitch L, McCulloch EA (1968) Cytological evidence for a relationship between normal and hematopoietic colony-forming cells and cells of the lymphoid system. J Exp Med 127:455–463

Zipori D, Duksin D, Tamir M, Argaman A, Toledo J, Malik Z (1985) Cultured mouse marrow stromal cells lines. II. Distinct subtypes differening in morphology, collagen types, myelopoietic factors and leukemic cell growth modulation activities. J Cell Physiol 122:81–90

Microenvironments for Hematopoiesis

Stem Cell-Stromal Cell Interactions

C. Chabannon and B. Torok-Storb

1 Introduction

Purification of stem cells has been a major goal over the past decades. By 1991, significant progress has been made, allowing us to work with highly enriched populations of stem cells and progenitors (Spangrude et al. 1988; Andrews et al. 1990). By contrast, progress in the isolation of components of the marrow microenvironment (ME) has been slower, although its importance has long been recognized. One contributing reason is that the stroma is a fixed tissue which limits experimental accessibility. Obtaining single cell suspensions needed to initiate in vitro studies results in the disruption of the general architecture and all of the possible cell–cell contacts that are believed to play an important role in regulating hematopoiesis. Currently no in vitro system is able to reproduce exactly the complexity of the in vivo myeloid ME; the best suitable model, long-term marrow cultures (LTMC), is a gross approximation as reflected by its limited ability to generate quite exclusively granulomonocytic cells over a few weeks. These limitations hampered the dissection of the different lineages within the heterogeneous stromal cell populations and the study of their relationship to hematopoietic cells.

Our understanding of the interactions between stromal cells and hematopoietic progenitors will benefit from an increased understanding of the

Clinical Research Division/Department of Transplantation Biology, Fred Hutchinson Cancer Research Center, 1124 Columbia Street, Seattle, WA 98104, USA

Current Topics in Microbiology and Immunology, Vol. 177

molecular mechanisms underlying cell communications. Interactions between cells in the marrow ME involve a ligand secreted or expressed on the regulatory cell and a receptor on the target cell. Recent years witnessed the discovery of an increasing number of soluble ligands, and the study of their receptors is expanding. Mechanisms underlying direct cell-cell interactions or interactions of cells with the extracellular matrix are also being investigated. The specific applications of these investigations to studies of the marrow ME hold great promise.

2 SL/SL and W/W^v Models

Some of the most original and compelling evidence for the role of cell–cell interactions in hematopoietic regulation comes from studies of the SL/SL or "steel" mutant mice (RUSSELL 1979). These mice die in utero with anemia even though they have competent stem cells. The defect in these mice was attributed to a defective stromal element that fails to provide appropriate signals. Allelic forms of the steel mutation have resulted in viable, but severely anemic SL/SL^d mice. Attempts to cure the anemia in these mice with bone marrow transplantation have been uniformly unsuccessful. However, implantation of an intact normal spleen into the peritoneum of the SL/SL^d mice results in hematopoietic recovery, with hematopoiesis occurring within the ME of the transplanted spleen. These studies indicated that the critical ME component probably did not circulate and function as a humoral factor, but acted locally within the architecture of the ME (BERNSTEIN 1970; MCCULLOCH et al. 1965; WOLF 1974).

Recently, a new murine cytokine, termed stem cell factor (SCF) (ZSEBO et al. 1990a, b; MARTIN et al. 1990) or mast cell growth factor (ANDERSON et al. 1990), was identified. The gene encoding this molecule was mapped to the SL locus (HUANG et al. 1990; COPELAND et al. 1990). Further, SCF was shown to be the ligand for c-*kit* (WILLIAMS et al. 1990; HUANG et al. 1990) which is the gene product of the W locus. The W locus, like the SL locus, gives rise to a nonviable embryo when it occurs in the homogenous state. In fact, the gross phenotype of the W/W or its viable allelic form W/W^v is remarkably similar to the SL/SL and SL/SL^d, respectively. However, whereas the steel defect results in a defective or missing signal normally provided by the ME, the W defect results from a defective or missing receptor for the SL gene product.

The W and SL mouse models provide a clear example of receptor ligand interactions between stem cells and the ME that are required for normal hematopoiesis. What needs to be determined is how the signals are exchanged, whether the cells must be in contact, and, if so, what other signals may play a role in bringing this contact about. Given the organized architecture of the ME, it is reasonable to hypothesize that specific positional effects exist and that they are established and maintained by products of the ME. These spatial relationships are probably not only critical for facilitating binding of SCF and other factors, but also for controlling the kinds and amounts of other cytokines that reach progenitor cells.

3 Stromal Cell Models

Studies of the ME rely heavily on LTMC. It was first described in 1977 in mice (DEXTER et al.) and adapted for human marrow cells in 1980 (GARTNER and KAPLAN). Briefly, when cultured in appropriate conditions, marrow cells give rise to a complex adherent layer of stromal cells on the bottom of the flask. Basic conditions include a mixture of horse and fetal calf serums with the addition of hydrocortisone. The cellular composition of the adherent layer reproduces the in vivo ME. Its maintenance is critical for long-term hematopoiesis in vitro; immature progenitors preferentially seed within the adherent layer (COULOMBEL et al. 1983) where they differentiate before being released in the supernatant. While the system proved to be efficient in mice (SAKAKEENY and GREENBERGER 1982), with production of hematopoietic cells lasting up to 1 year, human cultures are usually "exhausted" within 18 weeks or less. LTMC were also adapted in dogs (SCHUENING et al. 1989; CARTER et al. 1990) and in monkeys (TAKAHASHI et al. 1987). LTMC allow us to identify several components in the ME and investigate their roles in hematopoiesis. The reproduction of in vivo pathogenesis suggests that this culture system may be appropriate for addressing questions of hematopoietic regulation (DEXTER and MOORE 1977; MIGLIACCIO et al. 1990).

4 Composition of the Hematopoietic ME

Although the description of cells in the ME can differ depending on whether it is studied "in vivo" on marrow sections or in vitro in LTMC, there is general agreement on the following. *Fibroblastoid cells* represent a major component of the adherent layer in LTMC. The nature of these cells has been questioned since a proportion of them contain actin microfilaments specific for smooth muscle cells (CHARBORD et al. 1985). The uptake of horse serum proteins by stromal cells and smooth muscle cells, but not cultured endothelial cells or fibroblasts, also suggests similarities between stromal cells and smooth muscle cells (CHARBORD et al. 1987). Although many authors believe that fibroblastoid cells do not play an active role in the regulation of hematopoiesis, the demonstration by ROBERTS et al. (1987) that fixed NIH3T3 cells can stimulate proliferation and differentiation of early murine progenitors suggests that these cells execute some form of ctitical cell–cell interactions. *Reticular cells* are large cells with cytoplasmic extensions and are positive for alkaline phosphatase. The cytoplasmic extensions are in close contact with endothelial cells. They also contain actin microfilaments. *Adipocytes* are identified by their size, small nucleus, vacuolar cytoplasm, and the positivity with lipid-specific dyes (TAVASSOLI 1984). The appearance of adipocytes in LTMC depends upon the addition of hydrocortisone (in the 10^{-5}–10^{-7} *M* range) to the culture medium (GREENBERGER 1978). *Endothelial cells* line the marrow sinusoids in vivo and thus represent a critical component during the homing phenomenon of hematopoietic cells. Their existence in LTMC has been discussed (BENTLEY and TRALKA 1982), and immunologic studies proved that they are present in human (STROBEL et al. 1986) as well as mouse cultures (ZUCKERMAN and WICHA 1983). Endothelial cells express von Willebrand's factor, and very likely CD34 (FINA et al. 1990).

Macrophages are easily identified. Their ubiquity, along with the multiple signals they can deliver, explains why it is difficult to delineate their role within the stroma. *Lymphocytes*, although long considered a nonessential part of the ME, are present in the adherent layer (SHIBATA and INOUE 1986; BERNEMAN et al. 1989). Some T cells may be long-lived peripheral blood lymphocytes contaminating the initial marrow sample; however, immature T and B cells also appear to be present in LTMC (DORSHKIND and PHILLIPS 1982; SCHRADER and SCHRADER 1978).

Osteoblasts have not been detected in LTMC; however, the presence of *osteoclasts* was demonstrated in mouse (TAKAHASHI et al. 1988) and primate cultures (TAKAHASHI et al. 1987; KUKITA et al. 1989). Osteoclasts derive from monocyte-macrophages (UDAGAWA et al. 1990) or from progenitor cells (KURIHARA et al. 1989). There is no evidence that they exert a direct regulatory role in hematopoiesis.

The extracellular matrix is a complex mixture of molecules which provide a physical support as well as regulatory signals. Thirteen types of *collagens* have been described, three being abundantly produced in the marrow. Fibroblasts secrete types I and III, while endothelial cells produce type IV as a major protein of the basement membrane (BENTLEY et al. 1981). *Fibronectin* is a 450-kd heterodimeric glycoprotein with a disulfide bond near the carboxy terminus. The molecule contains several functional domains mediating cell attachment, adhesion to collagen, and laminin. At least two domains are active for cell attachment, one containing the RGDS sequence, the other involving an alternative peptide. *Laminin*, a 900-kd glycoprotein, is composed of three chains arranged in a cross-shaped tertiary structure and stabilized by disulfide bonds. It is part of the basal membrane secreted by endothelial cells. *Hemonectin* is a 60-kd protein, identified in the marrow only (CAMPBELL et al. 1987), which is believed to preferentially bind maturing granulocytes. *Proteoglycans* are high-molecular-weight molecules in which sulfated glycosaminoglycans are covalently attached to a core protein during posttranscriptional modifications. Bonds are either *O*-glycosidic bonds or *N*-glycosylamine bonds. The polysaccharide moieties (*glycosaminoglycans*) differ by their composition, usually a repetitive alternance of uronic acid and hexosamine residues, the glycosidic linkage, and the site for sulfatation. The marrow contains heparan sulfate, chondroitin sulfate, and hyaluronic acid (SPOONCER et al. 1983).

5 Origin of the Bone Marrow ME

The origin of stromal cells other than macrophages and lymphocytes has been controversial. Studies using G6PD as markers in LTMC established from transplanted patients suggested that stromal cells were transplantable (KEATING et al. 1982). Hematopoietic and stromal cells from LTMC established from chronic myelogenous leukemia patients heterozygous for G6PD were shown to share the same isoenzyme, again suggesting a clonal origin (SINGER et al. 1984). However, subsequent analysis suggested these studies were probably detecting donor macrophages in the LTMC adherent layer. In two patients undergoing sex-mismatched marrow transplantation, HOLLINGS et al. (1984) demonstrated that marrow fibroblasts were host-derived in chimeric subjects. Using in situ hybridization for the Y body, this laboratory showed that non-hemopoietic components of the stroma remain host in origin (SIMMONS et al. 1987). This observation was confirmed by LAVER et al. (1987). LENNON and MICKLEM (1986) and PERKINS and FLEISCHMAN (1988) reached similar conclusions in murine models. These studies prove that, with conditioning regimens currently in use for clinical marrow transplantation, the stroma is not transplanted. Whether this reflects an inadequate conditioning regimen or the inability of stromal cells to "home" to the marrow is still unknown.

6 The ME as a Source of Regulatory Signals

Stromal cells are a potential source of multiple cytokines. Whether or not all of these molecules are produced in vivo and play a physiologic role remains uncertain. LTMC supernatants contain activities stimulating granulomonocytic differentiation (SHADDUCK et al. 1983; TSURUSAWA et al. 1983, 1984). This activity is mainly due to granulocyte-macrophage colony-stimulating factor (GM-CSF), which is regulated by exogenous interleukin 1 (IL1) (LOVHAUG et al. 1986; YANG et al. 1988) and stimulated by irradiation (ALBERICO et al. 1987; GUALTIERI et al. 1984; IZUMI et al. 1984). Potential sources of GM-CSF include macrophages, lymphocytes, fibroblastoid cells (KAUSHANSKY et al. 1988; YANG et al. 1988), and endothelial cells (BAGBY et al. 1986; SIEFF et al. 1987). Granulocyte colony-stimulating factor (G-CSF) and macrophage colony-stimulating factor (M-CSF) are also upregulated by IL1 (KAUSHANSKY et al. 1988; FIBBE et al. 1988). Potential sources for G-CSF and M-CSF include macrophages and fibroblastoid cells. The presence of IL3 in LTMC, and therefore its physiologic role, is far more controversial. Attempts to detect the protein in supernatant or the mRNA in the adherent layer by Northern blotting failed (LI and JOHNSON 1985; ELIASON et al. 1988; GREENBERGER et al. 1989). The availability of the polymerase chain reaction allowed us to detect an IL3 transcript in stromal cells as well as fresh marrow cells. This was further cofirmed by the presence in some LTMC supernatants of an activity detected with an IL3-dependent cell line and neutralized by anti-IL3 antibodies. This activity was not produced by mature T cells, since the transcript was not present in peripheral blood, and a CD2 transcript was present in some but not all cultured marrows (BARGE et al. 1989). Two recent publications also suggest that IL3 production is not restricted to T cells (MORRIS et al. 1990; DALLOUL et al. 1991). Contrary to the three other CSF (GARNETT et al. 1982; LANOTTE et al. 1982; NAPARSTEK et al. 1986), to date no stromal cell line, either murine or human, has been shown to produce IL3 (KODAMA et al. 1986).

Macrophages also represent a source of IL1, IL6, tumor necrosis factor (TNF), and transforming growth factor β (TGFβ), while T lymphocytes may produce IL2, IL4, IL5, IL6, TNF and TGF-β. The production of these molecules within the marrow environment has not been clearly documented. TGFβ exerts both a positive action on mesenchymal cells (IGNOTZ and MASSAGUÉ 1986, 1987) and a negative effect on hematopoietic progenitors (KELLER et al. 1988). TNF is able to suppress colony formation, but also stimulates fibroblasts to produce CSF(ZUCALI et al. 1988). The presence of megakaryocyte-stimulating activity detected by GUALTIERI et al. (1984) may be attributed to high levels of GM-CSF in irradiated cultures (ROBINSON et al. 1987).

A number of arguments suggest that the role of stromal cells is not limited to the production of humoral factors. For example, the extracellular matrix plays a role as a "humoral factor presenting network," (CAMPBELL et al. 1985; ZUCKERMAN et al. 1985). Glycosaminoglycans trap GM-CSF (GORDON et al. 1987), thus creating an ME favoring proliferation of committed progenitors. This ability is limited to heparan sulfate (ROBERTS et al. 1988) and may partly account for the difficulties encountered by investigators trying to demonstrate production of CSF in LTMC (HEARD et al. 1982). It is still unknown if production and presentation of cytokines/inhibitors is the only function of the ME with regard to hematopoietic differentiation.

The discovery of a precise spatial relationship between stromal and hematopoietic elements stimulated the interest in possible direct interactions between these two components. For example, BESSIS (1958) described the erythroblastic island, where erythroblasts tightly cluster around a central macrophage, appearing more mature as they become more distal; survival of hematopoietic progenitors in vitro decreases when they are cultured in diffusion chambers rather than directly onto preestablished marrow-derived adherent cells (BENTLEY 1981); and a stromal preadipocyte cell line provides optimal support for colony-forming units (CFU) when close contact is allowed, while conditioned medium does not have this ability (KODAMA et al. 1982). Additional examples are provided by COULOMBEL et al. (1983) who demonstrated the perferential location of immature progenitors in the adherent layer of LTMC, YAMAZAKI et al. (1989) who described the morphologic interactions involved between NIH3T3 cells and a murine IL3-dependent cell line, and KIERNEY and DORSHKIND (1987) who showed that B cell differentiation required close contact with stromal cells.

Cell–cell interactions within the hematopoietic ME were the subject of a recent review by TAVASSOLI and HARDY (1990). In this model, homing of murine hematopoietic progenitors in the marrow can be mimicked with a hematopoietic cell line and a stromal cell line. The "homing" or binding was shown to be mediated by a lectin-like system with mannosyl and galactosyl specificities (AIZAWA and TAVASSOLI 1987). Further, the injection of neoglycoproteins with these specificities was shown to block engraftment in irradiated mice (AIZAWA and TAVASSOLI 1988a) and interfere with reinitiation of hematopoiesis in established LTMC. As spleen CFU (CFU-S) and CFU-GM but not stromal cells are selectively agglutinated by neoglycoproteins, the homing protein appears to be present on progenitors (AIZAWA and TAVASSOLI 1988b). Partial purification identified a heterodimeric glycoprotein of 110 kd with a disulfide bond (MATSUOKA and TAVASSOLI 1989). The structure recognized on stromal cells is a carbohydrate which has not been identified yet.

The lack of a suitable assay comparable to the STAMPER and WOODRUFF assay (STAMPER and WOODRUFF 1976) used for lymphocyte homing has hampered the elucidation of the homing phenomenon in the marrow. However, the availability of probes initially designed for the study of lymphocyte homing has provided an opportunity to investigate some stromal cell-hematopoietic cell interactions. Integrins are highly conserved proteins composed of two chains, α and β, associated in a noncovalent way. A common β chain allows them to be classified into three subfamilies. The very late antigens (VLA) (HEMLER et al. 1987) were so called because the first molecule discovered was expressed on lymphocytes late after activation, 2–4 weeks. Currently, six molecules have been identified with the same $\beta 1$ chain and most serve as receptors for components of the extracellular matrix: VLA4 can bind VCAM-1, an adhesion molecule induced on human endothelial cell (ELICES et al. 1990). VLA2 is a receptor for collagen, VLA3 binds to fibronectin, collagen, and laminin; VLA5 binds to the RGDS peptide of fibronectin, and VLA6 binds to laminin. The leukocyte adhesion molecule family (Leu CAM) includes three members with a common $\beta 2$ chain: LFA-1, the receptor for ICAM-1, as well as Mac-1 and p150-95, both of which are receptors for the third component of complement C3bi. Cytoadhesins with a common $\beta 3$ chain include the vitronectin receptor and the glycoprotein IIbIIIa which is expressed on platelets and megakaryocytes and is a receptor for fibronectin, von Willebrand's factor, and collagen.

The expression of integrins on bone marrow cells has been reported (MILLER et al. 1985; CAMPANA et al. 1986; SOLIGO et al. 1990). Currently, only VLA5 expression has been demonstrated to be related to differentiation. Erythroid cells bind to fibronectin (PATEL et al. 1985; TSAI et al. 1987; COULOMBEL et al. 1988) and lose this property when reticulocytes mature into erythrocytes. The adhesion of erythroid progenitors (CFU-E) is inhibited by antibodies directed against VLA- 5 (GIANCOTTI et al. 1986; VUILLET-GAUGLER et al. 1990) which, as mentioned above, binds the RGDS peptide of fibronectin. B cells also have a maturation-linked pattern of adherence. Rearrangement of the immunoglobulin genes in murine pre-B cell lines is coincident with the loss of the receptor for fibronectin (BERNARDI et al. 1987). The receptor expressed on these cell lines is similar to the receptor expressed on murine réticulocytes and erythroleukemia cell lines; it exhibits similarities with membranes of the integrin family and binds to the RGDS peptide (PATEL and LODISH 1986), which points to VLA-5 as the candidate receptor. Similarly, CAMPBELL et al. (1990) demonstrated that binding of human granulocytic cells and HL60 cells to hemonectin decreases with differentiation; however, the receptor for hemonectin is not known, and there is no evidence that it is related to the integrin family. Granulocytes also bind to fibronectin, but a relation with the differentiation process is unclear (TSAI et al. 1987; POMMIER et al. 1984). The observation that lymphoid precursors adhere to two different sites on fibronectin (BERNARDI et al. 1987), along with the characterization of VLA-4 and VLA-5, leads to the hypothesis that VLA-4 may provide a basis for a nonspecific adhesion of hematopoietic cells to stromal fibronectin, while binding through VLA-5 may be a regulated event during the differentiation process. As yet there is little evidence that other members of the integrin family play a role in stromal cell-stem cell interactions. Stromal cells express N-CAM (THOMAS et al. 1988), a molecule closely related to ICAM1, the ligand for LFA-1 (SIMMONS et al. 1988). Megakaryocytes interact with the extracellular matrix (LEVINE et al. 1985), but the role of GpIIbIIIa in this phenomenon has not been elucidated.

CD44 is a monomeric 90-kd glycoprotein which mediates the interaction between lymphocytes and mucosal high endothelial venules, but also functions as a receptor for the extracellular matrix (HAYNES et al. 1989). CD44 is brightly expressed on mature and immature hematopoietic cells, including the CD34-positive marrow subpopulation (LEWINSOHN et al. 1990; KANSAS et al. 1990). Antibodies against the canine analog of CD44 interfere with engraftment (SCHUENING et al. 1987). In this model α CD44 antibody given during a narrow window in relation to total body irradiation plus autologous marrow prevented long-term hematopoietic reconstitution. Additional studies (B.M. SANDMAIER, personal communication) suggested this was due to interference with cellular interactions within the ME. As the introduction of anti-CD44 monoclonal antibodies also blocks lymphohematopoiesis in murine LTMC (MIYAKE et al. 1990), these observations suggest that CD44 may mediate interactions between progenitor cells and stromal cells.

7 Conclusion

Our ability to isolate relatively pure populations of stem cells and progenitor cells, coupled with the application of molecular techniques for identifying ME functional components has allowed us to unravel the complexities of hematopoietic regulation. The emerging picture suggests that numerous cytokines, some with exclusive activities, others with redundant functions, can influence proliferation and differentiation of stem cells, as well as maturation and function of end-stage cells. These activities act in a concentration-dependent manner which may be controlled at the level of receptor/ligand gene transcription, but also by positional effects mediated by a fixed architecture of cells and extracellular matrix. The expression of relevant receptors and ligands may also be regulated by signals provided by juxtaposed cells. We have begun to appreciate that the spatial relationships of various cell types may influence the nature of their interactions, contributing to the complexity of hematopoietic regulation.

Acknowledgements. This work was supported by grants HL36444, CA18221, and DK34431 of the National Institutes of Health, DHHS.

References

Aizawa S, Tavassoli M (1987) In vitro homing of hemopoietic stem cells is mediated by a recognition system with galactosyl and mannosyl specificities. Proc Natl Acad Sci USA 8:4485–4489

Aizawa S, Tavassoli M (1988a) Molecular basis of the recognition of intravenously transplanted hemopoietic cells by bone marrow. Proc Natl Acad Sci USA 85:3180–3183

Aizawa S, Tavassoli M (1988b) Detection of membrane lectins on the surface of hemopoietic progenitor cells and their changing pattern during differentiation. Exp Hematol 16:325–329

Alberico TA, Ihle JN, Liang C-M, McGrath HE, Quesenberry PJ (1987) Stromal growth factor production in irradiated lectin exposed long-term murine bone marrow cultures. Blood 69: 1120–1127

Anderson DM, Lyman SD, Baird A, Wignall JM, Eisenman J, Rauch C, March CJ, Boswell HS, Gimpel SD, Cosman D, Williams DE (1990) Molecular cloning of mast cell growth factor, a hematopoietin that is active in both membrane bound and soluble forms. Cell 63:235–243

Andrews RG, Singer JW, Bernstein ID (1990) Human hematopoietic precursors in long term culture: single CD34+ cells that lack detectable T, B, and myeloid antigens produce multiple colony-forming cells when cultured with marrow stromal cells. J Exp Med 172:355–358

Bagby GC, Dinarello CA, Wallace P, Wagner C, Hefeneider S, McCall E (1986) Interleukin 1 stimulates granulocyte macrophage colony-stimulating activity release by vascular endothelial cells. J Clin Invest 78:1316–1323

Barge AJ, Johnson G, Simmons P, Torok-Storb B (1989) Interleukin 3 transcript, bioactivity, and protein is detectable in normal bone marrow stroma in the absence of mature lymphocytes. Blood 74:115a

Bentley SA (1981) Close range cell: cell interaction required for stem cell maintenance in continuous bone marrow culture. Exp Hematol 9:308–312

Bentley SA, Tralka TS (1982) Characterization of marrow-derived adherent cells. Evidence against an endothelial subpopulation. Scand J Haemtol 28:381–388

Bentley SA, Alabaster O, Foidart JM (1981) Collagen heterogeneity in normal human bone marrow. Br J Haematol 48:287–291

Bernardi P, Patel VP, Lodish HF (1987) Lymphoid precursor cells adhere to two different sites on fibronectin. J Cell Biol 105:489–498

Berneman ZN, Chen Z-Z, van Bockstaele D, Ramael M, Korthout M, Peetermans ME (1989) The nature of the adherent hemopoietic cells in human long-term bone marrow cultures (HLTBMCs): presence of lymphocytes and plasma cells next to the myelomonocytic population. Leukemia 3:648–661

Bernstein SE (1970) Tissue transplantation as an analytic and therapeutic tool in hereditary anemias. Am J Surg 119:448–451

Bessis M (1958) L'ilot érythroblastique, unite fonctionnelle de la moelle osseuse. Rev Hematol 13: 8
Campana D, Sheridan B, Tidman N, Hoffbrand, AV, Janossy G (1986) Human leukocyte function-associated antigens on lympho-hemopoietic precursor cells. Eur J Immunol 16: 537–542
Campbell A, Wicha MS, Long M (1985) Extracellular matrix promotes the growth and differentiation of murine hematopoietic cells in vitro. J Clin Invest 75: 2085–2090
Campbell AD, Long MW, Wicha MS (1987) Haemonectin, a bone marrow adhesion protein specific for cells of granulocyte lineage. Nature 329: 744–746
Campbell AD, Long MW, Wicha MS (1990) Developmental regulation of granulocytic cell binding to hemonectin. Blood 76: 1758–1764
Carter RF. Kruth SA, Valli VEO, Dubé ID (1990) Long-term culture of canine marrow: cytogenetic evaluation of purging of lymphoma and leukemia. Exp Hematol 18: 995–1001
Charbord P, Gown AM, Keating A, Singer JW (1985) CFA-7 and HHF, two monoclonal antibodies that recognize muscle actin and react wiht adherent cells in human long-term bone marrow cultures. Blood 66: 1138–1142
Charbord P, Tippens D, Wight TS, Gown AM, Singer JW (1987) Stromal cells from human long-term marrow cultures but not cultured marrow fibroblasts, phagocytose horse serum constituents: studies with a monoclonal antibody that reacts with a species-specific epitope common to multiple horse serum proteins. Exp Hematol 15: 72–77
Copeland NG, Gilbert DJ, Cho BC, Donovan PJ, Jenkins NA, Cosman D, Anderson D, Lyman SD, Williams DE (1990) Mast cell growth factor maps near the steel locus on mouse chromosome 10 and is deleted in a number of steel alleles. Cell 63: 175–183
Coulombel L, Eaves AC, Eaves CJ (1983) Enzymatic treatment of long-term human marrow cultures reveals the preferential location of primitive hemopoietic progenitors in the adherent layer. Blood 62: 291–297
Coulombel L, Vuillet MH, Leroy C, Tchernia G (1988) Lineage- and stage- specific adhesion of human hematopoietic progenitor cells to extracellular matrices from marrow fibroblasts. Blood 71: 329–334
Dalloul AH, Arock M, Fourcade C, Hatzfeld A, Bertho J-M, Debre P, Mossaylayi MD (1991) Human thymic epithelial cells produce interleukin-3. Blood 77: 69–74
Dexter TM, Moore MAS (1977) In vitro duplication and 'cure' of haemopoietic defects in genetically anaemic mice. Nature 269: 412–413
Dexter TM, Allen TD, Lajtha LG (1977) Conditions controlling the proliferation of haemopoietic stem cells in vitro. J Cell Physiol 91: 335–344
Dorshkind K, Phillips RA (1982) Maturational state of lymphoid cells in long-term bone marrow cultures. J Immunol 129: 2444–2450
Eliason JF, Thorens B, Kindler V, Vassalli P (1988) The roles of granulocyte-macrophage colony-stimulating factor and interleukin 3 in stromal cell-mediated hemopoiesis in vivo. Exp Hematol 16: 307–312
Elices MJ, Osborn L, Takada Y, Crouse C, Luhowsky S, Hemler ME, Lobb RR (1990) VCAM-1 on activated endothelium interacts with the leukocyte integrin VLA-4 at a site distinct from the VLA-4/fibronectin binding site. Cell 60: 577–584
Fibbe WE, Van Damme J, Billiau A, Goselink HM, Voogt PJ, Van Eeden G, Ralph R, Altrock BW, Falkenburg JHF (1988) Interleukin 1 induces human marrow stromal cells in long-term culture to produce granulocyte colony-stimulating factor and macrophage colony-stimulating factor. Blood 71: 430–435
Fina L, Molgaard HV, Robertson D, Bradley NJ, Monaghan P, Delia D, Sutherland DR, Baker MA, Greaves MF (1990) Expression of the CD34 gene in vascular endothelial cells. Blood 75: 2417–2426
Garnett HM, Cronkite EP, Harigaya K (1982) Regulation of the colony- stimulating activity produced by a murine marrow-derived cell line (H-1). Proc Natl Acad Sci USA 79: 1545–1548
Gartner S, Kaplan HS (1980) Long-term culture of human bone marrow cells. Proc Natl Acad Sci USA 77: 4756–4759
Giancotti FG, Comoglio PM, Tarone G (1986) Fibronectin-plasma membrane interaction in the adhesion of hemopoietic cells. J Cell Biol 103: 429–437
Gordon, MY, Riley GP, Watt SM, Greaves ME (1987) Compartmentalization of a haematopoietic growth factor (GM-CSF) by glycosaminoglycans in the bone marrow microenvironment. Nature 326: 403–405
Greenberger JS (1978) Sensitivity of corticosteroid-dependent insulin-resistant lipogenesis in marrow preadipocytes of obese-diabetic (db/db) mice. Nature 275: 752–754

Greenberger JS, Fitzgerald TJ, Anklesaria P (1989) Recent studies of the hematopoietic microenvironment in long-term bone marrow cultures. Immunol Res 8: 226–235

Gualtieri RJ, Shadduck RK, Baker DG, Quesenberry PJ (1984) Hematopoietic regulatory factors produced in long-term murine bone marrow cultures and the effect of in vitro irradiation. Blood 64: 516–525

Haynes BF, Telen MJ, Hale LP, Denning SM (1989) CD44- a molecule involved in leukocyte adherence and T-cell activation. Immunol Today 10: 423–428

Heard J-M, Fichelson S, Varet B (1982) Role of colony-stimulating activity in murine long-term bone marrow cultures: Evidence for its production and consumption by the adherent cells. Blood 59: 761–767

Hemler ME, Huang C, Schwarz L (1987) The VLA protein family. Characterization of five distinct cell surface heterodimers each with a common 130,000 molecular weight β subunit. J Biol Chem 262: 3300–3309

Hollings PE, Fitzgerald PH, Heaton DC, Beard MEJ (1984) Host origin of in vitro bone marrow fibroblasts after marrow transplantation in man. Int J Cell Cloning 2: 348–355

Huang E, Nocka K, Beier DR, Chu T-Y, Buck J, Lahm H-W, Wellner D, Leder P, Besmer P (1990) The hematopoietic growth factor KL is encoded by the Sl locus and is the ligand of the c-kit receptor, the gene product of the W locus. Cell 63: 225–233

Ignotz RA, Massagué J (1986) Transforming growth factor- β stimulates the expression of fibronectin and collagen and their incorporation into the extracellular matrix. J Biol Chem 261: 4337–4345

Ignotz RA, Massagué J (1987) Cell adhesion protein receptors as targets for transforming growth factor-β action. Cell 51: 189–197

Izumi H, Miyanomae T, Tsurusawa M, Fujita J, Mori KJ (1984) Effects of bacterial lipopolysaccharide and X-irradiation on the production of colony-stimulating factor and the maintenance of granulopoiesis in bone marrow culture. Leuk Res 8: 295–305

Kansas GS, Muirhead MJ, Dailey MO (1990) Expression of the CD11/CD18, leukocyte adhesion molecule 1, and CD44 adhesion molecules during normal myeloid and erythroid differentiation in humans. Blood 76: 2483–2492

Kaushansky K, Lin N, Adamson JW (1988) Interleukin 1 stimulates fibroblasts to synthesize granulocyte-macrophage and granulocyte colony-stimulating factors. Mechanism for the hematopoietic response to inflammation. J Clin Invest 81: 92–97

Keating A, Singer JW, Killen PD, Striker GE, Salo AC, Sanders J, Thomas ED, Thorning D, Fialkow PJ (1982) Donor origin of the in vitro haematopoietic microenvironment after marrow transplantation in man. Nature 298: 280

Keller JR, Mantel C, Sing GK, Ellingsworth LR, Ruscetti SK, Ruscetti FW (1988) Transforming growth factor β selectively regulates early murine hematopoietic progenitors and inhibits the growth of IL-3-dependent myeloid leukemia cell lines. J Exp Med 168: 737–750

Kierney PC, Dorshkind K (1987) B lymphocyte precursors and myeloid progenitors survive in diffusion chamber cultures but B cell differentiation requires close association with stromal cells. Blood 70: 1418–1424

Kodama H-A, Amagai Y, Koyama H, Kasai S (1982) A new preadipose cell line derived from newborn mouse calvaria can promote the proliferation of pluripotent hemopoietic stem cells in vitro. J Cell Physiol 112: 89–95

Kodama H-A, Hagiwara H, Sudo H, Amagai Y, Yokota T, Arai N, Kitamura Y (1986) MC3T3-G2/PA6 preadipocytes support in vitro proliferation of hemopoietic stem cells through a mechanism different from that of interleukin 3. J Cell Physiol 129: 20–26

Kukita T, McManus LM, Miller M, Civin C, Roodman GD (1989) Osteoclast-like cells formed in long-term human bone marrow cultures express a similar surface phenotype as authentic osteoclasts. Lab Invest 60: 532–538

Kurihara N, Suda T, Miura Y, Nakauchi H, Kodama H, Hiura K, Hakeda Y, Kumegawa M (1989) Generation of osteoclasts from isolated hematopoietic progenitor cells. Blood 74: 1295–1302

Lanotte M, Metcalf D, Dexter TM (1982) Production of monocyte/macrophage colony-stimulating factor by preadipocyte cell lines derived from murine marrow stroma. J Cell Physiol 112: 123–127

Laver J, Jhanwar SC, O'Reilly R, Castro-Malaspina H (1987) Host origin of the human hematopoietic microenvironment following allogeneic bone marrow transplantation. Blood 70: 1966–1968

Lennon JE, Micklem HS (1986) Stromal cells in long-term murine bone marrow culture: FACS studies and origin of stromal cells in radiation chimeras. Exp Hematol 14: 287–292

Levine RF, Eldor A, Hyam E, Gamliel H, Fuks Z, Vlodavsky I (1985) Megakaryocyte interaction with subendothelial extracellular matrix is associated with adhesion, platelet-like shape change, and thromboxane A_2 production. Blood 66: 570–576

Lewinsohn DM, Nagler A, Ginzton N, Greenberg P, Butcher EC (1990) Hematopoietic progenitor cell expression of the H-CAM (CD44) homing-associated adhesion molecule. Blood 75: 589–595

Li CL, Johnson GR (1985) Stimulation of multipotential, erythroid and other murine haematopoietic progenitor cells by adherent cell lines in the absence of detectable multi-CSF (IL-3). Nature 316: 633–636

Lovhaug D, Pelus LM, Nordlie EM, Boyum A, Moore MAS (1986) Monocyte- conditioned medium and interleukin 1 induce granulocyte-macrophage colony-stimulating factor production in the adherent cell layer of murine bone marrow cultures. Exp Hematol 14: 1037–1042

Martin FH, Suggs SV, Langley KE, Lu HS, Ting J, Okino KH, Morris CF, McNiece IK, Jacobsen FW, Mendiaz EA, Birkett, NC, Smith KA, Johnson MJ, Parker VP, Flores JC, Patel AC (1990) Primary structure and functional expression of rat and human stem cell factor DNAs. Cell 63: 203–211

Matsuoka T, Tavassoli M (1989) Purification and partial characterization of membrane-homing receptors in two cloned murine hemopoietic progenitor cell lines. J Biol Chem 264: 20193–20198

McCulloch EA, Siminovitch L, Till JE, Russel ES, Bernstein SE (1965) The cellular basis of the genetically determined hemopoietic defect in anemic mice of genotype SL/SL^d. Blood 26: 399–410

Migliaccio AR, Migliaccio G, Johnson G, Adamson JW, Torok-Storb B (1990) Comparative analysis of hematopoietic growth factors released by stromal cells from normal donors or transplanted patients. Blood 75: 305–312

Miller BA, Antognetti G, Springer TA (1985) Identification of cell surface antigens present on murine hematopoietic stem cells. J Immunol 134: 3286–3290

Miyake K, Medina KL, Hayashi S-I, Ono S, Hamaoka T, Kincade PW (1990) Monoclonal antibodies to Pgp-1/CD44 block lympho-hemopoiesis in long-term bone marrow cultures. J Exp Med 171: 477–488

Morris CF, Young IG, Hapel AJ (1990) Molecular and cellular biology of interleukin-3. In: Dexter TM, Garland JM, Testa NG (eds) Colony-stimulating factors: molecular and cell biology. Dekker, New York, pp 177–214

Naparstek E, Donnelly T, Shadduck RK, Waheed A, Wagner K, Kase KR, Greenberger JS (1986) Persistent production of colony-stimulating factor (CSF- 1) by cloned bone marrow stromal cell line D2XRII after X-irradiation. J Cell Physiol 126: 407–413

Patel VP, Lodish HF (1986) The fibronectin receptor on mammalian erythroid precursor cells: characterization and development regulation. J Cell Biol 102: 499–456

Patel VP, Ciechanover A, Platt O, Lodish HF (1985) Loss of adhesion of erythrocyte precursors to fibronectin during erythroid differentiation. In: Cronkite EP, Dainiak N, McCaffrey RP, Palek J, Quesenberry PJ (eds) Hematopoietic stem cell physiology vol 189. Liss, New York pp 355–368

Perkins S, Fleischman RA (1988) Hematopoietic microenvironment. Origin, lineage, and transplantability of the stromal cells in long-term bone marrrow cultures from chimeric mice. J Clin Invest 81: 1072–1080

Pommier CG, O'Shea J, Chused T, Takahashi T, Ochoa M, Nutman TB, Bianco C, Brown EJ (1984) Differentiation stimuli induce receptors for plasma fibronectin on the human myelomonocytic cell line HL-60. Blood 64: 858–866

Roberts RA, Spooncer E, Parkinson EK, Lord BI, Allen TD, Dexter TM (1987) Metabolically inactive 3T3 cells can substitute for marrow stromal cells to promote the proliferation and development of multipotent haemopoietic stem cells. J Cell Physiol 132: 203–214

Roberts R, Gallagher J, Spooncer E, Allen TD, Bloomfield F, Dexter TM (1988) Heparan sulphate bound growth factors: a mechanism for stromal cell mediated haemopoiesis. Nature 332: 376–378

Robinson BE, McGrath HE, Quesenberry PJ (1987) Recombinant murine granulocyte macrophage colony-stimulating factor has megakaryocyte colony-stimulating activity and augments megakaryocyte colony stimulation by interleukin 3. J Clin Invest 79: 1648–1652

Russell ES (1979) Hereditary anemias of the mouse: a review for geneticists. Adv Genet 20: 357–459

Sakakeeny MA, Greenberger JS (1982) Granulopoiesis longevity in continuous bone marrow cultures and factor-dependent cell line generation: significant variation among 28 inbred mouse strains and outbred stocks. J Natl Cancer Inst 68: 305–317

Schrader JW, Schrader S (1978) In vitro studies on lymphocyte differentiation. I. Long term in vitro culture of cells giving rise to functional lymphocytes in irradiated mice. J Exp Med 148: 823–828

Schuening FG, Storb R, Goehle S, Meyer J, Graham TC, Deeg HJ, Appelbaum FR, Sale GE, Graf L, Loughran TP (1987) Facilitation of engraftment of DLA-nonidentical marrow by treatment of recipients with mab directed against marrow cells surviving radiation. Transplantation 44: 607–613

Schuening FG, Storb R, Meyer J, Goehle S (1989) Long-term culture of canine bone marrow cells. Exp Hematol 17:411–417

Shadduck RK, Waheed A, Greenberger JS, Dexter TM (1983) Production of colony stimulating factor in long-term bone marrow cultures. J Cell Physiol 114:88–92

Shibata T, Inoue S (1986) Mature T cells are part of adherent cells in human long-term bone marrow cultures. Exp Hematol 14:234–240

Sieff CA, Tsai S, Faller DV (1987) Interleukin 1 induces cultured human endothelial cell production of granulocyte-macrophage colony-stimulating factor. J Clin Invest 79:48–51

Simmons D, Makgoba MW, Seed B (1988) ICAM, an adhesion ligand of LFA-1, is homologous to the neural cell adhesion molecule NCAM. Nature 331:624–627

Simmons PJ, Przepiorka D, Thomas ED, Torok-Storb B (1987) Host origin of marrow stromal cells following allogeneic bone marrow transplantation. Nature 328:429–432

Singer JW, Keating A, Cuttner J, Gown AM, Jacobson R, Killen PD, Najfeld V, Powell J, Sanders J, Striker GE, Fialkow PJ (1984) Evidence for a stem cell common to hematopoiesis and its in vitro microenvironment: Studies of patients with clonal hematopoietic neoplasia. Leuk Res 8: 535–545

Soligo D, Schiro R, Luksch R, Manara G, Quirici N, Parravicini C, Deliliers GL (1990) Expression of integrins in human bone marrow. Br J Haematol 76:323–332

Spangrude GJ, Heimfeld W, Weissman IL (1988) Purification and characterization of mouse hematopoietic stem cells. Science 241:58–62

Spooncer E, Gallagher JT, Krizsa F, Dexter TM (1983) Regulation of haemopoiesis in long-term bone marrow cultures. IV. Glycosaminoglycan synthesis and the stimulation of haemopoiesis by β-D-xylosides. J Cell Biol 96:510–514

Stamper HB, Woodruff JJ (1976) Lymphocyte homing into lymph nodes: in vitro demonstration of the selective affinity of recirculating lymphocytes for high-endothelial venules. J Exp Med 144:828–833

Strobel E-S, Gay RE, Greenberg PL (1986) Characterization of the in vitro stromal microenvironment of human bone marrow. Int J Cell Cloning 4:341–356

Takahashi N, Mundy GR, Kuehl TJ, Roodman GD (1987) Osteoclast-like cell formation in fetal and newborn long-term baboon marrow cultures is more sensitive to 1,25-dihydroxyvitamin D3 than adult long-term marrow cultures. J Bone Miner Res 2:311–317

Takahashi N, Yamana H, Yoshiki S, Roodman GD, Mundy GR, Jones SJ, Boyde A, Suda T (1988) Osteoclast-like cell formation and its regulation by osteotropic hormones in mouse bone marrow cultures. Endocrinology 122:1373–1382

Tavassoli M (1984) Marrow adipose cells and hemopoiesis: an interpretative review. Exp Hematol 12:139–146

Tavassoli M, Hardy CL (1990) Molecular basis of homing of intravenously transplanted stem cells to the marrow. Blood 76:1059–1070

Thomas PS, Pietrangeli CE, Hayashi S, Schachner M, Goridis C, Low M, Kincade PW (1988) Demonstration of neural cell adhesion molecules on stromal cells that support lymphopoiesis. Leukemia 2:171–175

Tsai S, Patel V, Beaumont E, Lodish HF, Nathan DG, Soeff CA (1987) Differential binding of erythroid and myeloid progenitors to fibroblasts and fibronectin. Blood 69:1587–1594

Tsurusawa M, Miyanomae T, Isumi H, Mori KJ (1983) Presence of colony promoting activity (CPA) in the supernatant of the long-term bone marrow cultures. Leuk Res 7:167–173

Tsurusawa M, Aizawa S, Miyanomae T, Mori KJ (1984) Colony promoting activity (CPA) produced in long-term culture of murine bone marrow cells. Leuk Res 8:377–386

Udagawa N, Takahashi N, Akatsu T, Tanaka H, Sasaki T, Nishihara T, Koga T, Martin TJ, Suda T (1990) Origin of osteoclasts: mature monocytes and macrophages are capable of differentiating into osteoclasts under a suitable microenvironment prepared by bone marrow-derived stromal cells. Proc Natl Acad Sci USA 87:7260–7264

Vuillet-Gaugler MH, Breton-Gorius J, Vainchenker W, Guichard J, LeRoy C, Tchernia G, Coulombel L (1990) Loss of attachment to fibronectin with terminal human erythroid differentiation. Blood 75:865–873

Williams DE, Eisenman J, Baird A, Rauch C, Van Ness K, March CJ Park LS, Martin U, Mochizuki DY, Boswell HS, Burgess GS, Cosman D, Lyman SD (1990) Identfication of a ligand for the c-kit proto-oncogene. Cell 63:167–174

Wolf NS (1974) Dissecting the hematopoietic microenvironment. I Stem cell lodgement and commitment, and the proliferation and differentiation of erythropoietic descendants in the SL/SL^d mouse. Cell Tissue Kinet 7: 89–98

Yamazaki K, Roberts TA, Spooncer E, Dexter TM, Allen TD (1989) Cellular interactions between 3T3 cells and interleukin-3-dependent multipotent haemopoietic cells: a model system for stromal-cell-mediated haemopoiesis. J Cell Physiol 139: 301–312

Yang Y-C, Tsai S, Wong GG, Clark SC (1988) Interleukin-1 regulation of hematopoietic growth factor production by human stromal fibroblasts. J Cell Physiol 134: 292–296

Zsebo KM, Williams DA, Geissler EN, Broudy VC, Martin FH, Atkins HL, Hsu R-Y, Birkett NC, Okino KH, Murdock DC, Jacobsen FW, Langley KE, Smith KA, Takeishi T, Cattanach BM, Galli SJ, Suggs SV (1990a) Stem cell factor is encoded at the Sl locus of the mouse and is the ligand for the c-kit tyrosine kinase receptor. Cell 63: 213–224

Zsebo KM, Wypych J, McNiece IK, Lu HS, Smith KA, Karkare SB, Sachdev RK, Yschenkoff VN, Birkett NC, Williams LR, Satyagal VN, Tung W, Bosselman RA, Mendiaz EA, Langley KE (1990b) Identification, purification, and biological characterization of hematopoietic stem cell factor from buffalo rat liver-conditioned medium. Cell 63: 195–201

Zucali JR, Broxmeyer HE, Gross MA, Dinarello CA (1988) Recombinant human tumor necrosis factors α and β stimulate fibroblasts to produce hemopoietic growth factors in vitro. J Immunol 140: 840–844

Zuckerman KS, Rhodes RK, Goodrum DD, Patel VR, Sparks B, Wells J, Wicha MS, Mayo LA (1985) Inhibition of collagen deposition in the extracellular matrix prevents the establishment of a stroma supportive of hematopoiesis in long-term murine bone marrow cultures. J Clin Invest 75: 970–975

Zuckerman KS, Wicha MS (1983) Extracellular matrix production by the adherent cells of long-term murine bone marrow culture. Blood 61: 540–547

The Stem Cell Compartment: Assays and Negative Regulators

E. G. WRIGHT[1] and I. B. PRAGNELL[2]

1 Introduction

The majority of cells in hematopoietic tissue are relatively short-lived and have to be replaced constantly. Indeed, the requirement for billions of new hematopoietic cells per day indicates that the hematopoietic system is highly active and extremely well regulated considering that aberrations resulting in neoplasia are very rare. The system is also capable of responding to extreme stress or cytotoxicity, that is, the sudden removal of large numbers of maturing and mature cells by a rapid regenerative process. This extremely efficient repopulation process is only possible by virtue of a series of overlapping populations of primitive cells which can become committed to proliferation and differentiation as required. It is likely that only in the most extreme conditions a pluripotent stem cell will be required for repopulation. Under normal steady-state conditions, only a very small part of the stem cell compartment is in active cell cycle. In conditions of stress, the stem cells are triggered into an active cell cycle as required for differentiation into mature cells and replenishment of the system. Once this has occurred, steady state is reached and the stem cell compartment returns to quiescence. Thus the stem cell compartment must come under the influence of both negative and positive locally acting stimuli. It is also becoming clear that the stem cell compartment is heterogeneous in

[1] M.R.C. Radiobiology Unit, Chilton, Didcot, Oxon, OX11 ORD, England, UK
[2] Cancer Research Campaign Beatson Laboratories, Beatson Institute for Cancer Research, Garscube Estate, Bearsden, Glasgow, G61 1BD, Scotland, UK

Current Topics in Microbiology and Immunology, Vol. 177

nature. To understand fully the mechanism involved in hematopoiesis, it is essential to consider the compartment as a whole as well as the most primitive pluripotent stem cells. The latter, which are characterized by marrow reconstitution ability, may be particularly important for gene therapy and transplantation studies. Studies on the whole hematopoietic stem cell compartment may well yield information and concepts of importance in other tissue systems and vice versa (BURGESS and NICOLA 1983; HALL and WATT 1989; POTTEN and LOEFFLER 1990). In this review we focus on the assays which have been used to define myeloid stem cells. We discuss a negative regulator of myeloid stem cell proliferation recently described (GRAHAM et al. 1990) using assay systems which detect a significant proportion of the myeloid stem cell compartment (PRAGNELL et al. 1988; LORIMORE et al. 1990).

The stem cell compartment comprises an overlapping series of cell types which display most, but not necessarily all, of the properties ascribed to a stem cell. Stem cells have been described as cells with extensive self-maintenance and proliferative capacity and are capable of repopulating the tissue after injury (LAJTHA 1979). In the mouse, the spleen colony-forming unit (CFU-S) has been widely regarded as the cell most closely fitting this definition (TILL and MCCULLOCH 1961; reviewed by TILL and MCCULLOCH 1980). More recent studies have shown that the CFU-S comprises a very heterogenous compartment (WORTON et al. 1969; CURRY and TRENTIN 1967; SCHOFIELD and LAJTHA 1973; ROSENDAAL et al. 1976; HARRIS et al. 1984; MOLINEUX et al. 1986; VISSER et al. 1984; BERTONCELLO et al. 1985; SPANGRUDE et al. 1988) with cells that have some stem cell properties, including the ability to generate secondary colonies (SIMINOVITCH et al. 1963), i.e., self-maintenance. However, cells with short-term marrow repopulating ability (MRA) can be separated from day 12 CFU-S and are considered to be pre-CFU-S (PLOEMACHER and BRONS 1989). Evidence for pre-CFU-S which may be regarded as *long-term* MRA cells has received support from experiments demonstrating that 5-fluorouracil (5-FU) spares cells responsible for long-term myeloid and lymphoid repopulation using a competitive repopulation assay (LERNER and HARRISON 1990). A competitive repopulation assay has been used to establish a quantitative in vivo assay for cells within the stem cell compartment capable of long-term reconstitution (SZILVASSY et al. 1990).The incidence of this primitive stem cell (competitive repopulation unit, CRU; $\sim 10/10^5$) is in the range of frequencies determined in a number of studies designed to estimate the numbers of primitive cells in marrow which are capable of long-term repopulation (BOGGS et al. 1982; MICKLEM et al. 1987) (Table 1).

JONES et al. (1989, 1990), principally by using an elutriator technique, demonstrated in an elegant series of experiments that short-term and long-term repopulating cells could be separated from murine bone marrow and that two phases of bone marrow engrafment could be observed. The initial engrafment (short-term repopulation) was due to cells which also produced spleen colonies. Long-term engraftment was attributed to a different cell fraction, but long-term repopulation was only possible if animals received short-term repopulating

Table 1. The incidence of clonogenic cells within the stem cell compartment

Cell nomenclature	Incidence[a]	Reference
Mix-CFC	150	JOHNSON (1980)
Day 14 BFU-E	5–10	HUMPHRIES et al. (1979a)
Blast CFC	4	NAKAHATA and OGAWA (1982)
CFU-A	150–200	PRAGNELL et al. (1988)
		LORIMORE et al. (1990)
HPP-CFC-1	250	BARTELMEZ et al. (1989)
HPP-CFC-2	110	BARTELMEZ et al. (1989)
CAFC	6	PLOEMACHER et al. (1989)
CFU-S	200–400	METCALF and MOORE (1971)
CRU	10	SZILVASSY et al. (1990)

[a] The incidence of the clonogenic cells derived from in vivo and in vitro assays is expressed per 10^5 normal bone marrow cells. The figures are approximate and cited in the references quoted

cells. It follows therefore that the stem cell compartment would be described as a continuum in which the most primitive members (long-term repopulating, pluripotent cells) have the greatest long-term repopulating capacity and are most resistant to proliferation and differentiation stimuli. These cells are therefore unlikely to be readily detected with in vitro clonogenic assays. By contrast, these cells, in subsequent divisions in vivo, give rise to daughter cells that become detectable as CFU-S with decreasing self-renewal and increasing capacity to become committed to hematopoietic lineages. *It follows, therefore, that the stem cell should be defined in the context of the properties under consideration,* and that the extremely low-frequency pluripotent stem cell is part of the compartment and is most likely critical only in abnormal stress conditions, but is important in gene therapy and transplantation studies.

2 Assays for the Stem Cell Compartment In Vitro

There is considerable literature documenting that cultures of murine bone marrow can be manipulated with appropriate growth factors to yield colonies derived from within the continuum of the stem cell compartment. These can be conveniently grouped into three classes: mixed-lineage colony-forming cells, (mix-CFC) high proliferative potential colony-forming cells (HPP-CFC) and blast colony-forming cells (blast CFC). It is most likely that the mix-CFC represent the least primitive part of the compartment (e.g., day 7 CFU-S), and that the blast CFC represent the most primitive. It should be noted, however, that clonogenic cells are only detected by virtue of the particular in vitro conditions used, and these cells may not necessarily normally contribute to hematopoiesis in normal marrow.

Mixed-Lineage Colony-Forming Cells. Mix-CFC have been described in a number of systems (JOHNSON and METCALF 1977; METCALF et al. 1978; JOHNSON 1980). They are generally mixed-erythroid in nature with an approximate incidence of $150/10^5$ (Table 1). These colonies contain cells with a varying capacity for secondary colony formation, with infrequent presence of CFU-S in the primary colonies. A more primitive mix-CFC has also been described (HUMPHRIES et al. 1979a) with an incidence in the range $5–10/10^5$ normal bone marrow cells (Table 1). Macroscopic bursts observed in this culture system contain cells capable of secondary colony formation of single lineage colonies. Subsequent experiments indicated that each macroscopic burst contained about one CFU-S (HUMPHRIES et al. 1979b). Another assay for multilineage CFC was described in which the incidence of the clonogenic cell was $3–4/10^5$, and the percentage of cells in DNA synthesis was 1.9 ± 3.5 (HARA and OGAWA 1978). The presence of secondary CFC and the infrequent presence of CFU-S in the mix-CFC described above suggests that this class of clonogenic cells overlaps marginally with the stem cell compartment.

Blast Colony-Forming Cells. An in vitro clonal assay for a cell that is ancestral to mix-CFC has been described in which the incidence of the clonogenic cell was approximately $4/10^5$ in normal marrow (NAKAHATA and OGAWA 1982). These colonies consisted of blast cells after extended incubation time (16 days); the colonies also contained progenitor cells and produced secondary multilineage colonies on replating. Most blast-CFC usually contained about two to four CFU-S per colony (range 0–16). The frequency in bone marrow and the properties of the blast-CFC suggest that they may be pre-CFU-S in nature, although long-term repopulating studies with these cells will be required to clarify this point.

High Proliferative Potential Colony-Forming Cells. HPP-CFC able to generate large colonies in agar culture (> 0.5 mm in diameter, containing an average of 5×10^4 cells) were first reported as primitive macrophage progenitors (BRADLEY and HODGSON 1979). More recently, HPP-CFC, usually assayed in marrow from 5-FU-treated mice, have been defined further, and three classes have emerged HPP-CFC-1, -2, and -3 in order of primitiveness (MCNIECE et al. 1986, 1987; see review by MCNIECE et al. 1990). HPP-CFC-1 generated zero to six day 13 CFU-S per day 4, developing colony and cells with short-term repopulating ability (MCNIECE et al. 1987; BARTELMEZ et al. 1989). The properties ascribed to the HPP-CFC series suggest that this compartment falls into and may well mostly include the CFU-S day 12 to day 8 population detected by in vivo assay (Table 1). The CFU-A clonogenic cell (see below) may also be part of this compartment since its properties most closely resemble day 12 CFU-S.

The relationship of HPP-CFC to blast-CFC is not yet entirely clear, but it is possible that the most primitive HPP-CFC may be similar to blast-CFC.

CFU-A Colony-Forming Cells: Interrelationships in the CFU-S Compartment. We have developed an assay system using conditioned media as a source of growth factors which trigger cells from within the CFU-S compartment into proliferation and differentiation (PRAGNELL et al. 1988; ECKMANN et al. 1989). It is most likely that the critical growth factors in this assay system are granulocyte-macrophage colony-stimulating factor (GM-CSF), M-CSF, and another activity, possibly the "stem cell factor" recently described (MARTIN et al. 1990). The conditioned medium from the rat fibroblast line AF-1-19T (FRANZ et al. 1985) we use in the CFU-A assay produces a cocktail of growth factors which includes those above and very small amounts of interleukin 1 (IL-1) and IL-6, but no IL-3 (L.B. Pragnell and D. Donaldson, unpublished results). The salient features of this assay system are as follows: (a) a low cell inoculum; (b) normal bone marrow cells; (c) 10–12-day incubation time; (d) heterogeneity in colony size, in which the largest (>2 mm diameter) are derived from clonogenic cells (termed CFU-A). CFU-A share properties with CFU-S but differ from the more mature GM-CFC progenitors (LORIMORE et al. 1990). The assay system detects quiescent clonogenic cells which give rise to macroscopic (>2 mm diameter) CFU-A, as well as cells in the active cell cycle which give rise to smaller colonies in the same dish. The incidence of CFU-A in normal bone marrow from various strains of mice is in the range of $150–200/10^5$ (Table 1). The data from our investigations, which include assessment of proliferative status and response to CFU-S proliferation regulators, recovery after 5-FU treatment in vivo, buoyant cell density, radial distribution in the femur, and response to ionizing radiation, are consistent with the CFU-A assay detecting part of the multipotential stem cell compartment which is detected by the in vivo CFU-S assay (LORIMORE et al. 1990; ECKMANN et al. 1988). Isolated day 12 CFU-S (LORD and SPOONCER 1986), when cultured in the CFU-A assay system, give rise almost exclusively to identical macroscopic colonies (>2 mm diameter) at high cloning efficiency (L.B. PRAGNELL, G.J. GRAHAM, and B.I. LORD, unpublished results). *It is important to note here that all the experiments described above were performed with the CFU-A assay using conditioned medium as a source of synergizing growth factors.* The combination of GM-CSF and M-CSF, when used in this assay system, most likely stimulates more mature clonogenic cell as evidenced by the fact that these clonogenic cells are in cycle in normal bone marrow and are killed by 5-FU treatment in vivo (KRIEGLER et al. 1990). In contrast, CFU-A, as described, are quiescent in normal bone marrow (PRAGNELL et al. 1988; LORIMORE et al. 1990). We conclude therefore that the CFU-A clonogenic cell closely resembles both the day 12 CFU-S and the HPP-CFC-1. This is further supported by the similar frequencies of HPP-CFC-1, CFU-A, and CFU-S (allowing for a spleen "seeding factor") (BARTELMEZ et al. 1989) (Table 1) We have also attempted to develop an equivalent human CFU-A assay using bone marrow as a source of primitive cells (G. KONWALINKA et al.

unpublished). The combination of recombinant human GM-CSF and M-CSF in otherwise identical conditions to the murine assay leads to macroscopic multilineage colony formation by day 23. The assay detects a high-incidence ($178/10^5$) clonogenic cell, and developing colonies can be replated at high efficiency. However, these clonogenic cells may not be as primitive as those detected in the murine CFU-A assay, as they have variable cycle status and a lower proliferative potential. At present it is assumed that the most primitive, long-term repopulating stem cell will not give rise to colony formation in vitro, although it is conceivable that, under appropriate conditions of extended culture, the long-term repopulating stem cell should give rise to undifferentiated and differentiated progeny in vitro. Recently, an assay system which was used to detect primitive cells (cobblestone area-forming cells, CAFC) at an incidence of $6/10^5$ has been described (Ploemacher et al. 1989) (Table 1). The cells that form CAFC are contained in a population of cells that have short-term MRA. It will be of interest to see, using appropriate markers, whether this population contains the very primitive pluripotent long-term repopulating stem cells and whether they are identical to competitive repopulation units (CRU) (Szilvassy et al. 1990).

Summary. A large number of in vitro assay systems have been described using a range of growth factor combinations which detect a series of overlapping clonogenic cell populations broadly defined by the in vivo CFU-S assay. It is most likely that this population of cells may be responsible for continued long-term hematopoiesis in vivo in normal physiological conditions. In conditions of extreme stress, however, such as post irradiation, post severe cytotoxic drug treatment, and after transplantation into lethally irradiated recipients, the more primitive pluripotent stem cells must play a crucial role in the regenerative process.

3 Inhibitors of Stem Cell Proliferation

3.1 Stem Cell Inhibitor/Macrophage Inflammatory Protein-1α

The proliferative status of the CFU-S compartment has been postulated to be controlled by locally acting regulatory factors present in the stromal microenvironment (Toksoz et al. 1980). However, the purification of these factors was hampered by the lack of suitable in vitro assay systems. We have shown that the CFU-A assay can be used to detect both an inhibitor (Lord et al. 1976) and a stimulator (Lord et al. 1977) of CFU-S proliferation (Pragnell et al. 1988). Bone marrow macrophages have been identified as sources of these regulators (Wright and Lorimore 1987). Direct addition of unpurified, normal bone marrow-derived preparations of this inhibitor to the CFU-A assay system results in suppression of macroscopic colony formation (G.J. Graham et al., in

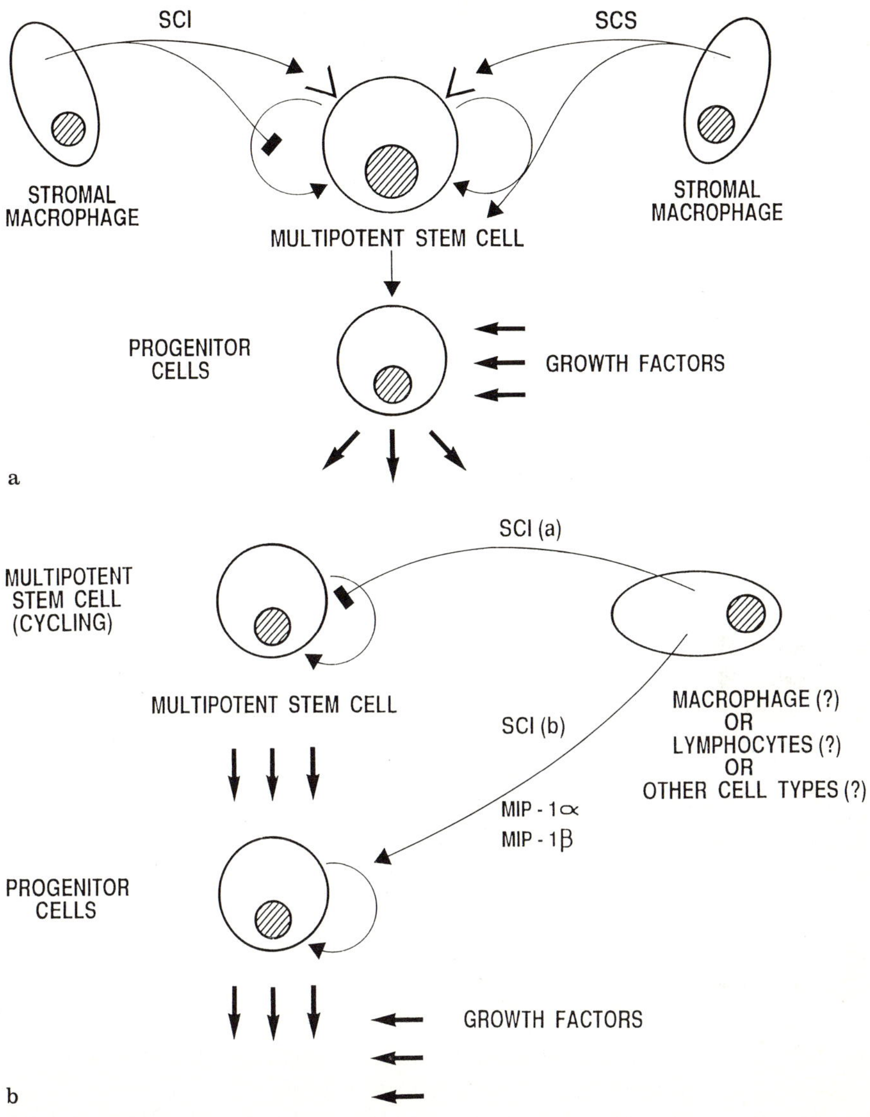

Fig. 1 a, b. A model for the control of stem cell proliferation. **a** Normal bone marrow: stem cell inhibitor/MIP-1α (SCI) and stem cell stimulator (SCS) (LORD et al. 1977) produced by macrophages in vivo modulate the proliferative status of the myeloid stem cell compartment, which is normally quiescent (less than 10% in cycle). **b** Regenerating marrow (including stem cell and inflammatory response). Feedback mechanisms, including induction of SCS, result in an active stem cell compartment in order to replace stem cells lost through differentiation or killing. Dampening of the stem cell proliferation response in the latter part of the regenerative phase could be attained by up-regulation of SCI gene expression to attain steady-state conditions. Induction of MIP-1α and MIP-1β can also take place as a result of activation of lymphocytes and monocyte/macrophages in peripheral blood, resulting in neutrophil activation and localized progenitor amplification mediated by MIP-1α and MIP-1β in synergy with other growth factors. Thus SCI/MIP-1α may act primarily as a stem cell proliferation regulator, but can also act as a biological response modifier within the mature hematopoietic compartment preparation). We therefore used this assay during biochemical purification of a macrophage cell line-derived inhibitor SCI (stem cell inhibitor, SCI). We found that SCI is antigenically and functionally identical to the normal bone marrow-derived inhibitor described above (GRAHAM et al. 1990) and is identical to a previously described macrophage inflammatory protein-1α (MIP-1α) (WOLPE et al. 1988). The closely related MIP-1β does not appear to function as an SCI at the concentration at which MIP-1α is active (GRAHAM et al. 1990), although it cannot be ruled out that relatively high concentrations of MIP-1β may be inhibitory. The fact that SCI is detectable in normal bone marrow and is active at picomolar concentration, in contrast to the inflammatory activities which require micromolar concentration, strongly suggests that the primary physiological role of SCI is to function as a modulator of the proliferative status of the CFU-S compartment. Further experimentation, using some of the in vitro and in vivo assays described in the previous section will be required to determine what proportion of the total stem cell compartment is sensitive to SCI/MIP-1α. The mechanisms involved in the inhibition are unknown at present, although it has been shown that unpurified SCI/MIP-1α acts by preventing cycling CFU-S from entering the phase of DNA synthesis and that the block in the cell cycle is close to the G1/S transition point (LORD 1981).

BROXMEYER et al. (1990) have shown that the MIP family of proteins (including MIP-1α and MIP-1β) will stimulate proliferation of progenitor cells such as GM-CFC. Our recent data confirm this (M. FRESHNEY, G.J. GRAHAM, and I.B. PRAGNELL, unpublished), but the effect can only be demonstrated in vitro in a reproducible manner when considerably suboptimal amounts of growth factor (GM-CSF) are used in an otherwise standard GM-CFC assay.

In view of the conclusions stated above, we suggest a modified model of the control of myeloid stem cell proliferation (Fig. 1) in which we propose that SCI/MIP-1α is an important modulator of CFU-S proliferation in normal bone marrow (Fig. 1a). MIP-1α could act by dampening the proliferation of the myeloid stem cell compartment in the later stages of recovery in regenerating bone marrow (Fig. 1b). An additional function of MIP-1α and MIP-1β would be as an inflammatory cytokine primarly inducible in peripheral blood lymphocytes and monocytes (Fig. 1b). This function could be ascribed perhaps to all members of the MIP family and other related cytokines (for review of these inflammatory mediators see WOLPE and CERAMI 1989).

MIP-1α mRNA sequences can be detected in RNA extracted from macrophages cultured from normal bone marrow, but there is only a very low level detectable in total normal bone marrow. Highly sensitive polymerase chain reaction (PCR) techniques can be used to detect MIP-1α mRNA in both systems (P. DAUBERSIES, A. REID, I.B. PRAGNELL et al., unpublished). In contrast, significant abundance of MIP-1α mRNA can be detected in total bone marrow in vivo during the regenerative period after cytotoxic drug treatment (A. Reid, unpublished). This preliminary data is supportive of the model proposed above, in which the levels of SCI/MIP-1α are either increased or replenished in the later stages of regeneration.

3.2 Other Inhibitors of Stem Cell Proliferation

A number of other potential physiologically important negative regulators of the stem cell compartment have been described (for recent brief review see GRAHAM and PRAGNELL 1991), including transforming growth factor β (TGFβ) (SING et al. 1988; OHTA et al. 1987; HAMPSON et al. 1989; RUSCETTI et al. 1989; CASHMAN et al. 1990), a pentapeptide, pyroGLU-GLU-ASP-CYS-LYS (pEEDCK) (LAERUM and PAUKOWITS 1984a) and a tetrapeptide, Acetyl-N-SER-ASP-LYS-PRO, (AcSDKP) (LENFANT et al. 1989).

Generally speaking, TGFβ is inhibitory to hematopoietic progenitor cells with a decreasing effect being observed as cells mature (SING et al. 1988). Colony formation by primitive cell populations such as CFU-S (HAMPSON et al. 1989), HPP-CFC (KELLER et al. 1990), and CFU-A (our unpublished observations) are strongly inhibited by TGFβ. The more mature GM-CFC are less sensitive (HAMPSON et al. 1989). The tetrapeptide AcSDKP was recently reported to be detectable in vivo and in long-term marrow culture (WDZIECZAK-BAKALA et al. 1990) and, in contrast to SCI-MIP-1α is probably only active at the GO/G1 transition or early in G1 rather than at the G1/S transition point in the cycle of CFU-S. This mechanism of action would predict that AcSDKP would have no effect on cycling cells in vivo or in vitro, as has been demonstrated (MONPEZAT and FRINDEL 1989; LAURET et al. 1989). It has been suggested that this tetrapeptide may derived in vivo by protelytic cleavage of thymosin B4 (GRILLON et al. 1990). The pentapeptide pEEDCK (or hemoregulatory peptide) has been shown to reduce the numbers of CFU-S and GM-CFC in vivo (LAERUM and PAUKOWITS 1984a) and will inhibit proliferation of GM-CFC but not BFU-E in vitro (LAERUM and PAUKOWITS 1984b). It is conceivable, therefore, that this peptide may primarily act on the CFU-S compartment in vivo by an indirect feedback mechanism. The dose-response kinetics of the action of both pEEDGK and AcSDKP on target cells approximate to an inverse bell-shaped curve. Furthermore, in most cases, pre-incubation of target cells with the peptides prior to plating out may be necessary (GUIGON et al. 1990; LAERUM et al. 1990). Thus the action of these peptides is complex, perhaps indirect, involving accssory cells and excess amounts of peptide may induce factors which abrogate their effect. LAERUM (1990) has recently identified sequences similar to pEEDCK in the G1 α subunit of G-proteins. The significance of this observation is not clear, especially as G-proteins are intracellular proteins. Similarly, Thymosin B4 is also intracellular in action. Thus both peptides may be part of an intracellular precursor protein in vivo, whereas TGFβ, normally secreted in a latent form (LYONS et al. 1988; BROWN et al. 1990) could be activated in certain physiological conditions.

The potential use of negative regulators, such as those indicated above, in clinical therapy for cancer is an exiciting prospect. In a number of animal models, preclinical trials have indicated that the inhibitory peptides (referred to above) can be used to increase survival of tumor-bearing mice (GUIGON 1987) and to alleviate cytotoxic effects of drug therapy (PAUKOWITS et al. 1991).

TGFβ has also been tested in preclinical trials with some success (RUSCETTI et al. 1991). Similar experiments are underway in Glasgow (UK) with SCI/MIP-1α (D. Dunlop, I.B. Pragnell et al., unpublished) and Manchester (UK) (B.I. Lord et al. unpublished) with encouraging results. It has not yet been established whether leukemic cells are less sensitive to SCI/MIP-1α than their normal counterparts, but, if they are, then drug purging of leukemic progenitors in the presence of SCI/MIP-1α has an exciting potential in therapy.

4 Conclusion

There has been a dramatic increase in our knowledge of the positive growth factors which interact with maturing progenitor cells in the hematopoietic system, many of which will act synergistically on target cells within the stem cell compartment. The availability of colony assays in vitro has been an essential element in this progress. We are now entering an exciting era in which the availability of cell separation techniques, in vivo and in vitro assays for the stem cell compartment and molecular approaches to the assessment of stem cell clonal contribution to hematopoiesis should lead to a more fundamental understanding of the control of hematopoiesis. This is particularly true for the molecular basis and physiological significance of stem cell heterogeneity. An important challenge in future studies on the control of stem cell proliferation will be analyses of the interaction of negative regulators with target cells in vivo and in vitro, and determination of how the balance of negative and positive regulatory factors is controlled in normal and neoplastic tissue.

Acknowledgements. The work of the authors is supported by grants from the Cancer Research Campaign, Medical Research Council, and the Leukaemia Research Fund. We thank Mrs Rae Fergusson for excellent secretarial assistance.

References

Bartelmez SH, Bradley TR, Bertoncello I, Mochizuky DY, Tushinsk RJ, Stanley ER, Hapel AJ, Young IG, Kriegler AB, Hodgson GS (1989) Interleukin 1 plus interleukin 3 plus colony-stimulating factor 1 are essential for clonal proliferation of primitive myeloid bone marrow cells. Exp Hematol 17: 240–245

Bertoncello I, Hodgson GS, Bradley TR (1985) Multiparameter analysis of transplantable hemopoietic stem cells. I. The separation and enrichment of stem cells homing to marrow and spleen on the basis of rhodamine-123 fluorescence. Exp Hematol 13: 999–1006

Boggs DR, Boggs SS, Sasce DS, Gress RA, Confield DR (1982) Hematopoietic stem cells with high proliferative potential. J Clin Invest 70: 242–253

Bradley TR, Hodgson GS (1979) Detection of primitive macrophage progenitor cells in mouse bone marrow. Blood 54: 1446–1450

Brown PD, Wakefield LM, Levinam AD, Sporn MB (1990) Physiochemical activation of recombinant latent transforming growth factor-beta's 1, 2 and 3. Growth Factors 3: 35–43

Broxmeyer HE, Sherry B, Lilu SC, Oh-Oh K, Tekamp-Olson P, Kwon BS, Cerami A (1990) Enhancing and suppressing effects of recombinant murine macrophage inflammatory proteins in colony formation in vitro by bone marrow myeloid progenitor cells. Blood 6: 1110–1116

Burgess A, Nicola N (1983) Growth factors and stem cells. Academic, New York

Cashman JD, Eaves AC, Raines EW, Ross R, Eaves CJ (1990) Mechanisms that regulate the cell cycle status of very primitive hematopoietic cells in long-term human marrow cultures I. Stimulatory role of a variety of mesenchymal cell activation and inhibitory role of TGFβ. Blood 75: 96–104

Curry JL, Trentin JJ (1967) Hemopoietic spleen colony studies. I. Growth and differentiation. Dev Biol 15: 395–413

Davatelis G, Tekampoison P, Wolpe SD, Hermsen K, Luedke C, Gallegos C, Coit D, Merryweather J, Cerami A (1988) Cloning and characterisation of a cDNA for murine macrophage inflammatory protein (MIP), a novel monokine with inflammatory and chemokinetic properties. J Exp Med 167: 1939–1944

Eckmann L, Freshney MG, Wright EG, Sproul A, Wilkie N, Pragnell IB (1988) A novel in vitro assay for murine haematopoietic stem cells. Br J Cancer [Suppl 9]: 36–40

Franz T, Lohler J, Fusco A, Pragnell I, Nobis P, Padua R, Ostertag W (1985) Transformation of mononuclear phagocytes in vivo and malignant histiocytosis caused by a novel murine spleen focus forming virus. Nature 315: 149–151

Graham GJ, Pragnell IB (1990) Negative regulation of haemopoiesis—current advances. Pro Growth Factor Res 2: 181–192

Graham GJ, Wright EG, Hewick R, Wolpe SD, Wilkie NM, Donaldson D, Lorimore S, Pragnell IB (1990) Identification and characterisation of an inhibitor of haemopoietic stem cell proliferation. Nature 344: 442–444

Grillon C, Rieger K, Bakala J, Schott D, Morgat J-L, Hannappel E, Voelter W, Lenfant M (1990) Involvement of thymosin β4 and endoproteinase Asp-N in the biosynthesis of the tetrapeptide AcSerAspLysPro, a regulator of the hematopoietic system. FEBS Lett 274: 30–33

Guigon M (1987) Biological properties of low molecular weight pluripotent stem cell (CFU-S) inhibitors. In: Najman A, Guigan M et al. (eds) The inhibitors of haematopoiesis. pp 241–251

Guigon M, Bonnet D, Lemoine F, Kobari L, Parmentier C, Mary JY, Najman A (1990) Inhibition of human bone marrow progenitors by the synthetic peptide AcSDKP. Exp Hematol 18: 1112–1115

Hall PA, Watt FM (1989) Stem cells: the generation and maintenance of cellular diversity. Development 106: 619–633

Hampson J, Ponting ILO, Cook L, Vodinelich S, Redmond S, Roberts AB, Dexter TM (1989) The effects of TGFβ on haemopoietic cells. Growth Factors 1: 193–202

Hara H, Ogawa M (1978) Murine hemopoietic colonies in culture containing normoblasts, macrophages and megakaryocytes. Am J Hematol 4: 23–24

Harris RA, Hogarth PM, Wadeson LJ, Collins P, McKenzie IFC, Pennington DG (1984) An antigenic difference between cells forming early and late haemopoietic spleen colonies. Nature 307: 638–641

Humphries RK, Eaves AC, Eaves CJ (1979a) Characterization of a primitive erythropoietic progenitor found in mouse marrow before and after several weeks in culture. Blood 53: 746–763

Humphries RK, Jacky PB, Dill FJ, Eaves AC (1979b) CFU-S in individual erythroid colonies in vitro derived from adult mouse marrow. Nature 279: 718–720

Johnson GR (1980) Colony formation in agar by adult bone marrow multipotential hemopoietic cells. J Cell Physiol 103: 371–383

Johnson GR, Metcalf D (1977) Pure and mixed erythroid colony formation in vitro stimulated by spleen conditioned medium with no detectable erythropoietin. Proc Natl Acad Sci USA 74: 3879–3882

Jones RJ, Celano P, Sharkis SJ, Sensenbrenner LL (1989) Two phases of engraftment established by serial bone marrow transplantation in mice. Blood 73: 397–401

Jones RJ, Wagner JE, Celano P, Zicha MS, Sharkis SJ (1990) Separation of pluripotent haematopoietic stem cells from spleen colony-forming cells. Nature 347: 188–189

Keller JR, Niece IK, Sill KT, Ellingsworth LR, Quesenberry PJ, Sing GK, Ruscetti FW (1990) Transforming growth factor β directly regulates primitive murine hematopoietic cell proliferation. Blood 75: 596–602

Kriegler AB, Bradley TR, Bertoncello I, Hamilton JA, Hart PH, Piccoli DS, Hodgson G (1990) Progenitor cells in murine bone marrow stimulated by growth factors produced by the AF1-19T rat cell line. Exp Hematol 18: 372–378

Laerum OD (1990) The sequence of the hemoregulatory peptide is present in G-1 alpha protein. FEBS Lett 269: 11–14
Laerum OD, Paukowits WR (1984a) Modification of murine haemopoiesis in vivo by a synthetic hemoregulatory peptide (HP5β). Differentiation 27: 106–112
Laerum OD, Paukowits WR (1984b) Inhibitory effects of a synthetic peptapeptide on haemopoietic stem cells in vivo and in vitro. Exp Hematol 12: 7–17
Laerum OD, Aakruag A, Frostad S, Kalland T, Langen P, Maurer HR (1990) Selectivity of homoregulatory peptide (HP5b) action in culture. Int J Cell Cloning 8: 431–444
Lajtha LG (1979) Haemopoietic stem cells: concepts and definitions. Blood Cells 5: 447–454
Lauret E, Dumersil D, Miyanomae T, Santeny F (1989) Further studies on the biological activities of the CFU-S inhibitory peptide AcSDKP.II. Unresponsiveness of isolated rat hepatocytes, 3T3, FDC-P2 and K562 cell lines to AcSDKP. Possible involvement of intermediary cells in the mechanism of AcSDKP action. Exp Hematol 17: 1081–1085
Lenfant M, Wolzieczak-Bakala J, Guittet E, Prome J-C, Solty D, Frindel E (1989) Inhibitor of haematopoietic pluripotent stem cell proliferation: purification and determination of its structure. Proc Natl Acad Sci USA 86: 779–782
Lerner C, Harrison DE (1990) 5 Fluorouracil spares haemopoietic stem cells responsible for long-term repopulation. Exp Hematol 18: 114–118
Lord BI (1981) The relationship of Go to the cell cycle of haemopoietic spleen colony-forming cells. Cell Tissue Kinet 14: 425–431
Lord BI, Spooncer E (1986) Isolation of haemopoietic spleen colony forming cells. Lymphokine Res 5: 59–63
Lord BI, Mori KJ, Wright EG (1977) A stimulator of stem cell proliferation in regenerating bone marrow. Biomedicine 27: 223–226
Lord BI, Mori KJ, Wright EG, Lajtha LG (1976) An inhibitor of stem cell proliferation in normal bone marrow. Br J Haematol 34: 441–445
Lorimore SA, Pragnell IB, Eckman L, Wright EG (1990) Synergistic interactions allow colony formation in vitro by murine haemopoietic stem cells. Leuk Res 14: 481–489
Lyons RM, Keski-Oja J, Moses HL (1988) Proleolytic activation of latent transforming growth factor-β from fibroblast-conditioned medium. J. Cell Biol 106: 1659–1665
Martin FH, Suggs SV, Langley KE et al. (1990) Primary structure and functional expression of rat and human stem cell factor DNAs. Cell 63: 203–211
McNiece IK, Kriegler AB, Bradley TR, Hodgson GS (1986) Subpopulations of mouse bone marrow high-proliferative potential colony-forming cells (HPP-CFC). Exp Hematol 14: 856–860
McNiece IK, Williams NT, Johnson GR, Kriegler AB, Bradley TR, Hodgson GS (1987) Generation of murine hematopoietic precursor cells from macrophage high-proliferative potential colony-forming cells. Exp Hematol 15: 972–977
McNiece IK, Bertoncello I, Kriegler AB, Quesenberry PJ (1990) Colony-forming cells with high proliferative potential (HPP-CFC). Int J Cell Cloning 8: 146–160
Metcalf D, Moore MAS (1971) Haemopoietic cells. North Holland, New York, pp 70–72
Metcalf D, Johnson GR, Mandel TR (1978) Colony formation in agar by multipotential hemopoietic cells. J Cell Physiol 98: 401–420
Micklem HS, Lennon JE, Ansell JD, Gray RA (1987) Numbers and dispersion of repopulating cell clones in radiation chimeras as functions of injected cell clone. Exp Hematol 15: 251–257
Migdalska A, Molineux G, Demuynck H, Evans GS, Ruscetti F, Dexter TM (1991) Growth inhibitory effects of transforming growth factor β-1 in vivo. Growth Factors 4: 239–245
Molineux G, Schofield R, Testa NG (1986) Development of spleen CFU-S colonies from day 8 to 11: relationship to self-renewel capacity. Exp Hematol 14: 710–713
Monpezat J-P, Frindel E (1989) Further studies on the biological activities of the CFU-S inhibitory peptide AcSDKP. 1. The precise point of the cell cycle sensitive to AcSDKP. Exp Hematol 17: 1077–1080
Nakahata T, Ogawa M (1982) Clonal origin of murine hemtopoietic colonies with apparent restriction to granulocyte-macrophage-megakaryocyte (GMM) differentiation. J Cell Physiol 111: 239–246
Ohta M, Greenberger JS, Anklesaria P, Bassols A, Massague J (1987) Two forms of transforming growth factor β distinguished by multipotential haematopoietic progenitor cells. Nature 329: 539–541
Paukovits WR, Moser MH, Binder KA, Paukovits JB (1990) The use of haemoregulatory peptides (pEEDCK monomer and dimer) for reduction of cytostatic drug-induced haemopoietic damage. Cancer Treat Rev 17: 347–354

Ploemacher RE, Brons RHC (1989) Separation of CFU-S from primitive cells responsible for reconstitution of the bone marrow hemopoietic stem cell compartment following irradiation: evidence for a pre-CFU-S cell. Exp Hematol 17: 263–266

Ploemacher RE, van der Sluijs JP, Voerman JSA, Brons NHC (1989) An in vitro limiting-dilution assay of long-term repopulating hematopoietic stem cells in the mouse. Blood 74: 2755–2763

Potten CS, Loeffler M (1990) Stem cells: attributes, cycles, spirals, pitfalls and uncertainties. Lessons for and from the Crypt. Development 110: 1001–1020

Pragnell IB, Wright EG, Lorimore SA, Adam J, Rosendaal M, DeLamarter JF, Freshney M, Eckmann L, Sproul A, Wilkie N (1988) The effect of stem cell proliferation regulators demonstrated with an in vitro assay. Blood 72: 196–201

Rosendaal M, Hodgson GS, Bradley TR (1976) Haemopoietic stem cells are organized for use on the basis of their generation age. Nature 264: 68–69

Ruscetti FW, Sing GK, Ellingsworth LR, Keller JR (1989) Transforming growth factor β: a negative regulatory of normal and leukemic haemopoietic progenitor cell growth. In: Baum JJ, Dicke KA, Lotzova E, Pluznik D (eds) Experimental hematology today. Springer, Berlin Heidelberg New York, pp 119–124

Schofield R, Lajtha LG (1973) Effect of isopropylmethane sulphonate (IMS) on haemopoietic colony-forming cells. Br J Haematol 25: 195–202

Siminovitch L, McCulloch EA, Till JE (1963) The distribution of colony-forming cells among spleen colonies. J Cell Comp Physiol 62: 327–336

Sing GK, Keller JR, Ellingsworth LR, Ruscetti FW (1988) Transforming growth factor β selectively inhibits normal and leukemic human bone marrow cell growth in vitro. Blood 72: 1504–1511

Spangrude GJ, Heimfield S, Weissman IL (1988) Purification and characterization of mouse hematopoietic stem cells. Science 241: 58–62

Szilvassy SJ, Humphries RK, Lansdorp PM, Eaves AC, Eaves CJ (1990) Quantitative assay for totipotent reconstituting hematopoietic stem cells by a competitive repopulation strategy. Proc Natl Acad Sci USA 87: 8736–8740

Till JE, McCulloch EA (1961) A direct measurement of the radiation sensitivity of normal mouse bone marrow cells. Radiat Res 14: 213–222

Till JE, McCulloch EA (1980) Hemopoietic stem cell differentiation. Biochim Biophys Acta 605: 431–459

Toksoz D, Dexter TM, Wright EG, Lajtha LG (1980) The regulation of haemopoiesis in long-term bone marrow cultures. II. Stimulation and inhibition of stem cell proliferation. Blood 55: 931–936

Visser JWM, Bauman JGJ, Mulder AH, Eliason JF, DeLeeuw AM (1984) Isolation of murine pluripotent hemopoietic stem cells. J Exp Med 50: 1576–1590

Wdzieczak-Bakala J, Fache M-P, Lenfant M, Frindel E, Santony F (1990) AcSDKP, an inhibitor of CFU-S proliferation, is synthesised in mice under steady state condition and secreted by bone marrow in long-term culture. Leukemia 4: 235–237

Wolpe SD, Cerami A (1989) Macrophage inflammatory proteins 1 and 2: members of a novel superfamily of cytokines. FASEB J 3: 2656–2573

Wolpe SD, Davatelis G, Sherry B, Beutler B, Hesse DG, Nguyen HT, Moldawer LL, Nathan CF, Lowry SF, Cerami A (1988) Macrophages secrete a novel heparinbinding protein with inflammatory and neutrophil chemokinetic properties. J Exp Med 167: 570–581

Worton RG, McCulloch EA, Till JE (1969) Physical separation of haemopoietic stem cells differing in their capacity for self renewal. J Exp Med 130: 91–103

Wright EG, Lorimore SA (1987) The production of factors regulating the proliferation of haemopoietic spleen colony-forming cells by bone marrow macrophages. Cell Tissue Kinet 20: 191–203

Stroma-Dependent Hematolymphopoietic Stem Cells

P. J. QUESENBERRY

1 Introduction

Hematolymphopoietic marrow stem cells are defined by their potential for extensive proliferation and differentiation. Self-renewal, a characteristic as yet not established for any cell type, has also been put forward as another stem cell characteristic. The colony-forming unit-spleen (CFUS) at days 8–10 was initially felt to represent an assay for pluripotent stem cells (TILL and MCCULLOCH 1961), but this was then supplanted by CFU-S forming at 12–14 days after transplantation (MAGLI et al. 1982). However, neither appears to fully mirror the most primitive stem cells that repopulate in vivo cell lineages. A number of in vitro clonal assays also have been developed including colony forming unit granulocyte-macrophage (CFU-GM) (BRADLEY and METCALF 1966; PLUZNIK and SACHS 1965), CFU-erythroid CFU-C and burst-forming unit erythroid (BFU-E) (AXELRAD et al. 1978; MCLEOD et al. 1974), CFU megakaryocyte (CFU-meg) (METCALF et al. 1975), and CFU-granulocyte-erythroid-macrophage-megakaryocyte (CFU-GEMM) (FAUSER and MESSNER 1978), all of which appear to monitor differentiated progeny of early hematopoietic stem cells. Assays which may mirror relatively early marrow stem cells include CFU blast (CFU-blast) first described by Ogawa and colleagues (OGAWA et al. 1985; SUDA et al. 1984; NAKAHATA and OGAWA 1982a, b), the high proliferative potential colony-forming cell (HPP-CFC) (HODGSON and BRADLEY 1979; KRIEGLER et al. 1984) and CFU diffusion chamber (CFU-D) (NISKANEN and WYANDT 1985). Harrison and

Div. of Hematology, University of Virgina School of Medicine, Box 502, Charlottesville, VA 22908, USA colleagues (STONE and HARRISON 1990; HARRISON et al. 1984; HARRISON 1980) have described a stem cell assay that assesses the capacity of congenic murine marrow populations to compete against genetically different marrow after transplantation in vivo, and this may in fact give the best measure of marrow stem cells. Studies on retrovirally marked marrow have suggested that individual clones of stem cells may dominate and contribute the majority of differentiated marrow cells (LEMISCHKA et al. 1986; JORDAN and LEMISCHKA 1990) although these observations may relate to the experimental model used in which a small number of stem cells may have been transplanted.

2 Hematopoietic Cells in Dexter Culture

Murine Dexter marrow culture represents a system in which long-term stroma-dependent stem cell and hematopoietic cell production occurs (DEXTER et al. 1977; DEXTER and LAJTHA 1974; DEXTER 1979). Initial studies established that CFU-GM and CFU-S could be maintained for relatively long culture periods (3–4 months) in plateau phase growth. Hematopoiesis in this culture system is characterized by a close association of predominantly granulocytic elements with the adherent cell population. Early observations also indicated that differentiated hematopoiesis and nonadherent supernatant CFU-GM were derived largely from a progenitor/stem cell population which was tightly adherent to the adherent stromal cells (MAUCH et al. 1980). Repetitive washing of this stromal system did not deplete hematopoiesis, but, in fact, over several weeks of culture appeared to augment it (P. J. QUESENBERRY, unpublished data).

To understand the nature of stroma-associated hematopoietic stem cells in Dexter cultures, one must consider the basic nature of the culture system in both human and mouse, in particular, the constituent cells and their organization in the adherent layer (QUESENBERRY 1989). There are distinct differences between the human and murine Dexter culture systems, separate from differences in media, serum, temperature, or other culture details. The murine systems evidence a prolonged plateau phase in which differentiated cells (largely granulocytes, macrophages, and megakaryocytes) and progenitor cells (including day 8 or 9 CFU-S and CFU-GM) are produced and maintained at least for several months. The human Dexter culture systems have not evidenced such a plateau growth phase despite extensive efforts to optimize the system. Instead, there is ordinarily a steady and fairly rapid decline of all cell types in human Dexter cultures (HOCKING and GOLDE 1980; DEXTER et al. 1984; CHARBORD et al. 1986). Active hematopoiesis is critically dependent upon the formation of a mixed adherent layer of cells in both human and murine Dexter cultures, but again there are marked differences between the two systems. In the murine system, up to 95% of the cells in the adherent layer appear to be either macrophages or alkaline phosphatase-positive preadipocytic fibroblasts with a small component of endothelial cells (QUESENBERRY 1989), while in human Dexter cultures adherent stromal vascular smooth muscle cells may make up an important component of the mixed cell population (CHARBORD et al. 1985, 1990). However, the differences between the human and murine Dexter cultures may be more apparent than real. HARRISON et al. (1987) have presented evidence, using in vivo stem cell competition as an assay, that there is an immediate and steady decline of stem cells in murine Dexter cultures, and thus both systems may be in "decline" from their initiation. This, of course, does not mean that important insights into progenitor/stem cells and their in vivo physiology cannot be gained by studies of these stromal support systems.

Studies in murine Dexter culture have now demonstrated support for a variety of progenitor/stem cells including CFU-GM, CFU-S (LEVITT and QUESENBERRY 1980), CFU-meg (MCGRATH et al. 1987), CFU-D (DOUKAS et al. 1986), and HPP-CFC (WADE et al. 1982) in both supernatant and adherent phases. Erythropoiesis can be induced by the addition of erythropoietin to the cultures or the transfer of erythropoietin cDNA into marrow stromal cells (MAYANI et al. 1990; COREY et al. 1990). Early multilineage progenitors are supported in Dexter culture, but differentiation in this system is relatively restricted to the myeloid pathways including granulocytes, monocyte-macrophages and megakaryocytes. However, stem cells with potential for in vivo repopulation of myeloid and lymphoid pathways are supported (DORSHKIND and PHILLIPS 1982; SHRADER, 1978). Lymphomyeloid-reconstituting cells are also maintained in these cultures for at least 4 weeks (DEXTER and SPOONCER 1978; SCHRADER and SCHRADER 1978; DORSHKIND and PHILLIPS 1983) and probably longer since switching Dexter to "lymphoid" Whitlock–Witte culture conditions after prolonged culture results in a switch from predominantly granulocyte, megakaryocyte, and macrophage production to that of pre-B and B cells (JOHNSON and DORSHKIND 1986). Modifying the culture system by omitting hydrocortisone, switching to fetal calf serum, increasing the temperature to 37°C, and adding 2-mercaptoethanol results in stroma-dependent cultures which support largely pre-B and B cell production (WHITLOCK and WITTE 1982; WHITLOCK et al. 1984; MÜLLER-SIEBURG et al. 1986), probably due to a shift in stroma cell type (RULAND et al. 1990) and/or colony-stimulating factor (CSF) production (JOHNSON and DORSHKIND 1986). Others have shown support of T cell precursors (TOUW and LOWENBERG 1984). The hematolymphopoietic support appears dependent upon direct cell–cell interaction between primitive hematopoietic stem cells and stromal cells, the latter probably being of fibroblast or endothelial origin. This adherence is probably mediated by yet to be defined adherence proteins, possibly with mannose or galactose specificity (AIZAWA and TAVASSOLI 1987, 1988).

3 Stromal Cytokine Regulation of Murine Stem Cells

The means whereby stromal populations nurture or support adherent hematopoietic stem cells represents a central issue and has been controversial. Early data suggested that the Dexter system was not a growth factor-dependent system (DEXTER 1979; N. WILLIAMS et al. 1978; DEXTER and SHADDUCK 1980), despite data on the induction of murine stromal growth factor or growth factor mRNA after exposure to various inducers (see below). As recently as 1990, DEXTER et al. (1990) stated "together the results indicate that although many of the myeloid cell growth factors so far characterized do play a role in the development of hematopoietic cells they may not be involved in stromal cell mediated hematopoiesis". These observations were based on a failure of many workers to detect constitutive cytokine production and on experiments in which anti-cytokine antibody or cytokine exposure failed to result in profound alteration in stem cell support in culture (HASS et al. 1989; COUTINHO et al. 1990).

The murine stromal system, however, appears in fact, to be a rich source of growth factors, and cell lines from these systems have been used to isolate new factors, for example, interleukin 7 (IL-7) (NAMEN et al. 1988) and hematolymphopoietic growth factor 1 (HLGF-1) or *c-kit* ligand (D.E. WILLIAMS et al. 1991; WOODWARD et al. 1990). Ongoing work in our laboratory established that irradiated stromal layers stimulated mixed lineage colonies and that a failure to detect CSF was probably due to utilization or degradation of the CSF in the system. Furthermore, adding back unirradiated Dexter culture stroma to irradiated Dexter stroma blocked the generation of cytokines (GUALTIERI et al. 1984). Subsequently we established that pokeweed mitogen (PWM) induced GM-CSF (ALBERICO et al. 1987), while CSF-1 was constantly produced by stromal cells. More recently, we have demonstrated that there is "constitutive" production from unirradiated or irradiated (1100 R) stroma of CSF-1, IL-6, GM-CSF, and G-CSF mRNA (QUESENBERRY et al. 1990). That these cytokines are in fact secreted has been shown by factor-dependent cell line stimulation experiments with selective antibody blockade. Various stromal cell lines also produce this pattern of cytokines (WHITLOCK et al. 1987; HUNT et al. 1987; RENNICK et al. 1987). Using polymerase chain reaction (PCR) technology IL-3 is also demonstrable, and preliminary data using a cell line-coculture technique suggests that this subliminal mRNA production results in IL-3 bioactivity. Thus, murine Dexter stroma appears to be a source of multiple cytokines.

The classical stromal cell deficiency of the $S1/S1^d$ mouse has been replicated in Dexter culture (DEXTER and MOORE 1977), as has the stem cell defect of the W/W^v mouse. The latter was shown to be due to defects in the product of the *c-kit* oncogene, a tyrosine kinase cell surface receptor (CHABOT et al. 1988; GEISSLER et al. 1988; NOCKA et al. 1990a,b; REITH et al. 1990; MAJUMDER et al. 1988; YARDEN et al. 1987). Since the W/W^v and $S1/S1^d$ defects could complement one another, it seemed likely that the $S1/S1^d$ defect might represent abnormalities of the ligand for the *c-kit* receptor. This, in fact, turns out to be the case. Several

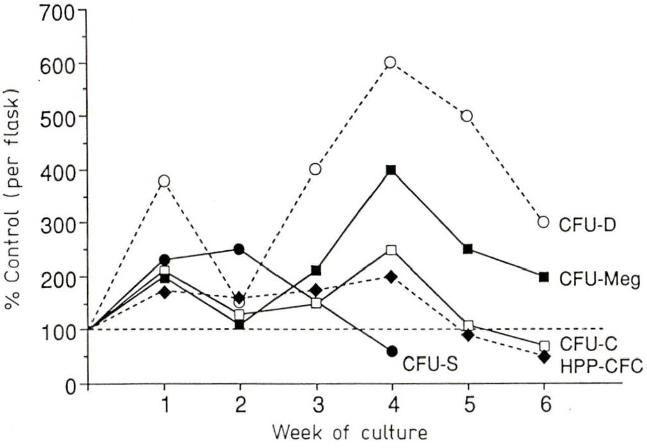

Fig. 1. Schematic model of effect of lithium on different progenitors in murine Dexter culture

groups have now described the *c-kit* ligand (also called stem cell factor, mast cell factor, or steel factor) (D.E. WILLIAMS et al. 1991; COPELAND et al. 1990; FLANAGAN and LEDER 1990; ZSEBO et al. 1990a, b; MARTIN et al. 1990; HUANG et al. 1990). This exists in a membrane and soluble form and is probably identical to a previously described factor from a stromal cell line which was termed HLGF-1 and which synergizes with CSF-1, IL-3, and other cytokines to stimulate HPP-CFC and also induces pre-B cells in liquid culture (WOODWARD et al. 1990; SONG et al. 1985; QUESENBERRY et al. 1987). The *c-kit* ligand is also easily demonstrable in Dexter stromal populations (P.J. QUESENBERRY, unpublished data) and may be a central factor for stem cell support in this system.

Lithium has been shown to stimulate granulocyte and probably platelet production when given to humans (MAYFIELD and BROWN 1966; ROTHSTEIN et al. 1978; LYMAN et al. 1980; STEIN et al. 1977). Exposure of Dexter cultures to 4 mEq/l lithium chloride results in a sustained stimulation of granulocyte, monocyte, and megakaryocyte production, and increases in CFU-S, day 9, CFU-GM (LEVITT and QUESENBERRY 1980), CFU-D (DOUKAS et al. 1986), CFU-meg (MCGRATH et al. 1987), and HPP-CFC (WADE et al. 1982) in both supernatant and adherent culture phases (Fig. 1). This effect was shown to be mediated by an indirect effect on stroma (QUESENBERRY et al. 1984), and more recently we have shown that lithium induces stromal GM-CSF mRNA and protein and mRNA for IL-6 and G-CSF without effecting the usual baseline CSF-1 production (Fig. 2).

These data on stroma-mediated stem cell support and cytokine production need to be considered in the light of recent investigations of the multifactor responsiveness of hematopoietic stem cells. Primitive stem/progenitors, represented by CFU-blast (OGAWA et al. 1985; SUDA et al. 1984; NAKAHATA and OGAWA 1982a,b) and HPP-CFC (HODGSON and BRADLEY 1979; KRIEGLER et al. 1984), are

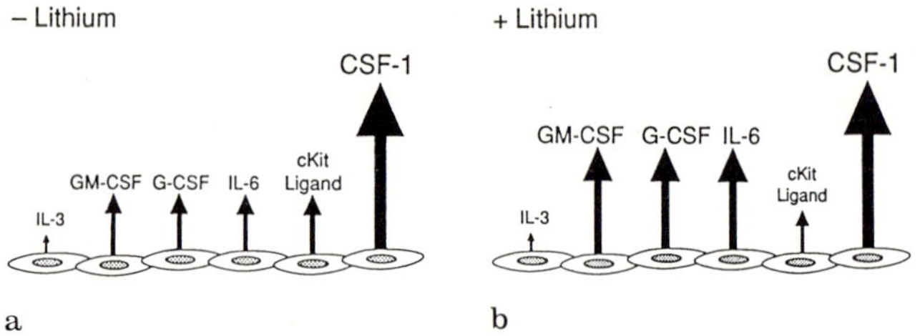

Fig. 2 a, b. Model of effect of lithium on stromal growth factor production

responsive to combinations of a variety of cytokines including IL-1, IL-3, GM-CSF, G-CSF, and IL-6. More recently, using a serum-free culture system and /or lineage-negative Sca^+ purified stem cells, we have shown that increasing cytokine numbers from two to seven stimulates increasing numbers of HPP-CFC; that *c-kit* ligand is a major synergizer with from one to six factors; and that, as one increases the number of factors in a combination, the level of at least some of the cytokines can be reduced to subliminal levels (LOWRY et al. 1991).

There is little question that inhibitors play an important role in stem cell regulation in Dexter culture. Transforming growth factor-β (TGF-β) inhibits hematopoiesis, and at higher doses lymphopoieses, and primitive stem cells. TGF-β also inhibits formation of osteoclast-like cells in Dexter culture (CHENU et al. 1988). Tumor necrosis factor-α also inhibits murine long-term marrow culture cell production possibly by an effect on stromal cells (ELIASON and VASSALLI 1988), while others have proposed that reactive oxygen intermediates (O_2^- and H_2O_2) derived from mature phagocytes might limit progenitor cell self-renewal (MEAGHER et al. 1988). The possible role of macrophage inflammatory protein 1-α (MIPI-α) as a stem cell inhibitory activity remains to be defined (GRAHAM et al. 1990).

4 Adherence of Stem Cells to Stroma

The adherence of progenitor/stem cells to stroma also appears to be an important feature of stromal support of hematolymphopoietic progenitors. Early studies have shown that murine erythroleukemia cells attach to fibronectin (FN), but not other extracellular matrix (ECM) proteins known to be present in marrow stroma (PATEL and LODISH 1984). Adhesion to FN was essential for differentiation into reticulocytes in vitro, and the capacity of erythroid precursors to attach to FN was lost with erythroid differentiation (PATEL and LODISH 1986). Loss of adhesion was due to loss of cell surface VLA5 integrin receptors. Both murine erythroleukemia (MEL) cells and pre-β cell lines attach to FN by binding to an Arg-Gly-Asp-Scr (RGDS) sequence and to the carboxy terminal heparin-binding domain of FN.

Furthermore, day 12 CFU-S attach to stromal cell-derived ECM by recognizing the CS-1 site of FN. As noted above, Tavassoli and colleagues have implicated galactosyl- and mannosyl-binding sites for adherence of primitive murine stem cells to stroma (AIZAWA and TAVASSOLI, 1987,1988), while late murine erythroid progenitors appear to bind to stroma by a fucosyl recognition mechanism (KONNA et al. 1990). Conversely, studies on human cultures have indicated that human $CD34^+$ normal myeloid or leukemic progenitors adhere to marrow stroma by mechanisms not involving fibronectin, laminin, RGDS-containing peptides, galactosyl or mannosyl specificities, or CD54 (LIESVELD et al. 1991). The importance of ECM proteins is further illustrated by the observations that addition of *cis*-hydroxyproline, which inhibits triple helical collagen formation, decreased stromal and hematopoietic cell production (ZUCKERMAN et al. 1985). Furthermore, addition of β-D-xylosides to murine Dexter cultures increased levels of chondroitin sulfate and hematopoiesis (SPOONCER et al. 1983). Other studies have indicated that stromal ECM glycosoaminoglycans might bind cytokines (GM-CSF) and present them to target cells (GORDON et al. 1987b).

The growth of adherent cells on irradiated stroma in Dexter conditions has been used as an assay for long-term repopulating cells measuring the frequency of cobblestone area-forming cells (CAFC) at varying intervals of culture (PLOEMACHER et al. 1989). Cells separated by a combination of centrifugal elutriation, depletion of lineage-defined cells, light scatter and dull rhodamine staining are enriched for marrow-repopulating ability (based on femoral day 12 CFU-S in an irradiated recipient 13 days post transplantation) and for precursors forming cobblestone areas at longer than 28-day culture intervals. These long-term CAFC probably correspond with reported Lin^- rhodamine-dull (BERTONCELLO et al. 1985, 1990), rhodamine dull/Hoechst 33342 low (WOLF et al. 1990) or $Thy\text{-}1^{lo}Lin^-Sca^+$ (MÜLLER-SIEBURG et al. 1988; MÜLLER-SIEBURG 1990; SPANGRUDE et al. 1988) in vivo repopulating cells. The day 7 CAFC may correspond to cells which provide only transient engraftment in vivo and are more closely correlated with day 12 CFU-S. In a similar vein, GORDON et al. (1985a,b, 1987a) have reported stroma-adherent blast colonies and proposed these as an assay for primitive progenitors.

5 Dexter Cultures as Stem Cell Assays

Much of the data on hematopoietic support in the human Dexter culture system fits the above hypothesis of stromal surface cytokine support of early stem cells. As noted above, relative long-term marrow growth has been obtained in human Dexter cultures although without the prolonged plateau phase seen with murine cultures (HOCKING and GOLDE 1980; DEXTER et al. 1984; CHARBORD et al. 1986). As with murine cultures, there is a continuous turnover and differentiation of granulocyte clonogenic cells in the nonadherent phase of human cultures

(CASHMAN et al. 1985). These cells appear to be released by cells from the adherents layers of the cultures (SLOVIK et al. 1984). In the mouse it appears that clonogenic cells do not correlate well with in vivo marrow repopulation. Cells that do repopulate tend to be resistant to 5-fluorouracil (5-FU) (HODGSON and BRADLEY 1979), lineage negative (SPANGRUDE et al. 1988), out of cycle (HELLMAN et al. 1978; ROSENDAAL et al. 1976), and to stain weakly with the dye rhodamine (BERONCELLO et al. 1985, 1990; PLOEMACHER and BRONS 1988). In the human, formal studies of marrow repopulation have not been possible, but data indicating that doses of 4-hydroperoxycyclophosphamide (4-HC), which kill up to 90% of clonogenic cells, spare the ability of at least some human marrow to function as protective autografts (SIENA et al. 1985; YEAGER et al. 1986; WINTON and COLENDA 1987) suggest a similar dichotomy between clonogenic cells and marrow renewal cells. Further studies have suggested that human marrow enriched for $CD34^+$ cells which are HLA-DR^- and rhodamine dull have few clonogenic cells, but adhere to and grow on stroma (VERFAILLE et al. 1990; SUTHERLAND et al. 1989; ANDREWS et al. 1986). $CD34^+$ cells may be further subdivided by expression of CD45RO and $CD45R^-$; $CD34^+$ $CD45RO^+$ cells may represent the more primitive long-term culture-initiating cells (LANSDORP et al. 1990). Further studies have suggested that marrow cells derived from long-term cultures can re-engraft upon autologous transplantation. Long-term culture-initiating cells have been defined by their ability to give rise to clonogenic units at 5–8 weeks of culture (SUTHERLAND et al. 1989, 1990). In general, these have been granulocyte clones, but erythroid clones have also been seen. The long-term culture (LTC)-initiating cell at 8 weeks also appears more resistant to 4-HC (WINTON and COLENDA 1987). The generation of colony-forming cells (CFC) from individual $CD34^+$ lineage-negative cells grown on irradiated human stroma has been demonstrated with 3.7% of these cells generating CFC by 4 weeks of culture and with each of these cells generating 6.3 ± 5.3 CFC (ANDREWS et al. 1990).

Human equivalents of CFU-blast (NAKAHATA and OGAWA 1982a) and HPP-CFC (MCNIECE et al. 1989a, b) have also been described—these are enriched in lineage-negative $CD34^+$ populations and are multifactor responsive. BRANDT et al. (1990) have also shown that $CD34^+$-, HLA-DR-, CD15-, and CD71-negative cells were supported by different cytokines for varying periods of time in liquid culture without an adherent layer. Cultures without cytokines produced clonogenic cells for 1 week, while IL-1 and IL-6 supported hematopoiesis for 2–3 weeks, GM-CSF for 3–4 weeks, and IL-3 for 4–5 weeks. IL-3 plus IL-α or IL-6 supported hematopoiesis for 8 weeks. The conclusion from these studies was "a major function of marrow adherent cells in long-term cultures is that of providing cytokines which promote the proliferation and differentiation of primitive hematopoietic stem cells" (BRANDT et al. 1990). Human stroma or stromal lines have also been found to be rich sources of cytokines, and recently IL-9 (YANG et al. 1989) and IL-11 (PAUL et al. 1990) were isolated from primate and human stromal lines, respectively. Human marrow fibroblast cells transfected with G-CSF or GM-CSF and G-CSF in a retroviral vector showed increased support of hematopoiesis. Stable induction of G-CSF or GM-CSF and G-CSF but not GM-CSF alone increased numbers of CFU-GM and BFU-E (HOGGE et al. 1991). In addition, G-CSF appeared to induce primitive adherent layer progenitors in long-term culture into S phase. When added exogenously GM-CSF and G-CSF appeared to act synergistically.

6 Model Stromal Hematopoiesis

These data form the basis for a model for stromal stem cell support. It seems likely that stromal cells produce most of the cytokines constitutively in low concentration and that stem cells are positioned next to stromal cells by binding to adherence proteins. This latter phenomena presumably may also modulate cytokine production. In this setting, the final outcome may be determined by the local concentration of certain critical stimulatory or inhibitory cytokines ("anchor factors"), the cytokine configuration or the specific mix of cytokines (Fig. 3). The geographic location of in vitro or in vivo hematopoiesis may also be explained by differential cytokine production. Rich and colleagues (VOGT et al. 1991; RICH 1986) have presented in situ hybridization evidence for the macrophage as a source of erythropoietin (epo) possibly explaining the erythroid nurse cells (BESSIS and BRETON-GORIUS 1959). These cells are also seen in in vitro murine Dexter cultures (DEXTER et al. 1984) following addition of epo and agitation. In a similar vein, OZAWA et al. (1990), using in situ hybridization, have presented evidence that only a small percentage of stromal cells in Dexter culture may produce G-CSF. These data suggest that the in vitro or in vivo location of various types of hematopoietic or lymphopoietic cells may relate to cell type-

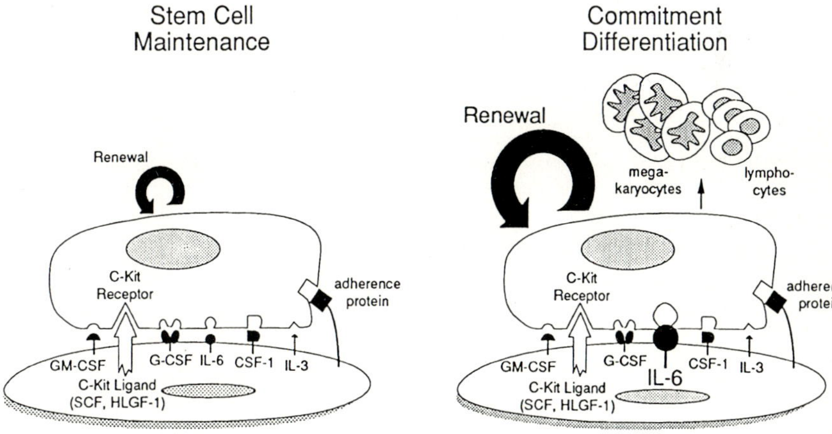

Fig. 3. Model of stromal cell regulation

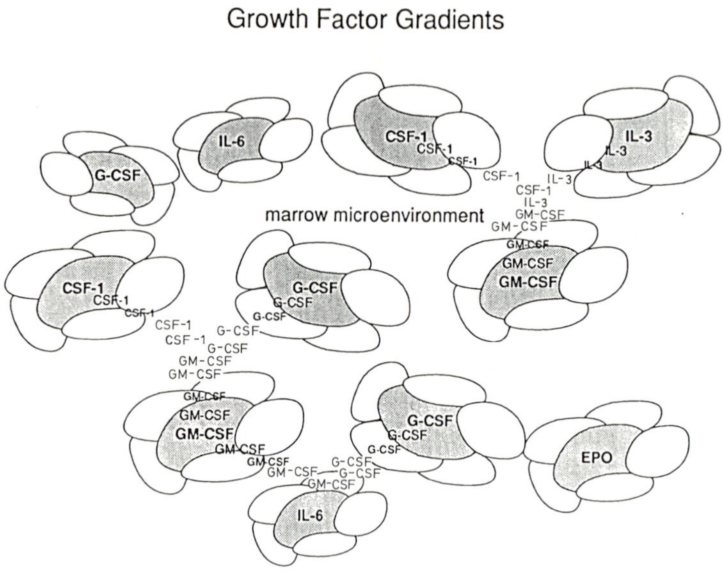

Fig. 4. Model of stromal growth factor gradients

specific cytokine production. Thus, the Dexter cultures, and by analogy the in vivo intramedullary environment, may represent a condition of growth factor gradients (Fig. 4).

References

Aizawa S, Tavassoli M (1987) In vitro homing of hemopoietic stem cells is mediated by a recognition system with galactosyl and mannosyl specificities. Proc Natl Acad Sci USA 84: 4485–4489

Aizawa S, Tavassoli M (1988) Detection of membrane lectins on the surface of hemopoietic progenitor cells and their changing pattern during differentiation. Exp Hematol 16: 325

Alberico T, Ihle JN, Quesenberry PJ (1987) Stromal growth factor production in irradiated lectin exposed long-term murine bone marrow cultures. Blood 69: 1120–1127

Andrews RG, Takahashi M, Segal GM, Powell JS, Bernstein ID, Singer JW (1986) The L4F3 antigen is expressed by unipotent and multipotent colony forming cells but not by their precursors. Blood 68: 1030

Andrews RG, Singer JW, Bernstein ID (1990) Human hematopoietic precursors in long-term culture: single $CD34^+$ cells that lack detectable T cell, B cell and myeloid cell antigens produce multiple colony-forming cells when cultured with marrow stromal cells. J Exp Med 172: 355–358

Axelrad AA, McLeod DL, Suzuki S, Shreeve MM (1978) Regulation of the population size of erythropoietic progenitor cells. In: Clarkson B, Marks PA, Till JE, Differentiation of normal and neoplastic hematopoietic cells, Cold Spring Harbor Laboratory, New York.

Bertoncello I, Hodgson GS, Bradley TR (1985) Multiparameter analysis of transplantable hemopoietic stem cells. I. The seperation and enrichment of stem cells homing to marrow and spleen on the basis of rhodamine-123 fluorescence. Exp Hematol 13: 999

Bertoncello I, Kriegler AB, Bradley TR, Dunlop JM, McNiece IK (1990) The effect of interleukin-6 on hemopoietic stem cells subsets resolved by rhodamine 123. Exp Hematol 18: 550 (abstract)

Bessis M, Breton-Gorius J (1959) Nouvelles observations sur piolot erythroblastique et la rhopheocytose de la ferritine. Rev Hematol 14: 165

Bradley TR, Metcalf D (1966) The growth of mouse bone marrow cells in vitro. Aust J Exp Biol Med Sci 44: 287–299

Brandt J, Srour EF, van Beisen K, Briddell RA, Hoffman R (1990) Cytokine-dependent long-term culture of highly enriched precursors of hematopoietic progenitor cells from human bone marrow. J Clin Invest 86: 932–941

Cashman J, Eaves AC, Eaves CJ (1985) Regulated proliferation of primitive hematopoietic progenitor cells in long-term human marrow cultures. Blood 66: 1002–1005

Chabot B, Stephenson DA, Chapman VM, Besmer P, Bernstein A (1988) The proto-oncogene c-kit encoding a transmembrane tyrosine kinase receptor maps to the mouse W locus. Nature 335: 88–89

Charbord P, Gown AM, Keating A, Singer JW (1985) CGA-7 and HHF, two monoclonal antibodies that recognize muscle actin and react with adherent cells in human long-term bone marrow cultures. Blood 66: 1138–1142

Charbord P, Fujiwara M, Singer JW (1986) Long-term cultures of human marrow cells. Nouv Rev Fr Hematol 28: 65–74

Charbord P, Lerat H, Newton I, Tamayo E, Gown AM, Singer JW, Herve P (1990) The cytoskeleton of stromal cells from human bone marrow cultures resembles that of cultured smooth muscle cells. Exp Hematol 18: 276–282

Chenu C, Pfeilschifter J, Mundy GR, Roodman GD (1988) Transforming growth factor β inhibits formation of osteoclast-like cells in long-term human marrow cultures. Proc Natl Acad Sci USA 85: 5683–5687

Copeland NG, Gilbert DJ, Cho BC, Donovan PJ, Jenkins NA, Cosman D, Anderson D, Lyman SD, Williams DE (1990) Mast cell growth factor maps near the steel locus on mouse chromosome 10 and is deleted in a number of steel alleles. Cell 63: 175–183

Corey CA, Desilva A, Williams DA (1990) Erythropoiesis in murine long-term marrow cultures following transfer of the erythropoietin cDNA into marrow stromal cells. Exp Hematol 18: 201–204

Coutinho LH, Will A, Radford J, Schiro R, Testa NG, Dexter TM (1990) Effects of recombinant human granulocyte colony stimulating factor (CSF), human granulocyte macrophage-CSF and gibbon interleukin-3 on hematopoiesis in human long term bone marrow culture. Blood 75: 2118–2129

Dexter TM (1979) Cell interactions in vitro. Clin Haematol 8: 453–468

Dexter TM, Lajtha LG (1974) Proliferation of haemopoietic stem cells in vitro. Br J Haematol 28: 525–530

Dexter TM, Moore MAS (1977) In vitro duplication and "cure" of haemopoietic defects in genetically anaemic mice. Nature 269: 412–414

Dexter TM, Shadduck RK (1980) The regulation of haemopoiesis in long-term bone marrow cultures. I. Role of L-cell colony stimulating factor. J Cell Physiol 102: 279–286

Dexter TM, Spooncer E (1978) Loss of immunoreactivity in long-term bone marrow culture. Nature 275: 135–136

Dexter TM, Allen TD, Lajtha LG (1977) Conditions controlling the proliferation of haemopoietic stem cells in vitro. J Cell Physiol 91: 335–344

Dexter TM, Spooncer E, Simmons P, Allen TD (1984) Long-term marrow culture: an overview of techniques and experiences. In: Wright DG, Greenberger JS (eds) Long-term bone marrow culture, Liss, New York, pp 57–96

Dexter TM, Coutinho LH, Spooner E, Heyworth CM, Daniel CP, Schiro R, Chang J, Allen TD (1990) Stromal Cells in Haemopoiesis. In: Block G, March J (eds) Moleculer control of haemopoiesis. Wiley, Chichester, pp 76–95

Dorshkind K, Phillips RA (1982) Maturational state of lymphoid cells in long term bone marrow cultures. J Immunol 129: 2444–2450

Dorshkind K, Phillips RA (1983) Characterization of early B lymphocyte precursors present in long-term bone marrow cultures. J Immunol 131: 2240–2245

Doukas M, Niskanen E, Quesenberry P (1986) The effect of lithium on stem cell and stromal cell proliferation in vitro. Exp Hematol 14: 215–221

Eliason JF, Vassalli P (1988) Inhibition of hemopoiesis in murine marrow cell cultures by recombinant murine tumor necrosis factor α: evidence for long-term effects on stromal cells. Blood Cells 14: 339–354

Fauser AA, Messner HA (1978) Granuloerythropoietic colonies in human bone marrow, peripheral blood and cord blood. Blood 52: 1243–1248

Flanagan JG, Leder P (1990) The kit ligand: A cell surface molecule altered in steel mutant fibroblasts. Cell 63: 185–194

Geissler EN, Ryan MA, Housman DE (1988) The dominant-white spotting (W) locus of the mouse encodes the c-kit proto-oncogene. Cell 55: 185–192

Gordon MY, Hibbin JA, Dowding C, Gorden-Smith EC, Goldman JM (1985a) Separation of human blast progenitors from granulocytic, erythroid, megakaryocytic and mixed colony-forming cells by "panning" on cultured marrow-derived stromal layers. Exp Hematol 13: 937–940

Gordon MY, Hibbin JA, Kearney LU, Gordon-Smith EC, Goldman JM (1985b) Colony formation by primitive haemopoietic progenitors in cocultures of bone marrow cells and stromal cells. Br J Haematol 60: 129–136

Gordon MY, Dowding CR, Riley GP, Greaves MF (1987a) Characterization of stroma-dependent blast colony forming cells in human marrow. J Cell Physiol 130: 150–156

Gordon MY, Riley GP, Watt SM, Greaves MF (1987b) Compartmentalization of a hematopoietic growth factor (GM-CSF) by glycosaminoglycans in the bone marrow microenvironment. Nature 326: 403–405

Graham GJ, Wright EG, Hewick R, Wople SD, Wilkie NM, Donaldson D, Larimore S, Prognall IB (1990) Identification and characterization of an inhibitor of haemopoietic stem cell proliferation. Nature 344: 442–444

Gualtieri RJ, Shadduck RF, Baker DG, Quesenberry PJ (1984) Hematopoietic regulatory factors produced in long-term bone marrow cultures and the effect of in vitro irradiation. Blood 64: 516–525

Harrison DE (1980) Competitive repopulation: a new assay for long-term stem cell functional capacity. Blood 55: 77–81

Harrison DE, Astle CM, Lerner C (1984) Ultimate erythropoietic repopulating abilities of fetal, young adult and old adult cells compared using repeated irradiation. J Exp Med 160: 759–771

Harrison DE, Lerner CP, Spooncer E (1987) Erythropoietic repopulating ability of stem cells from long-term marrow culture. Blood 69: 1021–1025

Hass R, Ogniben E, Kiesel S, Hohaus S, Baumann M, Korbling M, Dorken B, Hunstein W (1989) Enhanced myelopoiesis in long-term cultures of human bone marrow pretreated with recombinant qranulocyte-macrophage colony stimulating factor. Exp Hematol 17: 235–239

Hellman S, Botnick LE, Hannan EP, Vignevlle RM (1978) Proliferative capacity of murine hematopoietic stem cells. Proc Natl Acad Sci USA 75: 490

Hocking WG, Golde DW (1980) Long-term human bone marrow cultures. Blood 56: 118–124

Hodgson GS, Bradley TR (1979) Properties of haemopoietic stem cells surviving 5-fluorourascil treatment: evidence for a pre-CFU-S cell? Nature 281: 381

Hogge DE, Cashman JD, Humphries RK, Eaves CJ (1991) Differential and synergistic effects of human granulocyte-macrophage colony-stimulating factor and human granulocyte colony stimulating factor on hematopoiesis in human long-term marrow cultures. Blood 77: 492–499

Huang E, Nocka K, Beier DR, Chu T-Y, Buck J, Lahm H-W, Wellner D, Leder P, Besmer P (1990) The hematopoietic growth factor KL is encoded by the Sl locus and is the ligand of the c-kit receptor, the gene product of the W locus. Cell 63: 225–233

Hunt P, Robertson D, Weiss D, Rennick D, Lee F, Witte ON (1987) A single bone marrow derived stromal cell type supports the in vitro growth of early lymphoid and myeloid cells. Cell 48: 997

Johnson A, Dorshkind K (1986) Stromal cells in myeloid and lymphoid long-term bone marrow cultures can support multiple hemopoietic lineages and modulate their production of hemopoietic growth factors. Blood 68: 1348–1354

Jordan CT, Lemischka IR (1990) Clonal and systemic analysis of long-term hematopoiesis in the mouse. Genes Dev 4: 220–232

Konna M, Hardy L, Tavassoli M (1990) Interaction of late murine erythroid progenitors and stroma involves a recognition mechanism with fucosyl specificity. Exp Hematol 18: 185–189

Kriegler AB, Bradley TR, Giap KH, Hodgson GS (1984) Sources of murine macrophage colony-stimulating factor also contain growth factors for primitive macrophage progenitor cells. Exp Hematol 12: 844–849

Lansdorp PM, Sutherland HJ, Eaves CJ (1990) Selective expression of CD45 isoforms on functional subpopulations of $CD34^+$ hemopoietic cells from human bone marrow. J Exp Med 172: 363–366

Lemischka IR, Raulet D, Mulligan RC (1986) Developmental potential and dynamic behavior of hematopoietic stem cells. Cell 45: 917–927

Levitt L, Quesenberry PJ (1980) The effect of lithium on murine hematopoiesis in a liquid culture system. N Eng J Med 302: 713–719

Liesveld JL, Winslow JM, Kempski MC, Ryan DH, Brennan JK, Abboud CN (1991) Adhesive interactions of normal and leukemic human CD34$^+$ myeloid progenitors: role of marrow stromal fibroblast, and cytomatrix components. Exp Hematol 19: 63–70

Lowry PA, Zsebo KM, Deacon DH, Fullen A, Quesenberry PJ (1991) Multifactor stimulation of the murine HPP-CFC. International Society of Experimental Hematology Meeting, July 1991, Parma (abstract)

Lyman GH, Williams CC, Preston D (1980) The use of lithium carbonate to reduce infection and leukopenia during systemic chemotherapy. N Engl J Med 302: 257

Magli MC, Iscove NN, Odartchenko N (1982) Transient nature of early haematopoietic spleen colonies. Nature 295: 527–529

Majumder S, Brown K, Qiu F-H, Besmer P (1988) c-kit protein, a transmembrance kinase: identification in tissues and characterization. Mol Cell Biol 8: 4896–4903

Martin FH, Suggs SV, Langley KE, Lu HS, Ting J, Okino KH, Morris CF, McNiece IK, Jacobsen FW, Mendiaz EA, Birkett NC, Smith KA, Johnson MJ, Parker VP, Flores JC, patel AC, Fisher EF, Erjavec HO, Herrera CJ, Wypych J, Sachdev RK, Pope JA, Leslie I, Wen D, Lin C-H, Cupples RL, Zsebo KM (1990) Primary structure and functional expression of rat and human stem cell factor DNAs. Cell 63: 203–211

Mauch P, Greenberger JS, Botnick L, Hannon E, Hellman S (1980) Evidence for structure variation in self-renewal capacity within long-term bone marrow cultures. Proc Natl Acad Sci USA 77: 2927

Mayani H, Guilbert LJ, Janowska-Wieczorek A (1990) Modulation of erythropoesis and myelopoiesis by exogenous erythropoietin in human long-term marrow cultures. Exp Hematol 18: 174–179

Mayfield D, Brown RG (1966) The clinical laboratory and electroencephalographic effects of lithium. J psychiatr Res 4: 207

McGrath HE, Liang C, Alberico T, Quesenberry PJ (1987) The effects of lithium on growth factor production in long term bone marrow cultures. Blood 70: 1136–1142

McLeod DL, Shreeve MM, Axelrad AA (1974) Improved plasma culture system for production of erythrocytic colonies in vitro: quantitative assay method for CFU-E. Blood 44: 517–534

McNiece IK, Andrews R, Stewart FM, Clark S, Souza L, Boone T, Quesenberry PJ (1989a) Action of IL-3, G-CSF and GM-CSF on highly enriched human hematopoietic progenitor cells: synergistic interaction of GM-CSF plus G-CSF. Blood 74: 110–114

McNiece IK, Stewart FM, Deacon DH,Temeles DS, Zsebo K, Clark S, Quesenberry PJ (1989b) Detection of a human CFC with a high proliferative potential. Blood 74: 609–612

Meagher RC, Salvado AJ, Wright DG (1988) An analysis of the multilineage production of human hematopoietic progenitors in long-term bone marrow culture: evidence that reactive oxygen intermediates derived from mature phagocytic cells have a role in limiting progenitor cell self-renewal. Blood 72: 273–281

Metcalf D, McDonald HR, Odartchenko N, Surdat B (1975) Growth of mouse megakaryocyte colonies in vitro. Proc Natl Acad Sci USA 72: 1744–1748

Müller-Sieburg CE (1990) Isolation and characterization of hematopoietic stem cells and progenitors. Prog Immunol 7 (in press)

Müller-Sieburg CE, Whitlock CA, Weissman IL (1986) Isolation of two early B lymphocyte progenitors from mouse bone marrow: a commited pre-pre-B cell and a clonogenic Thy-1^{lo} hematopoietic stem cell. Cell 44: 653–662

Müller-Sieburg CE, Townsend K, Weissman IL Rennick D (1988) Proliferation and differentiation of highly enriched mouse hematopoietic stem cells and progenitor cells in response to defined growth factors. J Exp Med 167: 1825–1840

Nakahata T, Ogawa M (1982a) Hemopoietic colony-forming cells in unbilical cord blood with extensive capability to generate mono- and multipotential hemopoietic progenitors. J Clin Invest 70: 1324

Nakahata T, Ogawa M (1982b) Identification in culture of a new class of hemopoietic colony-forming units with extensive capability to self-renew and generate multipotential colonies. Proc Natl Acad Sci USA 79: 3843

Namen AE, Schmierer AE, March CJ, Overell RW, Park LS, Urdal DL, Mochizuki DY (1988) B cell precursor growth promoting activity. Purification and characterization of a growth factor active on lymphocyte precursors. J Exp Med 167: 988–1002

Niskanen E, Wyandt HE (1985) Potentiality of murine colony forming cells detected by an in vivo diffusion chamber culture system. Blood 66: 686

Nocka K, Majumder S, Chabot B, Ray P, Cervone M, Bernstein A, Besmer P (1990a) Expression of c-kit gene products in known cellular targets of W mutations in normal and W mutant mice—evidence for an impaired c-kit kinase in mutant mice. Genes Dev 3: 816–826

Nocka K, Tan JC, Chiu E, Chu TY, Ray P, Traktman P, Besmer P (1990b) Molecular basis of a dominant negative and loss of function mutations at the murine c-kit/white spotting locus: W37, Wv, W41 and W. EMBO J 9: 1805–1813
Ogawa M, Pharr PN, Suda T (1985) Stochastic nature of stem cell functions in culture. In: Cronkite E, Dainiak N, McCaffrey R, Palek J, Quesenberry (eds) Hematopoietic stem cell physiology. Liss, New York
Ozawa K, Takahashi K, Nishikawa M, Yoshikubo T, Shimane M, Takahashi T, Takahashi S, Horiguchi N, Tsuruta T, Watari K, Tojo A, Tani K, Asano S (1991) Regulatory production of G-CSF by stromal cells. In: Quesenberry PJ, Asano S, Socito K (eds) Hematopoietic Growth Factors. Excerpta Medica, Tokyo, pp 55–81
Patel VP, Lodish HF (1984) Loss of adhesion of murine erythroleukemia cells to fibronection during erythroid differentiation. Science 244: 996–998
Patel VP, Lodish HF (1986) The fibronectin receptor on mammalian erythroid precursor cells: characterization and developmental regulation. J Cell Biol 102: 449–456
Paul SR, Bennett F, Caluetti JA, Kelleher K, O'Hara RM Jr, Leary AC, Clark SC, Williams DA, Yang YC (1990) Molecular cloning of a cDNA encoding interleukin-11: a novel stromal cell derived lymphopoietic and hematopoietic cytokine. Proc Natl Acad Sci USA 87: 7512
Polemacher R, Brons N (1988) Cells with marrow and spleen repopulating ability and forming spleen colonies on day 16, 12 and 8 are sequentially ordered on the basis of increasing rhodamine-123 retention. J cell Physiol 136: 531
Ploemacher RE, Van der Slijs JP, Uoerman JSA, Brans NHC (1989) An in vitro limiting dilution assay of long-term repopulating hematopoietic stem cells in the mouse. Blood 74: 2755-2763
Pluznik DH, Sachs L (1965) The cloning of normal "mast" cells in tissue culture. J Cell Comp Physiol 66: 319–324
Quesenberry PJ, Coppola MA, Gualtieri RJ, Wade PM, Song Z, Doukas MA, Shideler CE, Baker DG, McGrath HE (1984) Lithium stimulation of murine hematopoiesis in liquid culture: an effect mediated by marrow stromal cells. Blood 63: 121–127
Quesenberry PJ, Song ZX, McGrath HE, McNiece IK, Shadduck RK, Waheed A, Baber G, Kleeman E, Kaiser D (1987) Multilineage synergistic activity produced by a murine adherent marrow cell line. Blood 69: 827–835
Quesenberry PJ (1989) Stromal cells in long-term bone marrow cultures. In: Tavassoli M (ed) Handbook of the hematopoietic microenvironment, Humana, Clifton
Quesenberry P, McGrath E, Buxton K, Kittler E, Crittenden R, Temeles D (1990) Constitutive production of multiple hemopoietic growth factors by murine Dexter stroma. Blood 76: 161a (abstract)
Reith AD, Rottapel R, Giddens E, Brady C, Forrester L, Bernstein A (1990) W mutant mice with mild or severe developmental defects contain distinct point mutations in the kinase domain of the c-kit receptor. Genes Dev 4: 390–400
Rennick D, Yang G, Gemmell L, Lee F (1987) Control of hemopoiesis by a bone marrow stromal cell clone: lipopolysac-charide and interleukin-1 inducible production of colony stimulating factors. Blood 69: 682–691
Rich IN (1986) A role for the macrophage in normal hemopoiesis. Effect of verying physiological oxygen tensions on the release of hemopoietic growth factors from bone marrow derived macrophages in vitro. Exp Hematol 14: 746
Rosendaal M, Hodgson GS, Bradley TR (1976) Haemopoietic stem cells are organized for use on the baisis of their generation age. Nature 264: 68
Rothstein G, Clarkson DR, Larsen W, Grosser BI, Athens JW (1978) Effect of lithium on neutrophil mass and production. N Engl J Med 298: 178
Ruland LJ III, Quesenberry PJ, Balian G (1970) A 17 Kilodalton polypeptide, sensitive to bacterial collagenase, is synthesized by bone marrow macrophages in myeloid and lymphoid cultures. Exp Hematol 18: 969–973
Schrader JW, Schrader S (1978) In vitro studies on lymphocyte differentiation. I. Long term in vitro culture of cells giving rise to functional lymphocytes in irradiated mice. J Exp Med 148: 823–828
Siena S, Castro-Malaspina H, Gulati SC, Lu L, Colvin MO, Clarkson BD, O'Reilly RJ, Moore MAS (1985) Effects of in vitro purging with 4-hydroperoxycyclophosphamide on the hematopoietic and microenvironmental elements of human bone marrow. Blood 65: 655–662
Slovik FT, Abboud CN, Brennan JK, Lichtman MA (1984) Survival of granulocytic progenitors in the nonadherent and adherent compartments of human long-term marrow cultures. Exp Hematol 12: 327–338

Song ZX, Shadduck RK, Innes Jr DJ, Washeed A, Quesenberry PJ (1985) Hematopoietic factor production by a cell line (TC-1) derived from adherent murine marrow cells. Blood 66: 273–281

Spangrude GJ, Heimfeld S, Weissman IL (1988) Purification and characterization of mouse hematopoietic stem cells. Science 241: 58–62

Spooncer E, Gallagher JT, Krizsa F, Dexter TM (1983) Regulation of haemopoiesis by G-D-xylosides. J Cell Biol 96: 510–514

Stein RS, Beaman C, Ali MY, Hansen R, Jenkins DD, Jume'an HG (1977) Lithium carbonate alternation of chemotherapy-induced neutropenia. N Engl J Med 297: 430

Stone M, Harrison DE (1990) A stratified binonial marker model for bone marrow repopulation experiments. J Theo Biol 144: 267–273

Suda J, Suda T, Ogawa M (1984) Analysis of differentiation of mouse hemopoietic stem cells in cultures by sequential replating of paired progenitors. Blood 64: 393

Sutherland HJ, Eaves CJ, Eaves AC, Dragowska W, Lansdorp PM (1989) Characterization and partial purification of human marrow cells capable of initiating long-term hematopoietic in vitro. Blood 74: 1563–1570

Sutherland HJ, Lansdorp PM, Henkelman DH, Eaves AC, Eaves CJ (1990) Functional characterization of individual human hametopoietic stem cells cultured at limiting dilution on supportive marrow stromal layers. Proc Natl Acad Sci USA 87: 3584–3588

Till JE, McCulloch EA (1961) A direct measurement of the radiation sensitivity of normal mouse bone marrow. Radiat Res 14: 213–218

Touw I, Lowenberg B (1984) Production of T lymphocyte colony forming units from precursors in human long-term bone marrow cultures. Blood 64: 656–661

Verfaille C, Blakolmer K, McGlove P (1990) Purified primitive human hematopoietic progenitor cells with long-term in vitro repopulating capacity adhere selectively to irradiated bone marrow stroma. J Exp Med 172: 509–520

Vogt C, Noe G, Rich IN (1991) The role of the blood island during normal and 5-Fluorouracil perturbed hemopoiesis. Blood Cells (in press)

Wade PM Jr, Stewart FM, Doukas MA, Quesenberry PJ (1982) The effect of lithium chloride (LiCl) on murine Dexter culture high proliferative potential colony forming cells (HPP-CFC) and stromal colony stimulating activity (CSA) production. Blood 60: 105a (abstract)

Whitlock CA, Witte ON (1982) Long-term culture of B lymphocytes and their precursors from murine bone marrow. Proc Natl Acad Sci USA 79: 3608–3612

Whitlock CA, Robertson D, Witte ON (1984) Murine B cell lymphopoiesis in long-term culture. J Immunol Methods 67: 353–369

Whitlock CA, Tidmarsh GF, Müller-Sieburg C, Weissman IL (1987) Bone Marrow stromal cell lines with lymphopoietic activity espress high levels of a pre-B neoplasia associated molecule. Cell 48: 1009

Williams DE, Eisenman J, Baird A, Rauch C, VanNess K, March CJ, Park LS, Martin V, Mochizuki DY, Boswell HS, Burgess GS, Cosman D, Lyman SC (1991) Identification of a ligand for the c-kit protooncogne. Cell 63: 167–174

Williams N, Jackson H, Sheridan APC, Murphy MJ, Elste A, Moore MAS (1978) Regulation of megakaryopoiesis in long-term murine bone marrow cultures. Blood 51: 245–255

Winton EF, Colenda KW (1987) Use of long-term human marrow cultures to demonstrate progenitor cell precursors in marrow treated with 4-hydroperoxycyclophosphamide. Exp Hematol 15: 710–714

Wolf NS, Kone A, Priestely GV (1990) Primitive stem cells obtained with combined Hoechst 33342 and rhodamine 123 sorting: selection and capabilities. Exp Hematol 18: 549 (abstract)

Woodward TA, McNiece IK, Witte PL, Bender P, Crittenden R, Temeles DS, Robinson BE, Baber GB, Deacon DH, Isakson PC, Quesenberry PJ (1990) Further studies on growth factor production by the TC-1 stromal cell line: pre-B stimulating activity. Blood 75: 2130–2136

Yang Y-C, Ricciardi S, Ciarletta A, Caluetti J, Kelleher K, Clark SC (1989) Expression cloning of a cDNA encoding a novel human hematopoietic growth factor: human homologue of murine T cell growth factor p 40. Blood 74: 1880

Yarden Y, Kuang W-J, Yang-Feng T, Coussens L, Munemitsu S, Dull TJ, Chen E, Schlessinger J, Francke U, Ullrich A (1987) Human Proto-oncogene c-kit: a new cell surface receptor tyrosine kinase for an unidentified ligand. EMBO J 6: 3341–3351

Yeager AM, Kaizer H, Santos GW, Saral R, Colvin OM, Stuart RK, Braine HG, Burke PJ, Ambinder RF, Burns WH, Fuller DJ, Davis JM, Karp JE, May WS, Rowley SD, Sensenbrenner LL, Vogelsang GB, Wingard JR (1986) Autologous bone marrow transplantation in patients with acute nonlymphocytic leukemia, using ex vivo marrow treatment with 4-hydroperoxycyclophosphamide. N Engl J Med 315: 141–147

Zsebo KM, Wypych J, McNiece IK, Lu HS, Smith KA, Karkare SB, Sachdev RK, Yuschenkoff YN, Birkett NC, Williams LR, Satyagal VN, Tung W, Bosselman RA, Mendiaz EA, Langley KE (1990a) Identification, Purification and biological characterization of hematopoietic stem cell factor from buffalo rat liver-conditioned medium. Cell 63: 195–201

Zsebo KM, Williams DA, Geissler EN, Broudy VC, Martin FH, Atkins HL, Hsu R-Y, Birkett NC, Okino KH, Murdock DC, Jacobsen FW, Langley KE, Smith KA, Takeishi T, Cattanach BM, Galli SJ, Suggs SV (1990b) Stem cell factor is encoded at the Sl locus of the mouse and is the ligand for the c-kit tyrosine kinase receptor. Cell 63: 213–224

Zuckerman KS, Rhodes RK, Goodrum DD, Patel VR, Sparks B, Wells J, Wicha MS, Mayo LA (1985) Inhibition of collagen deposition in the extracellular matrix prevents the establishement of a stroma supportive of hematopoiesis in long term murine bone marrow cultures. J Clin Invest 75: 970–975

Human Stem Cell Transplantation

Marrow Transplantation

R. STORB

1 Introduction

During the past 20 years marrow transplantation has been used with increasing success as treatment for patients with acute and chronic leukemias, malignant lymphomas, severe aplastic anemia, thalassemia major, congenital immunologic deficiency diseases, and inborn errors of metabolism (reviewed in SANTOS 1983; THOMAS et al. 1975; GALE and CHAMPLIN 1987; KRIVIT and PAUL 1986; O'REILLY

Fred Hutchinson Cancer Research Center, Clinical Research Division, 1124 Columbia St., Seattle, WA 98104, USA

Current Topics in Microbiology and Immunology, Vol. 177

1983; LUCARELLI et al. 1990; LOUGHRAN and STORB 1990; VAN ROOD and ZWAAN 1984; WITHERSPOON et al. 1989; STORB 1989; MARMONT et al. 1986; SPECK and GRATWOHL 1987; GORIN et al. 1989; BURNETT and MORRISON 1989; PHILLIPS et al. 1989; ARMITAGE and BIERMAN 1989). The International Bone Marrow Transplantation Registry estimated that, as of 1987, 20 000 transplants had been carried out, more than 10 000 of them between 1985 and 1987 and most of them for patients with hematologic malignancies. This manuscript will review the general principles and clinical marrow transplant results obtained to date and will define the principal problems that need to be solved to improve results.

2 General Principles

2.1 The Source of Marrow

Most patients have received allogeneic marrow grafts from donors who are siblings—genetically different, but genotypically identical for the HLA antigens of the major histocompatibility complex coded for by genes located on chromosome 6. Syngeneic grafts are from identical twins, completely identical for all genetic loci. Recent interest has focused on identifying alternative donors who are not HLA identical. These include siblings and other relatives who are genotypically identical for one HLA haplotype and phenotypically matched for one or more of the HLA loci on the nonshared haplotype. Acceptable relatives are found for approximately 10% of patients. The establishment of donor registries has allowed the identification of phenotypically HLA-identical unrelated volunteer donors. Over the past 5 years transplants have been increasingly performed using marrow grafts from such unrelated donors. Finally, "autologous" marrow grafts can be carried out. Marrow can be harvested and cryopreserved at $-196\,^{\circ}C$, either while the patient is in a marrow remission induced by chemotherapy or because the marrow is not involved in the malignant process. Often, the marrow is treated with monoclonal antibodies plus complement, immunotoxins, or certain chemicals (e.g., 4-hydroxyperoxycyclophosphamide) to remove T lymphocytes or clonogenic malignant cells.

2.2 Marrow Harvest and Infusion

Between 400 and 1000 ml marrow (10 ml/kg of donor body weight) is obtained by multiple aspirations from the iliac crests under anesthesia. Marrow is collected in heparinized tissue culture medium, screened through wire mesh to remove tissue particles, and given to the recipient by intravenous infusion, usually in an amount $1–8 \times 10^8$ nucleated cells per kilogram. Stem cells derived from peripheral blood have recently been used in patients in whom marrow is difficult to harvest because of preceding radiotherapy. Another development has been successful transplantation of marrow maintained in long-term culture. This technique is attractive because it may permit propagation of desirable, and elimination of undesirable, cells.

2.3 Conditioning of the Recipient Before Transplantation

In patients with an inherited lack of an immune system but normal marrow, e.g., severe combined immunodeficiency disease, marrow grafts provide cells that repair the immune system without replacing the myeloid and erythroid components. Such grafts may not need immunosuppressive conditioning regimens.

The preparative regimen often used for patients with aplastic anemia is cyclophosphamide (Cy), given intravenously at a dose of 200 mg/kg body weight over 4 days, followed 36 h later by marrow infusion.

In patients with genetically determined disorders, e.g., Wiskott–Aldrich syndrome or thalassemia major, who have normal or greater than normal marrow cellularity, Cy (200 mg/kg) is combined with busulfan (4 mg/kg/day orally for 4 days) to increase immunosuppression and to eradicate affected marrow, thereby providing "space" for the donor marrow.

In patients with leukemia, regimens combining high doses of chemotherapy and/or total body irradiation (TBI) serve both to destroy malignant cells and to suppress host immunity in order to prevent graft rejection. A common regimen consists of Cy (60 mg/kg on each of 2 days) followed by TBI. TBI may be administered as a single dose or in fractions with total doses ranging from 5.0 to 15.75 Gy. TBI is effective because it has an antitumor effect that is not as dependent on cell cycle as most chemotherapeutic agents; it penetrates privileged sites, e.g., central nervous system and testicles, and is immunosupressive. Regimens not containing TBI include the combination of busulfan and Cy and bischloroethylnitrosourea, Cy, and etoposide.

3 Clinical Results

3.1 Congenital Diseases

Marrow transplantation has been the treatment of choice for severe combined immunodeficiency disease since the first success in 1968 (reviewed in GALE and CHAMPLIN 1987; KRIVIT and PAUL 1986; O'REILLY 1983); 55% of patients can be cured by marrow grafts from HLA-identical siblings. Graft-versus-host disease (GVHD) is rare, and failures are often related to pre-existing infections. Transplants from HLA-nonidentical relatives have also been successful, but immune reconstitution is often incomplete. Marrow grafting constitutes a definitive form of therapy for patients with Wiskott–Aldrich syndrome and Fanconi's anemia. In the latter condition, an unusual sensitivity to the toxicity of the conditioning programs has been noted, perhaps due to impaired repair mechanisms of somatic cells, and lower doses of immunosuppressive agents should be used.

Other genetic diseases that have been cured with marrow grafts include: ataxia telangiectasia, infantile agranulocytosis, chronic granulomatous disease, osteopetrosis, Blackfan-Diamond anemia, Chediak-Higashi disease, chronic mucocutaneous candidiasis, congenital red cell aplasia, mucopolysaccharidosis, cartilage-hair hypoplasia, Gaucher's and other lipid storage diseases. Existing central nervous system complications of some of the lipid storage diseases may not be improved, and early transplants before nervous system damage should be considered.

Cure of thalassemia major has now been reported in 70% of patients given HLA-identical marrow grafts. Patients who have regular transfusions and good chelation have a survival rate of 94% at 3 years without transplantation (LUCARELLI et al. 1990). However, those with one or more poor-risk factors have survival rates of 60%–80%. Unfortunately, these same factors probably also predict for poor success with transplantation. At present the choice of transplantation over conventional therapy is a personal one. The same applies to patients with sickle cell disease whose natural history is variable with very few longitudinal studies of survival being reported. In both diseases, prevention through genetic counseling should receive a high priority. Correcting the genetic defects by in vivo fetal to fetal transplantation is feasible in principle but untested (see J.-L. TOURAINE, this volume). If successful, this technique could correct not only β- but also α-thalassemia where homozygotes invariably die in utero. In the future, modification of the genome of hematopoietic cells may avert the clinical manifestations of genetically determined hematologic diseases. Until then, marrow grafting offers the only hope for cure for these disorders.

3.2 Severe Aplastic Anemia

Aplastic anemia may most frequently be due to a stem cell defect. Infusion of marrow from a monozygotic twin has often been successful in reconstituting the marrow without immunosuppression of the recipient. However, some syngeneic grafts have been successful only after preparation with Cy, suggesting other mechanisms, perhaps an autoimmune etiology, which can be overcome by Cy.

Allogeneic transplantation is effective therapy for aplastic anemia (reviewed in LOUGHRAN and STORB 1990; VAN ROOD and ZWAAN 1984). Graft rejection was originally the major problem limiting success, and only 45% of patients became long-term survivors. Two factors predict rejection: positive in vitro tests of cell-mediated immunity of host lymphocytes against donor cells and a low number of grafted marrow cells ($<3 \times 10^8$/kg). Transfusion-induced immunity is the major cause of graft rejection. When transplants are carried out before blood transfusions, graft rejection is rare; 80% of untransfused patients are alive between 4 and 18 years after grafting.

Rejection can be avoided in transfused patients by more intensive immunosuppression in addition to Cy, such as TBI, total lymphoid irradiation, or thoracoabdominal irradiation. Viable donor buffy coat cells infused along with the marrow are also helpful in overcoming rejection. With decreasing rates of rejection, survival has increased. Survival rate for transfused patients is now close to 70%.

Most regimens utilized to decrease rejection have associated risks. The addition of buffy coat cells leads to an increase in chronic GVHD. Radiation regimens have early toxic effects as well as the potential for causing malignancies later (WITHERSPOON et al. 1989). More recently, a combination of antithymocyte globulin and Cy promises to reduce the risk of graft rejection while avoiding the need for buffy coat cell infusions. Nevertheless, emphasis should be placed on measures to prevent rather than overcome sensitization by blood transfusion. If an HLA-identical family member is available, early transplant, before transfusions, is the therapy of choice, Newer techniques of blood filtration to remove leukocytes, which are the immunogen, may also decrease immunization in multiply transfused patients. Animal studies have shown that gamma irradiation of blood products almost eliminates the risk of sensitization to marrow grafts which are matched at the major histocompatibility complex.

3.3 Acute Nonlymphoblastic Leukemia

Allogeneic Marrow Grafts in First Remission. Transplantation in first remission offers improved survival since the body burden of leukemic cells is small, leukemic cells may be less resistant to therapy, and patients are clinically well and better able to tolerate a transplant. Twelve of the first 22 patients in whom transplants were performed between 1976 and 1978 are surviving between 13 and 15 years later. Studies comparing the outcome of transplantation to conventional chemotherapy showed either statistically significant or suggestive advantages for transplantation. Most transplant teams now report 50% disease-free survival with transplantation (reviewed in GALE and CHAMPLIN 1987; O'REILLY 1983; VAN ROOD and ZWAAN 1984; STORB 1989; MARMONT et al. 1986; SPECK and GRATWOHL 1987). The relapse rate is about 25%. Patients under age 20 years have a 70% survival rate as opposed to 40% for those over age 20.

Allogeneic Marrow Grafts after Relapse. Until the mid-1970s, marrow grafts were carried out only after other therapy had failed. Between 1970 and 1975, 54 patients in relapse were given HLA-identical grafts in Seattle, and six are living in remission 16–20 years later (THOMAS et al. 1975). The relapse rate was 65%. An improvement was seen in 132 more recent patients (reviewed in STORB 1989). Survival at 3 years was 34% in untreated first relapse and second remission and 24% in resistant first relapse. This suggests that patients should be transplanted in early first relapse rather than after another remission induction, as some patients may die during chemotherapy.

Allogeneic Versus Autologous Marrow Grafts. Recent reports have shown survival at 2 years after autologous transplants to be 50% for patients in first remission and 30% for those in second remission. These results were seen regardless of whether the marrow was treated to remove clonogenic leukemic cells (reviewed in MARMONT et al. 1986; SPECK and GRATWOHL 1987; GORIN et al. 1989; BURNETT and MORRISON 1989). In contrast to data from marrow allografts, the interval from remission induction to autologous transplant seems important. Autotransplants immediately after remission induction result in lessened survivals comparable to those with chemotherapy alone, while autotransplants at 6 months or greater produced improved results similar to allografts. Perhaps, there is an element of patient selection in many reported studies on autografts. In one study, 25% of patients scheduled to receive autografts in first remission did not actually get the transplant due to early relapse, poor marrow function, or patient refusal. These same biases are present in studies of autotransplants in second remission. The true effectiveness of autografting in acute nonlymphoblastic leukemia (ANL) can only be answered by prospective randomized trials comparing autografts to chemotherapy in early first remission or after first relapse, with registration of patients at diagnosis and again after remission induction. Another unanswered question concerns the effectiveness of methods of purging the marrow of leukemic cells, since there are no controlled studies (GORIN et al. 1989).

3.4 Acute Lymphoblastic Leukemia

Allogeneic Marrow Grafts in Remission. Results are better with transplants in remission (reviewed in GALE and CHAMPLIN 1987; O'REILLY 1983; VAN ROOD and ZWAAN 1984; STORB 1989; MARMONT et al. 1986; SPECK and GRATWOHL 1987) than in relapse. Survival rates for patients transplanted in first remission vary from 25% to 60% with relapse rates of 20%–40%. Differences may reflect patient selection or variations in follow-up. Most transplants are done in second remission. Of the first 22 Seattle patients grafted in second or third remission, 27% are surviving 13–15 years later. Children do better than adults. Leukemic relapse continues to be the major problem with an incidence of 60%.

Allogeneic Marrow Grafts in Relapse. As for ANL, marrow transplantation was first employed to treat patients with advanced disease (THOMAS et al. 1975). Of 56 patients transplanted between 1970 and 1975, six are currently living 16–20 years later. The actuarial relapse rate was 75%.

Autologous versus Allogeneic Transplantation. Long-term disease-free survival has been reported in a proportion of patients given autologous transplants (reviewed in MARMONT et al. 1986; SPECK and GRATWOHL 1987; GORIN et al.1989;

BURNETT and MORRISON 1989). In many cases, attempts have been made to purge the marrow of clonogenic leukemic cells with monoclonal antibodies and complement or immunotoxins. Autologous transplants are safer, but allogeneic transplants result in fewer relapses. Results with autografts in first remission are similar to those of allografts. However, it is unknown whether the patient populations in the two settings are truly comparable. In two studies of second remission transplants, autografts resulted in survival of 20% as compared to 30% with allografts, and these differences were not statistically significant. The major cause of failure in both groups was relapse.

When to Carry Out Marrow Grafts. Given the grim prognosis of patients with acute lymphoblastic leukemia (ALL) resistant to chemotherapy, there is little doubt that they should be transplanted. A prospective study showed transplants to be superior to chemotherapy in patients who had relapsed at least once. Patients who relapse more than 18–24 months after chemotherapy has been stopped, however, may do as well with chemotherapy as with transplantation. Transplantation in first remission is debatable in view of the good results with current chemotherapy for normal risk children and adults. However, those with high risk for early relapse, e.g., with Philadelphia-chromosome-positive ALL and L3 morphology with high white blood counts, may benefit from transplantation in first remission. Allogeneic or autologous marrow grafts are the treatment of choice for patients in first marrow remission who have an extramedullary relapse.

3.5 Chronic Myelocytic Leukemia

Allogeneic Marrow Transplants in Chronic Phase. By 1976 the Seattle team had begun a trial in 19 patients who had healthy identical twins to serve as marrow donors. Twelve of the 19 are alive and in remission without the Philadelphia chromosome, 5–15 (median 12) years later. These results have encouraged studies with marrow grafts from HLA-identical siblings (reviewed in GALE and CHAMPLIN 1987; O'REILLY 1983; VAN ROOD and ZWAAN 1984; STORB 1989; MARMONT et al. 1986; SPECK and GRATWOHL 1987); disease-free survival in this group has been 49%–56%, and the risk of relapse is 12%–20%. Currently, marrow grafts are the only form of cure for chronic myelocytic leukemia (CML). Any patient below age 65 years who has a suitable donor should be transplanted.

Allogeneic Grafting in Accelerated Phase and Blastic Transformation. The experience with transplants in patients with CML in blast crisis parallels that in patients with acute leukemia in advanced relapse (reviewed in SANTOS 1983; THOMAS et al. 1975; GALE and CHAMPLIN 1987; O'REILLY 1983; VAN ROOD and ZWAAN 1984; STORB 1989; MARMONT et al. 1986; SPECK and GRATWOHL 1987). Long-term disease-free survival is about 15%, and the risk of relapse is 50%–75%. Attempts to improve results by transplanting in the accelerated phase have been somewhat successful, with 15%–40% disease-free survival, and relapse rates of 30%–56%.

3.6 Lymphoma

As of 1985 the Seattle group had carried out 100 transplants for recurrent malignant lymphoma with a 22% disease-free survival at 5 years and a likelihood of relapse of 60%. Patients grafted either in first relapse or in second remission did significantly better with a 42% 2-year disease-free survival and a relapse rate of 41%. Neither the source of marrow (autologous, allogeneic or syngeneic) nor disease histology (high-grade versus intermediate-grade non-Hodgkin's lymphoma versus Hodgkin's disease) were predictive of outcome. These results are similar to those obtained by others and indicate that marrow transplants for recurrent malignant lymphoma yield better results than salvage chemotherapy (GALE and CHAMPLIN 1987; STORB 1989; MARMONT et al. 1986; SPECK and GRATWOHL 1987; PHILLIPS et al. 1989; ARMITAGE and BIERMAN 1989).

3.7 Multiple Myeloma

Syngeneic transplants established the feasibility of curing patients with multiple myeloma by marrow transplant, and the longest surviving recipient is now 12 years after transplant. The longest survivor from an allogeneic transplant is now 7 years after transplant. In general, transplantaion of refractory patients produces results similar to those seen for resistant leukemia with survival around 20%. Earlier transplant should improve the survival of patients with definable poor-risk characteristics, i.e., probability of failure to conventional therapy in the 1st year. More recently, autografts have been shown to have a palliative role for patients with refractory myeloma, and progress with posttransplant biologic modifiers and lymphokine therapy could improve these results.

3.8 Myelodysplastic Syndromes

Patients with myelodysplastic syndromes, i.e., preleukemia and refractory anemia with and without excess blasts, have an invariably fatal prognosis. Allogeneic marrow transplantation offers the best treatment. Survival for patients without excess blasts is similar to that in aplastic anemia, 60%–70%, while patients with excess blasts have a survival of 35% due to posttransplant relapse.

3.9 Other Hematologic Malignancies

Marrow grafting has also been successful in small numbers of patients with myelofibrosis, hairy cell leukemia, and histiocytic medullary reticulosis.

3.10 Solid Tumors

Pilot studies have been reported with autologous transplants, usually after failure of other therapy, for radiosensitive solid tumors such as neuroblastoma, ovarian, testicular, and small cell lung cancer. Results have been poor except for breast cancer and neuroblastoma, where long-term disease-free survival and probable cures have been seen.

4 Alternative Marrow Donors

Extensive data from experimental animals support the concept that best results occur with marrow from genotypically HLA-identical siblings. HLA-nonidentity may negatively affect engraftment, GVHD, and immunologic reconstitution. However, only 35%–40% of patients have an HLA-identical sibling. To overcome this limitation, alternative sources of marrow have been explored. Searches of the patients' families show 10%–11% to have either phenotypically matched or one HLA-locus mismatched relatives. Additionally, either fully HLA-matched or one-locus-mismatched unrelated donors can be identified in 25%–30% of cases.

4.1 HLA-Nonidentical Family Member Donors

Results of transplantation with marrow from relatives other than HLA-identical siblings have shown that transplant complications increase in direct proportion to the degree of HLA-disparity (reviewed in GALE and CHAMPLIN 1987; ANASETTI et al. 1989, 1990; BEATTY et al. 1985) Complications include delayed engraftment, low granulocyte and platelet counts, increased risk of graft rejection, GVHD, and infections. Despite the higher incidence of graft failure and GVHD, survival of phenotypically HLA-identical and one-HLA-locus mismatched patients is not significantly different from that of genotypically HLA-identical patients, since the risk of transplant-related complications is offset by lower leukemic relapse rates. Patients mismatched for two or more HLA loci on the nonshared haplotype fare worse, however, and actuarial survival is only 10%–20%. In these patients, autologous transplants should be considered first.

4.2 Unrelated Donors

Experimental canine studies have suggested that marrow grafts from phenotypically DLA-identical unrelated donors, while less successful than those from genotypically DLA-identical littermates, are better than those from DLA-haploidentical littermates. The human experience to date concurs with those observations (reviewed in GALE and CHAMPLIN 1987; BEATTY et al. 1991). More than 200 unrelated transplants have by now been carried out at the Fred Hutchinson Cancer Research Center. Increased risks of acute and chronic GVHD have been observed; however, long-term survival does not appear to differ from that seen after HLA-identical sibling transplants, most likely because the risk of leukemic relapse is decreased.

A national marrow donor registry is now being developed in the United States which includes data on 150 000 potential marrow donors. Similar registries exist in Europe. With the continued growth of these registries and exchange of marrows between the two continents, the number of unrelated marrow grafts has steadily increased.

5 Currrent Problems and Future Approaches

5.1 Hematopoietic and Immunologic Recovery

Following an uncomplicated marrow graft, granulocyte and platelet levels usually return to normal within 40–50 days, and hematocrit levels within 60–90 days (reviewed in MARMONT et al. 1986; VAN BEKKUM and LÖWENBERG 1985). In most cases, hematopoiesis is entirely of donor origin although some patients show a persistent mixture of donor and host cells. Hematopoiesis following transplant is stable, as seen in patients up to 21 years after transplant. The recovery of monocytes and bronchoalveolar and hepatic macrophages is equally prompt. Nevertheless, certain prolonged functional impairments may remain, including poor granulocyte chemotaxis, a reduced number of clonogenic precursors in the marrow, and diminished immunologic function. Clonal dominance suggestive of reconstitution of marrow from small numbers of pluripotent hematopoietic stem cells has been reported in several experimental situations and two clinical patients. Our own studies in 20 allogeneic marrow graft recipients followed for more than 1 year after transplant, using restriction fragment length polymorphisms on the X chromosome, failed to show evidence of clonal dominance and are consistent with polyclonal reconstitution.

The recipient's immunologic status after transplant consists of three components, all of which may provide protection against infections. First, vestiges of the host's oiginal immune system may still be operative, as documented by transient production of host-type isohemagglutinin titers in ABO-incompatible recipients for as long as 12 months. Second, immune cells from the donor are transplanted along with the marrow, thereby conveying "instant" immune function to the recipient. For example, transfer of immunity to a number of recall and neoantigens has been documented in 85% of patients without GVHD and in 45% of those with chronic GVHD. Finally, the most important component is the "rebirth" of the new immune system originating from the grafted marrow. All recipients, regardless of the kind of transplant, have profound impairment of most immune functions during the first 4–5 months, except for monocyte-macrophage function and cytotoxic effector cells mediating natural killing, antibody-dependent killing, and lecitn-dependent killing, all of which are normal within 30 days. Beyond 4–5 months, most patients show a return to normal of most other immunologic parameters. Nevertheless, even among healthy, long-term survivors, a few fail to produce appropriate humoral antibodies, and 30%–40% of these show persistent in vitro B cell defects and decreased T cell activity. These immune deficits are even more pronounced in patients with chronic GVHD.

It would be attractive to accelerate the recovery of immunity and restore a normal immunologic state in marrow graft recipients. Attempts to do so using grafts of thymic tissue or thymic epithelial monolayers, and injection of thymosin fraction V or thymopoietin, however, have as yet been unsuccessful.

5.2 Opportunistic Infections

Bacterial and Fungal Infections. Five percent of HLA-identical marrow graft recipients may die from bacterial or fungal infection during the early granulocytopenic period (reviewed in Bowden and Meyers 1985). This figure is increased in patients with alternative donors. Shortening the period of granulocytopenia would help to reduce both the incidence of infections and avoid the use of potentially toxic antibiotics needed to treat them. Recombinant human hematopoietic growth factors such as granulocyte colony stimulating factor (G-CSF) and granulocyte-macrophage (GM)-CSF have been tested in marrow graft recipients and been shown to hasten granulocyte recovery and decrease infection (Nemunaitis et al. 1990). Approximately 50% of marrow graft recipients who have graft failure will respond to GM-CSF. In two separate randomized trials high-dose IV gammaglobulin appears to be effective in decreasing bacterial infections.

Viral and Protozoan Infections. The most important infections during the first 3–4 months are viral (reviewed in Bowden and Meyers 1985). Activation of herpes simplex virus immediately post grafting used to be common but is now prevented with acyclovir. Varicella zoster infections seen in approximately 40% of all patients are also treatable and preventable with acyclovir. Pneumocystis carinii pneumonias have now been prevented with trimethoprim sulfamethoxazole. By far the most important infection is cytomegalovirus (CMV). Evidence of CMV activation is seen in about 75% of patients who have antibody to CMV pretransplant. While often asymptomatic, activation can become serious in the form of CMV pneumonia. CMV pneumonia is seen in about 15% of patients grafted for malignancy and has a case fatality rate of 50%–85%. Patients who are CMV seronegative pretransplant can be protected by the exclusive use of CMV seronegative blood products or blood products rendered leukopoor by filtration. Patients who are seropositive at the time of transplant have a lower reactivation rate if given prophylactic acyclovir. Immunoprophylaxis using CMV immunoglobulin has been controversial. Currently, an acyclovir derivative, ganciclovir, combined with CMV-immune globulin appears to be effective therapy for established CMV infection.

5.3 Graft Failure, GVHD, Recurrence of Malignant Disease

The long-term success of marrow transplants for the treatment of hematologic malignancies depends on avoiding three interrelated problems: graft failure, GVHD, and recurrence of underlying disease. Graft failure and recurrence of leukemia are both indicative of the inability of current conditioning programs to consistently destroy all host immune and malignant cells. For complete eradication to occur, a graft-versus-leukemia effect may be needed. Rodent studies suggest that a graft-versus-leukemia effect could be augmented without increasing GVHD.

Graft Failure. Graft failure is rare (<5%) in leukemic recipients of unmodified HLA-identical sibling marrow (reviewed in Beatty et al 1985; Anasetti et al. 1989; van Bekkum and Löwenberg 1985; Storb and Deeg 1986). Occasionally poor graft function is seen, which may be due to drug toxicity, e.g., methotrexate, trimethoprim sulfamethoxazole, interferon, etc., or related to certain viruses, e.g., CMV infection. Deletion of the incriminating drug or antiviral therapy may improve marrow function. Approximately 50% of poorly functioning grafts will respond to GM-CSF. Failure of engraftment and graft rejection have been recognized to occur in two settings. The first is with marrow from donors who are HLA nonidentical. Failure is seen in 5% of phenotypically HLA-identical pairs, 7%–10% of one-HLA-locus-mismatched pairs, and 15%–25% of two–three-HLA-loci-mismatched pairs. The second setting occurs with T-cell-depleted marrow grafts. Presumably, donor T cells are needed to destroy those host cells which cause rejection of the graft. Graft failure from rejection has a high case fatality rate and is not usually reversible by growth factors. Increased understanding of the mechanisms of graft failure is needed to develop ways to overcome the problem. Animal studies have suggested that these mechanisms vary depending upon the particular donor-recipient combination. Canine studies have implicated radioresistant Ia-positive non-T cells in destruction of DLA-nonidentical grafts and T cells in rejection of DLA-identical grafts. They suggest further that an increase in the dose of TBI from 9.2 to 18.0 Gy or pretreatment of the recipient with anti-CD44 antibodies is capable of overcoming resistance to mismatched grafts, while treatment with antithymocyte globulin is effective in histocompatible grafts. In humans, the incidence of graft rejection decreases with intensified conditioning regimens, but survival is not improved due to concomitant increases in regimen-related toxicities. The efficacy of adding antibodies directed at immune cells to the conditioning regimen is being explored. Also, it is possible that the T cells in the marrow inoculum which are needed to overcome graft rejection may be different from those which cause GVHD. Animal studies have yet to provide definite answers to this question. Human trials are in progress, however, using selective T cell depletion.

Pathophysiology of Acute GVHD. Acute GVHD is thought to be caused by T cells in the donor marrow which attack the skin, liver, and gastrointestinal tract (reviewed in BEATTY et al. 1991, VAN BEKKUM and LÖWENBERG 1985; DEEG et al. 1984; MÖLLER 1985; SALE and SHULMAN 1984; STORB 1986). It is manifested by skin rash, liver function abnormalities, abdominal pain, and diarrhea. Of patients given marrow from HLA-identical siblings, 20%–50% develop clinically significant acute GVHD. Acute GVHD is associated with a 2.5 times higher risk of fatality. Death in these patients is most often due to infections.

Relationship Between GVHD and Leukemic Recurrence. Some transplant recipients may have been cured of their leukemia due to a "graft-versus-leukemia effect." This hypothesis is based on murine studies and retrospective clinical analyses. At 150 days following transplant, the likelihood of remaining in remission is significantly higher in allogeneic recipients with GVHD than in those without GVHD or with syngeneic grafts. Similarly, recipients of T-cell-depleted marrow have less GVHD but more relapses than those given unmanipulated marrow. Attempts at inducing GVHD to decrease the relapse rate and improve survival have not met with success, since the lowered risk of relapse is offset by a greater risk of transplant complications (SULLIVAN et al. 1989). Certain murine studies, yet to be verified in humans, suggest that a graft-versus-leukemia effect can be distinguished from GVHD. Certain cells in the donor inoculum may specifically react with antigens expressed on leukemic cells while others may more broadly react with histocompatibility antigens on host tissue cells. Perhaps it would be possible to clone T cells sensitized to tumor antigens and expand them in vitro with the help of interleukin 2 (IL-2) before infusion into the tumor-bearing host along with IL-2.

Prevention of Acute GVHD. Most prevention of GVHD has involved the use of immunosuppressive drugs after grafting. Recipients of unmanipulated marrow who did not receive these drugs experienced an unacceptably high incidence of acute GVHD. Methotrexate and cyclosporine are effective in preventing GVHD, while Cy is less so. A combination of methotrexate and cyclosporine is significantly better than either drug alone and may lead to improved survival (STORB et al. 1986). Unfortunately, both drugs have associated toxicities often necessitating dose reductions, thereby impairing their efficacy in GVHD prevention. Another way to prevent GVHD has been the removal of T cells from the donor marrow with monoclonal antibodies used either for immunoadsorption or for targeting lytic activities of complement and toxins. T cell depletion has also been achieved by their selective agglutination to soybean lectin or through E-rosetting and elutriation. Any of these techniques reduces the number of T cells by 1–3 logs. T cell depletion appears to reduce substantially the risk of acute and possibly also chronic GVHD, but at the price of increases in graft rejection and leukemic relapse. Since rejection and leukemic relapse nearly always result in death, no improvement in survival has been seen to date in patients given T-cell-depleted marrow.

Treatment of Established Acute GVHD. Based mainly on animal studies, acute GVHD has been treated with prednisone, antithymocyte globulin, cyclosporine, and various monoclonal antibodies directed against T cells, and results have been variable. As the repertoire of monoclonal anti-T cell antibodies increases, reagents may be found that specifically interact with those cells that are involved in acute GVHD.

Chronic GVHD. Chronic GVHD develops in 25%–50% of allograft recipients between 3 and 15 months after transplantation (reviewed in GALE and CHAMPLIN 1987; LOUGHRAN and STORB 1990; VAN ROOD and ZWAAN 1984; VAN BEKKUM and LÖWENBERG 1985; MÖLLER 1985; SALE and SHULMAN 1984; STORB 1986). The clinical manifestations resemble those seen in systemic collagen vascular diseases, including skin lesions, keratoconjunctivitis, buccal mucositis, esophageal and vaginal strictures, intestinal involvement, chronic liver disease, generalized wasting, and pulmonary insufficiency. Patients with acute GVHD and older patients are at higher risk for developing chronic GVHD. If untreated, most patients become disabled or die of infection. Prednisone given alone or combined with cyclosporine is the most effective drug to treat chronic GVHD (SULLIVAN et al. 1988a, b). With therapy, approximately one-half of the patients survive with Karnofsky performance scores of 100%, and an additional one-fourth with scores of between 80% and 90%. In 50% of the patients, therapy can be stopped after 9–12 months. Nevertheless, the management of patients with chronic GVHD remains unsatisfactory, since 25% of affected patients die with infections, mainly from encapsulated gram-positive bacteria, resulting from their poor immunologic status.

Incidence and Origin of Recurrent Malignant Disease. Recurrent malignancy remains a major cause of failure of marrow transplantation (reviewed in THOMAS et al. 1975; GALE and CHAMPLIN 1987; O'REILLY 1983; VAN ROOD and ZWAAN 1984; STORB 1989; MARMONT et al. 1986; SPECK and GRATWOHL 1987; ANASETTI et al. 1990; BEATTY et al. 1991; DEEG et al. 1984). In more than 95% of the cases, the original malignant cell population reappears as determined by blood genetic markers. This demonstrates the inability of current pretransplant conditioning programs to eradicate all clonogenic leukemic cells in every patient. Less than 5% of recurrent leukemia involves donor type cells. It is not clear whether this is due to transfection of donor cells with DNA derived from host cells, or influences of the marrow microenvironment. A few relapses have been in the form of immunoblastic lymphosarcoma in cells of donor origin, usually in patients given T-depleted marrow or treated with monoclonal antibodies for steroid-resistant GVHD.

Epstein-Barr virus has been detectable in all these lymphosarcomas. The proliferation of Epstein-Barr-virus-transformed cells suggests normal immune surveillance mechanisms are impaired in these patients.

Modification of Conditioning Programs to Improve Tumor Cell Kill. For most approaches involving systemic chemotherapy with and without TBI, the limits of nonhematopoietic toxicity may have been reached and barring unexpected developments, no quantum improvements can be expected from these approaches. The most effective way to eradicate cancer would be through the use of agents which specifically interact with the malignant cells. Such an approach is the use of monoclonal antibodies directed against antigens on the tumor. Monoclonal antibodies injected in vivo can concentrate on tumor cells, but their anti-tumor effect is limited. This is due to the fact that some tumor cells lack target antigens, and others, though coated by antibody, are not killed by it. Antibodies linked to toxins such as the ricin-A chain are currently being explored for more effective tumor cell kill. Another way to use monoclonal antibodies is to attach them to short-lived radioactive isotopes (reviewed in APPELBAUM et al. 1988). In this way, not only would cells expressing the target antigens be killed, but also neighboring antigen-negative cells. If applied to hematologic malignancies, this approach would ablate normal marrow cells as well as malignant cells, and subsequent "rescue" by a transplant would still be necessary.

Studies in dogs have shown appropriate antibody isotope conjugates to localize preferentially in marrow and spleen, and to a lesser extent also in lymph nodes, with the amount of isotope in the marrow as compared to other organs achieving a ratio of 28:1 or better. Radiolabeled antibodies are capable of producing fatal marrow aplasia; aplasia can be reversed by infusion of cryopreserved autologous marrow 8 days later when little radioactivity is left. These studies, done initially with radioactive iodine, are continuing, and various combinations of chemotherapy, TBI, and radiolabeled antibody are being explored for their ability to condition dogs for grafts of T-cell-depleted marrow. The experiments have set the stage for the human application of this technique. Refinements of this approach, particularly the use of high energy β-emitting isotopes with short linear energy transfer, are likely to result in less toxic and more efficient pretransplant conditioning programs, not only providing for better eradication of malignant disease but also lessening the problem of graft failure.

6 Conclusions and Outlook

Twenty years ago marrow grafts were restricted to patients with advanced hematologic diseases. Now, marrow transplantation is the treatment of choice for patients with aplastic anemia, immunodeficiency disease, certain genetic disorders of hematopoiesis, myelodysplasia, CML, and any acute leukemia or lymphoma that has recurred at least once. Thalassemia major is the most prevalent disease curable by marrow transplantation. Despite impressive advances in marrow transplantation, major problems remain to be solved. These include infections associated with prolonged immunodeficiency, graft failure in patients given T-cell-depleted or HLA-nonidentical grafts, acute and chronic GVHD, recurrence of malignant disease, and late complications resulting from the conditioning programs. The development of more effective and less toxic conditioning programs is crucial. At present, approaches such as the use of monoclonal antibodies linked to short-lived radioisotopes seem promising. More effective conditioning programs should reduce the problems of recurrence of malignant disease and of graft failure. Most likely they will also permit a broader application of T cell depletion for GVHD prevention, thus extending marrow transplants to include more HLA-nonidentical and unrelated patients. Improved conditioning programs should also enlarge the scope of autologous marrow grafting for malignancy, especially if combined with better methods of marrow purging. It is hoped that recombinant hematopoietic growth factors will reduce the incidence of infections. Finally, gene transfer into stem cells might find clinical application, conferring drug resistance to marrow cells and inhibiting or altering the function of genes involved in the generation of disease.

Acknowledgements. This work was supported by grants HL36444, CA18221, CA18029, CA15704, and CA31787 of the National Heart, Lung and Blood Institute and the National Cancer Institute of the National Institutes of Health (U.S. Department of Health and Human Services).

References

Anasetti C, Amos D, Beatty PG, Appelbaum FR, Bensinger W, Buckner DC, Clift R, Dony K, Martin PJ, Mickelson E, Nisperos B, O'Quigley J, Ramberg R, Sanders JE, Stewart P, Storb R, Sullivan KM, Witherspoon RP, Thomas ED, Hansen JA (1989) Effect of HLA compatibility on engraftment of bone marrow transplants in patients with leukemia or lymphoma. N Engl J Med 320: 197–204

Anasetti, C, Beatty PG, Storb R, Martin PJ, Mori M, Sanders JE, Thomas ED, Hansen JA (1990) Effect of HLA incompatibility on graft-versus-host disease, relapse and survival after marrow transplantation for patients with leukemia or lymphoma. Hum Immuno 29: 79–91

Appelbaum FR, Brown PA, Graham TC, Sandmaier BM, Schuening FW, Storb R (1988) Characterization of malignant lymphoma in dogs and use as a model for the development of treatment strategies. In: Baum SJ, Santos GW, Takaku F (eds) Experimental hematology today 1987. Recent advances and future directions in bone marrow transplantation. Springer, Berlin Heidelberg New York, pp 31–35

Armitage JO, Bierman PJ (1989) Is there an optimum conditioning regimen for patients with lymphoma undergoing autologous bone marrow transplantations? In: Dicke KA, Spitzer G, Jagannath S, Evinger-Hodges MJ (eds) Autologous bone marrow transplantation. University of Texas, Houston, MD Anderson, pp 299–303

Beatty PG, Clift RA, Mickelson EM, Nisperos B, Flournoy N, Martin PJ, Sanders JE Stewart P, Buckner CD, Storb R, Thomas ED, Hansen JA (1985) Marrow transplantation from related donors other than HLA-identical siblings. N Engl J Med 313: 765–771

Beatty PG, Hansen JA, Thomas ED, Sanders JE, Martin PJ, Bearman SI, Longton GM, Anasetti C, Petersdorf EW, Mickelson EM, Pepe MS, Appelbaum FR, Buckner CD, Clift RA, Petersen FB, Stewart PS, Storb RF, Sullivan KM, Tesler, MC, Witherspoon RP (1991). Marrow transplantation from HLA-matched unrelated donors for treatment of hematologic malignacies. Transplantation 51: 443–447

Bowden RA, Meyers JD (1985) Infectious complications following marrow transplantation. Plasma Ther Transfus Technol 6: 285–302

Burnett AK, Morrison A (1989) The status of autograft for acute myeloid leukemia. In: Dicke KA, Spitzer G, Jagannath S, Evinger-Hodges MJ, Autologous bone marrow transplantation University of Texas, Houston, MD Anderson, pp 93–99

Deeg HJ, Storb R, Thomas ED (1984) Bone marrow transplantation: a review of delayed complications. Br J Haematol 57: 185–208

Gale RP, Champlin R (eds) (1987) Progress in bone marrow transplantation. Liss, New York

Gorin NC, Laporte J-P, Douay L, Lopez M, Najman A(1989) Autologous bone marrow transplantation for consolidation of acute leukemia in remission: purging or not purging? Influence of pretransplant intervals. In: Dicke KA, Spitzer G, Jagannath S, Evinger-Hodges MJ (eds) Autologous bone marrow transplantation. . University of Texas, Houston, MD Anderson, pp 41–53

Krivit W, Paul NW (eds) (1986) Bone marrow transplantation for treatment of lysosomal storage diseases. Liss, New York

Loughran TP Jr, Storb R (1990) Treatment of aplastic anemia. In: Forman SJ (ed) Hematology/oncology clinics of North America, vol 4 Saunders, Philadelphia, pp 559–575

Lucarelli G, Galimberti M, Polchi P, Angelucci E, Baronciani D, Giardini C, Politi P, Durazzi SMT, Muretto P, Albertini F (1990) Bone marrow transplantation in patients with thalassemia. N Engl J Med 322: 417–421

Marmont AM, Bacigalupo A, Van Lint MT (eds) (1986) Bone marrow transplantation. Proceedings of the 12th Annual Meeting of the European Cooperative Group, vol 1, Macmillan, New York

Möller G (ed) (1985) Graft-versus-host reaction. Munksgaard, Copenhagen (Immunological Reviews, vol 88)

Nemunaitis J, Singer J, Buckner CD, Durnam D, Epstein C, Hill R, Storb R, Thomas ED, Appelbaum FR (1990) The use of recombinant human granulocyte-macrophage colony-stimulating factor in graft failure after bone marrow transplantation. Blood 76: 245–253

O'Reilly RJ (1983) Allogeneic bone marrow transplantation: current status and future directions. Blood 62: 941–964

Phillips G, Fay JW, Herzig RH, Lazarus H, LeMaistre FC, Wolf SN, Herzig G (1989) Autologous bone marrow transplant conditioning regimens for non-Hodgkin's lymphoma. In: Dicke KA, Spitzer G, Jagannath S, Evinger-Hodges MJ (eds) Autologous bone marrow transplantation. University of Texas, Houston, MD Anderson, pp 305–310

Sale GE, Shulman HM (eds) (1984) The pathology of bone marrow transplantation. Masson, New York

Santos GW (1983) History of bone marrow transplantation. Clin Haematol 12: 611–639

Speck B, Gratwohl A (eds) (1987) Bone marrow transplantation. Proceedings of the 13th Annual Meeting of the European Cooperative Group. Macmillan, New York

Storb R (1986) Graft-versus-host disease after marrow transplantation. In: Meryman HT (ed) Transplantation: approaches to graft rejection. Alan R. Liss, New York, pp 139–157

Storb R (1989) Bone marrow transplantation. In: DeVita VT Jr, Hellman S, Rosenberg SA (eds) Cancer: principles and practice of oncology, 1989. Lippincott, Philadelphia, pp 2474–2489

Storb R, Deeg HJ (1986) Failure of allogeneic canine marrow grafts after total body irradiation: Allogeneic "resistance" vs transfusion induced sensitization. Transplantation 42: 571–580

Storb R, Deeg HJ, Whitehead J, Appelbaum F, Beatty P, Bensinger W, Buckner CD, Clift R, Doney K, Farewell V, Hansen J, Hill R, Lum L, Martin P, McGuffin R, Sanders J, Stewart P, Sullivan K, Witherspoon R, Yee G, Thomas ED (1986) Methotrexate and cyclosporine compared with cyclosporine alone for prophylaxis of acute graft versus host disease after marrow transplantation for leukemia. N Engl J Med 314: 729–735

Sullivan KM, Witherspoon RP, Storb R, Deeg HJ, Dahlberg S, Sanders JE, Appelbaum FR, Doney KC, Weiden P, Anasetti C, Loughran TP, Hill R, Shields A, Yee G, Shulman H, Nims J, Strom S, Thomas ED (1988a) Alternating-day cyclosporine and prednisone for treatment of high-risk chronic graft-v-host disease. Blood 72: 555–561

Sullivan KM, Witherspoon RP, Storb R, Weiden P, Flournoy N, Dahlberg S, Deeg HJ, Sanders JE, Doney KC, Appelbaum FR, McGuffin R, McDonald GB, Meyers J, Schubert MM, Gauvreau J, Shulman HM, Sale GE, Anasetti C, Loughran TP, Strom S, Nims J, Thomas ED (1988b) Prednisone and azathioprine compared with prednisone and placebo for treatment of chronic graft-v-host disease: prognostic influence of prolonged thrombocytopenia after allogeneic marrow transplantation. Blood 72: 546–554

Sullivan KM, Storb R, Buckner CD, Fefer A, Fisher L, Weiden PL, Witherspoon RP, Appelbaum FR, Banaji M, Hansen J, Martin P, Sanders JE, Singer J, Thomas ED (1989) Graft-versus-host disease as adoptive immunotherapy in patients with advanced hematologic neoplasms. N Engl J Med 320: 828–834

Thomas ED, Storb R, Clift RA, Fefer A, Johnson FL, Neiman PE, Lerner KG, Glucksberg H, Buckner CD (1975) Bone-marrow transplantation. N Engl J Med 292: 832–843, 895–902

van Bekkum DW, Löwenberg B (eds) (1985) Bone marrow transplantation. Biological machanisms and clinical practice. Dekker, New York

van Rood J, Zwaan F (eds) (1984) Bone marrow transplantation. I. Seminars in Hematology, vol 21: 1–63

Witherspoon RP, Fisher LD, Schoch G, Martin P, Sullivan KM, Sanders J, Deeg HJ, Doney K, Thomas D, Storb R, Thomas ED (1989) Secondary cancers after bone marrow transplantation for leukemia or aplastic anemia. N Engl J Med 321: 784–789

The Fetal Liver as a Source of Stem Cells for Transplantation into Fetuses In Utero

J.-L. TOURAINE

1 Introduction

Postnatal transplantation of fetal liver cells, alone or in association with syngeneic fetal thymic (epithelial) cells, leads to full immunologic reconstitution of patients with severe combined immunodeficiency disease (SCID) and to improvement of patients with a variety of other congenital or hematologic disorders. When no compatible donor is available for a bone marrow transplant to patients with these diseases, fetal liver transplantation (FLT) represents a very valuable therapeutic solution and it cures a significant proportion of the patients (TOURAINE et al. 1987).

Over the last 16 years, 213 fetal tissue transplants (especially FLT associated with syngeneic fetal thymus) have been performed in our institution to treat 58 patients with severe immunodeficiency diseases (IDD), severe aplastic anemia, or inborn errors of metabolism (IEM). Seventy-seven percent of patients with IEM and 50% of patients with IDD are alive and well with a follow-up of 1–16 years. They are presently either completely cured or significantly improved. Despite HLA mismatch between donor and recipient, in these cases there is no noticeable restriction of T cell functions in vivo, ex vivo, and in vitro, provided that investigations are carried out several years after the transplant (TOURAINE 1983; RONCAROLO et al. 1986).

FLT therefore represents an efficient postnatal therapy in severe hematologic, immunologic, or metabolic disturbances, and it does not require any HLA matching between donor and recipient. As exemplified below, it can also be very

Transplantation and Clinical Immunology, Pavillon P, Hôpital E, Herriot, 69437 Lyon Cedex 03, France

Current Topics in Microbiology and Immunology, Vol. 177

effective when performed prenatally, even in cases in which postnatal FLT does not usually result in graft take. In 1988 and 1989 we treated three sick fetuses with prenatal FLT. Two of these in utero transplants were carried out for inherited immunodeficiency, the third one for thalassemia major. Recommendations of the French National Committee for Bioethics were carefully followed for tissue procurement as well as for transplant conditions, both in postnatal and in prenatal FLT. The three children are the first three patients with engraftment of donor cells following in utero transplantation. They confirm the basic knowledge on the possibility of inducing immunologic tolerance provided that allogeneic cells are injected before immune maturation of the recipient fetus.

2 In Utero Transplantation of Stem Cells in IDD

The case reports of the first two patients treated by in utero FLT are summarized below. Both of them were fetuses with severe IDD diagnosed in mid-gestation. The first patient suffered from bare lymphocyte syndrome (BLS), i.e., a genetically transmitted form of combined IDD due to a lack of expression of HLA antigens (TOURAINE et al. 1987). Infections, especially with opportunistic micro-organisms, are responsible for the death of these infants unless they grow up in isolation in a fully sterile atmosphere while they are successfully reconstituted with stem cell transplants (SCT). When carried out postnatally, however, such a SCT—in the form of either a bone marrow transplant or an FLT—is usually associated with graft failure due to the presence of allogeneic reactions in the host (persisting transplant immunity) and a high susceptibility to infections (defective immunity to infectious antigens). Prenatal diagnosis of BLS can be performed by HLA analysis of fetal blood lymphocytes (DURANDY et al. 1987).

The mother had a first child who had died of BLS. When she became pregnant again, she asked for prenatal diagnosis. At 19 weeks after fecundation,

Table 1. Conditions and results of the first three in utero FLT. (Reproduced from TOURAINE et al. 1991)

Diseases	Age of fetal patients[a]	Age of fetal donors[a]	Date of transplant	Date of birth	Outcome
BLS	28	7 and 7.5	30.6.1988	7.8.1988	Graft take (HLA) and immunologic reconstitution
SCID	26	7.5	8.6.1989	7.8.1989	Some graft take (Y chromosome), no immunologic reconstitution yet
Thalassemia	12	9.5	10.10.1989	25.3.1990	Some graft take (Y chromosome), thalassemia still present

[a] Weeks after fecundation

a fetal blood sample (1.5 ml) was obtained by means of a direct umbilical cord puncture with a 20G spinal needle under ultrasound visualization. The placenta was posteriorly situated and the puncture of the umbilical vein was performed 2 cm ahead of the cord insertion on the placenta. The purity of the fetal blood sample was ascertained by means of Ii typing. Fetal blood lymphocytes were analyzed for HLA antigen expression using fluorescent monoclonal antibodies and a cytofluorometer. The virtually complete lack of expression of both class I and class II HLA antigens at the cell surface, contrasting with findings in immunologically normal fetuses, led to the diagnosis of type III BLS. Three choices were offered to this woman: (a) therapeutic abortion; (b) no treatment before birth and SCT after birth; (c) in utero SCT followed by postnatal SCT. She was informed that the latter treatment was unprecedented and that its results were therefore uncertain.

The mother and father decided for the earliest possible transplant. In June 1988, at 28 weeks after fecundation, a second fetal blood puncture was performed under direct ultrasound visualization at the insertion of the umbilical vein on the placenta; following withdrawal of 5 ml fetal blood, SCT was carried out by infusion in the umbilical vein of 7 ml culture medium containing a suspension of 16×10^6 fetal liver cells and fetal thymic epithelial cells (TOURAINE et al. 1989). The technique for intravenous infusion was comparable to that used for intravascular intrauterine transfusions (BERKOWITZ et al. 1987). Liver and thymic cells were obtained from two dead fetuses (whose ages were 7 and 7.5 weeks after fecundation). Over the months following the in utero transplant, no adverse effects were observed in the pregnant woman or in her baby, and the child was born normally in August 1988 (Table 1).

At birth, the diagnosis of BLS was again confirmed, but some cells with class I HLA antigens became progressively detectable (Fig. 1): 10% of lymphocytes had a normal expression of class I HLA antigens at the age of 1 month (TOURAINE et al. 1989), and these cells were of donor origin since their HLA specificities were not inherited from the child's parents. In particular, these cells expressed the HLA A9 specificity of donor origin, which made transplanted cells readily detectable in the initial test, at birth, and in the subsequent tests performed since. These results demonstrated the persisting engraftment of the fetal liver cells infused into the sick fetus. Based on past experience, graft take without extensive preconditioning is known to be very difficult in BLS after birth. The engraftment can therefore be attributed to the state of immaturity of the immune system of the fetus at the time of the transplant. Investigations carried out on peripheral blood lymphocytes (PBL) of the fetus at this same time (i.e., 28 weeks following fecundation) indeed showed significant immaturity of T cells compared to that found in normal fetuses of identical age.

The expression of class II HLA antigens remained comparatively low at the surface of resting lymphocytes. This finding confirms that B cell development does not occur as rapidly as T cell differentiation from stem cell transplants in immunodeficient patients (TOURAINE et al. 1987).

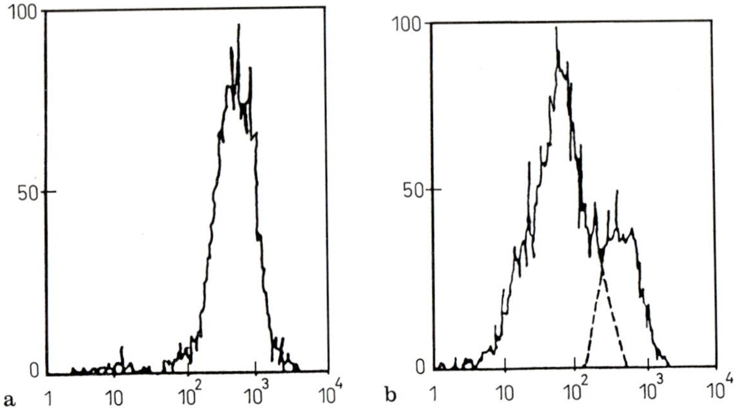

Fig.1 a, b. Cytofluorimetric analysis of PBL with W6/32 anti-HLA-ABC monoclonal antibody; *x axis*, fluorescence intensity (log scale); *y axis*, cell number; **a** Normal cells; **b** Cells of the patient 1 month after birth (i.e., 3 months after fetal liver transplantation and before additional cell infusion). Two populations of cells have been identified, a large one with HLA deficiency (patient's cells) and another with normal HLA expression (donor-derived cells). (Reproduced from TOURAINE et al. 1989)

As initially scheduled, the newborn was placed in a sterile bubble and, to accelerate reconstitution, he further received seven additional SCT from nine fetal donors. This complementary treatment was carried out after the tests demonstrating engraftment of in utero transfused stem cells. No engraftment of the cells infused after birth could be demonstrated, confirming the "resistance" to transplantation in these patients. However, the number of cells deriving from the in utero transplant increased and was found to be 26% among PBL at 1 year. In parallel, T cell maturity and immunologic reconstitution progressed. Although this reconstitution cannot yet be considered to be absolutely complete, the proliferative responses to antigens (candida antigens, cytomegalovirus antigens, and tetanus toxoid) developed well. Immunoglobulin levels are still relatively low, a finding consistent with the previous observation of slow reconstitution of humoral immunity following FLT (TOURAINE et al. 1987).

Because of this T cell reconstitution from in utero transplanted cells and in view of his good health, the patient was allowed to leave the isolator after 16 months. The child's present condition is excellent. He has not experienced infections and he lives a normal life at home; he is only receiving immunoglobulins every 3 weeks until he produces sufficient amounts of IgM and IgG.

After this first in utero FLT, and in view of the most encouraging results observed, a second fetus with IDD was treated prenatally in 1989. This second patient was a younger fetus with a complete form of SCID. She was treated by FLT in June 1989, at the age of 26 weeks following fecundation, after the prenatal diagnosis of severe IDD (complete lack of CD2+, CD3+, CD4+, and CD8+ lymphocytes in the peripheral blood of this fetus, while B cells were present; the findings were superimposable with those obtained in the first child of the family who also had SCID).

The transplant was carried out in the same conditions as in the first fetus treated in utero. It involved the intravenous infusion of 37×10^6 fetal liver and thymic cells in the umbilical vein of the fetal patient, under ultrasound control. The age of the donor was 7.5 weeks after fecundation, the age of the recipient at the time of FLT was 26 weeks after fecundation. There were no side effects and the child was born normally in August 1989 (Table 1).

At birth, the girl still had immunologic manifestations of SCID. She has therefore been maintained in sterile isolation and has received additional infusions of fetal liver and thymus cells, with the aim of accelerating development of the in utero transplanted stem cells. Using amplification techniques, DNA fragments specific of the Y chromosome have been repeatedly demonstrated to be present in the DNA material prepared from PBL of this girl. Engraftment has therefore been obtained, but the number of donor cells in this patient is considered to be still insufficient to sustain life in an unprotected environment. Isolation will be prolonged until transplanted stem cells have further proliferated and differentiated into T lymphocytes with adequate immune functions.

3 In Utero Transplantation of Stem Cells in β^0 Thalassemia

The results obtained in IDD prompted us to attempt in utero FLT in fetal patients with severe nonimmune hematologic disorders. In such conditions, however, graft take may not be facilitated by immune incompetence of the fetal host, and we assumed that FLT had to be carried out during the first trimester of pregnancy, at a time when normal fetuses have not yet developed cell-mediated immunity (Royo et al. 1987).

A pregnant woman with a family history of thalassemia requested early prenatal diagnosis. By molecular biology techniques, the fetus was demonstrated to have β^0-thalassemia major. The mother refused abortion and asked for in utero FLT. She was informed of the uncertainty of the results in such a disease.

FLT was carried out in the fetal patient, whose age was 12 weeks after fecundation. The transplant consisted of an intraperitoneal injection of 3×10^8 viable cells from a fetus aged 9.5 weeks after fecundation. There was no adverse effect in the mother nor in the fetus. Birth occurred in March 1990 (Table 1). The site of puncture could hardly be seen on the abdominal skin of the newborn, and ultrasound investigations demonstrated no modification of the abdominal wall.

Studies performed after birth showed the presence of thalassemia. However, there were a few cells of donor origin: amplification techniques revealed Y-chromosome-specific DNA fragments in the PBL of this girl. In addition, hemoglobin A was found to account for 0.9% of all hemoglobin (Table 2). No further transplant was done in this infant who is presently in very good general condition. New

Table 2. Postnatal hematologic data in girl with β^{0}-thalassemia who received FLT during her fetal life. (Reproduced from Touraine et al. 1991)

Hematologic criteria	Results
Adult hemoglobin (%)	0.9
Hemoglobin (g/dl)	10
Mean corpuscular hemoglobin (pg)	23.8
Mean corpuscular volume (fl)	68.1
White blood cells (no. per µl)	14 000
Y chromosome (of donor origin)	Present

investigations will be carried out to follow donor cells and to determine whether these few cells are likely to proliferate and improve the thalassemic condition.

4 Discussion

The results obtained in these three patients have demonstrated the feasibility of in utero FLT (before the end of the first trimester in one case). They confirm experimental data in animals (Harrison et al. 1989) and show that the procedure can be effective in humans: engraftment is obtained and can result in the cure of inherited immunodeficiency. However, it has not yet been established whether the present conditions are sufficient to improve significantly other IEM such as thalassemia. It may well be that the number of donor stem cells will have to be increased and the age of the fetal host to be lowered to treat efficiently fetal patients with disorders that do not involve immunodeficiency.

The main reasons for developing in utero FLT are the following:

1. Increased probability of graft take and chimerism, especially in diseases such as BLS in which a residual immunity can induce rejection, and even more so in diseases without immunodeficiency (provided that FLT is performed in very young fetuses).
2. Improved isolation at the time of transplant, since the uterus is even better than a sterile bubble.
3. A more optimal environment for fetal cell development when transplanted in the fetus rather than in the infant. The earlier the transplant is performed, the greater will be the chance of full and rapid development of the transplanted cells. Whether or not additional therapy in fetuses with normal immunity might be safe and beneficial has not been investigated. This will obviously have to be studied carefully in animal models before it is used in humans.

Fetal cell transplants can completely cure half of the patients with severe congenital immunodeficiencies treated after birth (Touraine et al. 1987). Immune immaturity of the fetal donor prevents acute graft-versus-host disease. When these cells are transplanted in utero, before any immune maturation in the recipient (in conditions such as BLS, adenosine deaminase deficiency, or even inherited diseases without immunodeficiency), rejection is prevented. In most cases, prenatal diagnosis of these various diseases has resulted in the decision to terminate pregnancy. This new opportunity of in utero transplantation may represent a more satisfying solution than abortion of the sick fetus. In inherited hematologic diseases and in other IEM it also offers very great promise and, in the future, it will share with in utero gene therapy the privilege of curing many of these severe diseases prenatally.

Acknowledgements. We are grateful to D. Raudrant, C. Vullo, D. Frappaz, F. Freycon, A. Rebaud, F. Barbier, M.G. Roncarolo, L. Gebuhrer, H. Bétuel, M.T. Zabot, E. Goillot, S. Laplace, N. Philippe, F. Rezzoug, C. Royo, K. Sandadji, G. Souillet, and F. Touraine for their contribution to the treatment and the immunologic investigations in these patients.

References

Berkowitz RL, Chitkara U, Wilkins I, Lynch L, Mehalek KE (1987) Technical aspects of intravascular intra-uterine-transfusions; lessons learned from 33 procedures. Am J Obstet Gynecol 157: 4–9

Durandy A, Cerf-Bensussan N, Dumez Y, Griscelli C (1987) Pre-natal diagnosis of severe combined immunodeficiency with detective synthesis of HLA molecules. Prenat Diagn 7: 27–31

Harrison MR, Slotnick RN, Crombleholme TM, Globus MS, Tarantal AF, Zanjani ED (1989) In utero transplantation of fetal liver haemopoietic stem cells in monkeys. Lancet ii: 1425–1427

Roncarolo MG, Touraine J-L, Banchereau J (1986) Cooperation between major histocompatibility complex mismatched mononuclear cells from a human chimera in the production of antigen-specific antibody. J Clin Invest 77: 673–680

Royo C, Touraine J-L, de Bouteiller O (1987) Ontogeny of T lymphocyte differentiation in the human fetus: acquisition of phenotype and functions.Thymus 10: 57–73

Touraine J-L (1983) Bone marrow and fetal liver transplantation in immunodeficiencies and inborn errors of metabolism: Lack of significant restriction of T-cell function in long-term chimeras despite HLA-mismatch. Immunol Rev 71: 103–121

Touraine J-L, Bétuel H, Souillet G, Jeune M (1978) Combined immunodeficiency disease associated with absence of cell-surface HLA A and B antigens. J Pediatr 93: 47–51

Touraine J-L, Raudrant D, Royo C, Rebaud A, Roncarolo MG, Souillet G, Philippe N, Touraine F, Bétuel H (1989). In utero transplantation of stem cells in a patient with the bare lymphocyte syndrome. Lancet i: 1382

Touraine J-L, Raudrant D, Vullo C, Frappaz D, Feycon F, Rebaud A, Barbier F, Roncarolo MG, Gebuhrer L, Btuel H, Zabot MT (1991) New developments in stem cell transplantation with special reference to the first in utero transplants in humans. Bone Marrow Transplantation, Vol. 7, Suppl. 3, pp 92–97

Touraine J-L, Roncarolo MG, Royo C, Touraine F (1987) Fetal tissue transplantation, bone marrow transplantation and prospective gene therapy in severe immunodeficiencies and enzyme defi ciencies. Thymus 10: 75–87

Human Umbilical Cord Blood as a Source of Transplantable Hematopoietic Stem and Progenitor Cells

H. E. BROXMEYER[1,2,4], S. COOPER[1,4], M. YODER[3], and G. HANGOC[1,4]

1 Introduction and Background

1.1 Hematopoietic Stem and Progenitor Cells

Mature blood cells found in the circulation are produced from rare populations of hematopoietic stem and progenitor cells (BROXMEYER 1982, 1983). Hematopoietic stem cells are operationally defined by their unique capacity for self-renewal, the ability to make more of themselves (BROXMEYER 1982). A hierarchy of stem cell populations exists beginning with cells that have the most "stemness" and ending with those cells with the least "stemness" (BROXMEYER 1982, 1983; BROXMEYER and WILLIAMS 1988). Little is known about self-renewal (BROXMEYER 1991a), but this property is most directly measured by the capability of clonal progeny to form other clones of cells as measured by secondary colony formation in vivo (in a mouse spleen colony-forming unit, CFU-S, assay) or in vitro (in a colony assay performed in a semisolid support culture medium). It would appear that stem cells with the least "stemness" give rise to more mature progeny, the progenitor cell.

[1] Department of Medicine (Hematology/Oncology),
[2] Department of Microbiology/Immunology,
[3] Department of Pediatrics,
[4] The Walther Oncology Center,
Indiana University School of Medicine, Indianapolis, IN 46202-5121, USA

Current Topics in Microbiology and Immunology, Vol. 177

Progenitor cells are operationally defined by their proliferative and differentiative capacities and until recently were considered to have minimal or no self-renewal ability (BROXMEYER 1983). However, it is apparent now that under the appropriate growth stimuli, colonies believed to derive from hematopoietic progenitor cells can form secondary colonies (CAROW et al. 1991), suggesting that the concept of lack of stemness of progenitor cells may need some modification. It is mainly hematopoietic stem cells, but also hematopoietic progenitor cells, that are involved in allogeneic and autologous bone marrow transplantation to correct hematopoietic disorders. Bone marrow is used because in the adult it is the primary source of hematopoietic stem/progenitor cells (BROXMEYER 1982). Circulating blood has also be used as a source of hematopoietic stem/progenitor cells, almost entirely thus far in an autologous setting, but adult blood contains 10–100-fold fewer of these cells than adult bone marrow (BROXMEYER 1983). Stem/progenitor cells are first found ontologically in the yolk sac, later they migrate to the fetal liver, and then to the fetal bone marrow (TAVASSOLI 1991). Practically nothing is known regarding the mechanisms involved in this migratory pattern in which these cells leave one tissue for ultimate residence in another tissue (BROXMEYER 1991a), but clinical medicine is already the beneficiary of this migration phenomenon. Recently, umbilical cord blood, a source of cells which is usually discarded, has been used in allogeneic stem/progenitor cell transplantation (GLUCKMAN et al. 1989; BROXMEYER et al. 1990a, 1881b; WAGNER et al. 1991).

1.2 Umbilical Cord Blood Transplantation

At present, there have been five HLA-matched sibling cord blood transplants performed. In each case, single collections were sent to our laboratory for testing and freezing prior to delivery to the site of the transplant. Four transplants were done in patients with Fanconi's anemia (three at the Hôpital St. Louis, Paris, France, and one at the Children's Hospital, Cincinnati, Ohio) (GLUCKMAN et al. 1989; BROXMEYER et al. 1990a, 1991b), and one transplant was done in a patient with juvenile chronic myelogenous leukemia (at Johns Hopkins Medical Center, Baltimore, MD) (WAGNER et al. 1991). Thus far, four of the transplants have been successful (three for Fanconi's anemia and the one for juvenile chronic myelogenous leukemia). At the time of writing (March 1991), it has been longer than 29, 16 and 10 months, respectively, for the three patients with Fanconi's anemia since the time of transplant and the patients remain healthy. The patient with leukemia relapsed after 8 months.

These encouraging results have suggested the possibility of establishing cord blood banks for the storage of cells for use in an autologous setting but also for access to allogeneic cells (MORRIS 1991; MCGLAVE 1991; BROXMEYER 1991b). Such banks would, out of necessity, differ from the bone marrow registry, which is essentially a computer listing of potentially available donor HLA types. Collection of bone marrow from adults can be done at any time if the donor is willing. In contrast, cord blood can only be collected at birth and must be stored in frozen form immediately after collection, or it will be lost. It is obvious that the storage of cord blood from many donors offers logistic problems not encountered, especially with regards to storage, if only a computer registry is required, but the problems are not insurmountable. In fact, it is believed that access to HLA-typed cryopreserved and stored cord blood will allow transplantation for minorities and other HLA types which are vastly under represented in the bone marrow registry (MORRIS 1991). It is principal author's (HEB) feeling that in the future most newborns will have their cord blood collected and stored for their own use, since the only perfectly matched cells available are one's own cells. This is especially underscored by a recent study (FLEISCHHAUER et al. 1990; PARHAM 1991) in which rejection of a bone marrow graft was associated with a single amino acid substitution in an MHC class I HLA molecule.

Human cord blood progenitors can be maintained in culture for many weeks (SALAHUDDIN et al. 1981; SMITH and BROXMEYER 1986). The ability to expand the numbers of stem/progenitor cells outside the body is a technology which is not yet refined sufficiently to obtain the large numbers of such cells needed for successful transplantation. Yet even better culture conditions will not necessarily obviate the need to collect and store cord blood. One can envision a time when very small numbers of adult bone marrow or blood cells can be withdrawn from a donor and exponentially expanded in culture for use. However, adult blood may not contain the earliest stem cells, or stem cells of the quality found in bone marrow. Furthermore, the withdrawal of bone marrow still involves an invasive, although relatively minor, procedure that donors still consider seriously before undergoing. The collection of cord blood offers no problems to the mother or baby, is not an invasive procedure, and more importantly, the blood contains populations of stem/progenitor cells at a frequency at least as good as bone marrow and superior in number to adult blood. There is also the distinct possibility that stem/progenitor cells in cord blood will have been subjected to far fewer insults than similar cells in the adult. This latter possibility will no doubt be an area of investigation in the near future and will include assessment of the broadness of applicability of cord blood (including types of disorders that can be treated with cord blood, and the use of cord blood for adults, as well as children) and the immunologic reactivity of cord verses adult bone marrow or blood stem/progenitor cells (BROXMEYER et al. 1991b).

The use of human unbilical cord blood as source of transplantable hematopoietic stem and progenitor cells began with a multi-institutional national laboratory study (BROXMEYER et al. 1989a) that addressed the following questions: (a) could one estimate the reconstituting cellular contents of cord blood by measurement of hematopoietic progenitor cell numbers and comparing these with numbers in adult bone marrow that have been associated with successful engraftment [others had already estimated such numbers in adult bone marrow (SPITZER et al. 1980; FAILLE et al. 1981; DOUAY et al. 1986; MA et al. 1987)]; (b) could one collect large enough samples of cord blood routinely and could these samples be transported in viable condition; (c) could cord blood cells be stored frozen and thawed in viable from for usage? The answers to all three questions turned out to be yes (BROXMEYER et al. 1990a), and this led to the first (GLUCKMAN et al. 1989) and subsequent (BROXMEYER et al. 1990a, 1991b; WAGNER et al. 1991) successful cord blood transplants (referred to above) that involved a multi-institutional national and international collaborative effort.

2 Estimation of Hematopoietic Stem and Progenitor Cells in Human Umbilical Cord Blood

2.1 Collection and Cryopreservation

The detailed procedures for collecting and processing cord blood for preservation and hematopoietic transplantation have been described (BROXMEYER et al. 1989a; ENGLISH et al. 1991). It became apparent early on that any attempts to separate cord blood cells even by the simplest maneuvers, such as removal of erythrocytes by sedimentation, lysis, density cut, or preparation of a buffy coat, resulted in unacceptable losses of progenitor cells, and attempts to wash the thawed cryopreserved cells also resulted in large losses (BROXMEYER et al. 1989a). This is why the cells used thus far for cord blood transplantation have been stored in unseparated form and infused into the conditioned recipients without washing the cells.

2.2 Variables Involved in Estimating Stem/Progenitor Cell Numbers

Assays for hematopoietic progenitor cells require colony assays in vitro that include specific growth factors (BROXMEYER 1983). Prior to our own study, there had been a number of reports on the presence in human umbilical cord blood of granulocyte-macrophage, progenitor cells (CFU-GM), erythroid progenitor (BFU-E) and precursor (CFU-E) cells, megakaryocytic (CFU-MK), and multipotential (CFU-GEMM) progenitors (KNUDTZON 1974; GABUTTI et al. 1975; FAUSER and MESSNER 1978; PRINDULL et al. 1978; VAINCHENKER et al. 1979; HASSON et al. 1979; TCHERNIA et al. 1981; LYNCH et al. 1982; GEISSLER et al. 1986; BODGER 1987). In addition, evidence for stem cells with replating capabilities were found in cord blood (NAKAHATA and OGAWA 1982; LEARY et al. 1984).

These studies demonstrate that the frequency (number of colonies formed per number of cells plated) of human umbilical cord blood hematopoietic progenitor cells assayable in vitro are equal to or greater than the frequency of these cells in adult human bone marrow. It is important to emphasize that each laboratory has its own optimized system for assessing numbers of hematopoietic progenitor cells, and the number of such cells detected in umbilical cord blood collections may vary between laboratories, not only because of the collection procedure, but also because of differences in the in vitro cultures. Differences in such cultures include: (a) the semisolid support medium (e.g., agar, methycellulose, plasma clot, fibrin clot, or agarose); (b) the incubation conditions (e.g., amounts of carbon dioxide or oxygen); (c) culture nutrients and serum; (d) growth factors, such as those present in medium conditioned by various cells or cell lines or purified colony stimulating factors [CSF: GM-CSF, G-CSF, M-CSF, interleukin 3 (IL-3)] or non-CSF molecules that enhance the action of CSF [e.g., IL-1,4,6,9,11, macrophage inflammatory proteins, (MIP)] (BROXMEYER and WILLIAMS 1988; BROXMEYER, 1990) and the ligand for the c-*kit* proto-oncogene protein variously termed mast cell growth factor (MGF) (WILLIAMS et al. 1990; BROXMEYER et al. 1991a), stem cell factor (SCF) (ZSEBO et al. 1991), *kit* ligand (KL) (HUANG et al. 1991), and steel factor (SF); (e) the cell populations assessed (unfractionated or fractionated) and the detection criteria used (e.g., designation of what constitutes a colony or cluster and the day at which colonies/clusters are scored). As long as the conditions for stem/progenitor cell assessment are clarified, it should still be possible to compare, to some degree, the data presented by different laboratories.

2.3 Culture Conditions Used in Our Laboratory

The culture conditions originally used to estimate numbers of hematopoietic progenitors in single collections of cord blood (BROXMEYER et al. 1989a) utilized 0.3% agar culture medium for assessment of CFU-GM colonies (>50 cells/aggregate) and clusters (3–50 cells/aggregate). This included 10% v/v heat-inactivated (56 °C/30 min) prescreened fetal bovine serum (FBS) with either 10% v/v medium conditioned by the human urinary bladder carcinoma cell line, 5637 (5637CM) or 100 units/ml recombinant human (rhu) GM-CSF (specific activity, 10^8 units/mg). Cultures were scored after 7 and 14 days of incubation as these scoring days appear to detect CFU-GM at different levels of maturity. For example, colonies scored after 14 days of incubation detect an earlier progenitor than colonies scored after 7 days of incubation (JACOBSEN et al. 1979; FERRERO et al. 1983). Since assessment of total aggregates seems to more accurately reflect the CFU-GM compartment (JACOBSEN et al. 1979), colonies plus clusters, as well as colonies, were scored. The exact culture ingredients used for our CFU-GM assays have been reported elsewhere (COOPER and BROXMEYER 1991). The BFU-E/CFU-GEMM compartments were assessed in 0.9% methylcellulose culture medium which included 30% v/v non-heat-inactivated prescreened FBS with erythropoietin (epo, 1 unit/ml), −/+ either 10% v/v 5637 CM, rhuGM-CSF (100 units/ml), or rhuIL-3 (100 units/ml, specific activity 10^8 units/mg), −/+ 0.1 mM hemin. The cultures were scored after 14 days of incubation.

The use of epo alone appears to detect a more mature population of BFU-E, while the addition of 5637CM, GM-CSF, or IL-3 with epo reveals an earlier form of

BFU-E (Broxmeyer et al. 1990b). CFU-GEMM are mainly detected with the combination of epo plus 5637CM, GM-CSF, or IL-3. Addition of hemin under certain conditions usually, but not always, enhances the detection of CFU-GEMM colonies (Lu and Broxmeyer 1983). CFU-GM, BFU-E, and CFU- GEMM cultures were incubated at 5% carbon dioxide at lowered (5%) oxygen tension. Use of lowered oxygen tension allows increased detection of progenitor cell-derived colonies (Broxmeyer et al. 1989b, 1990c).

The above culture conditions were used to estimate CFU-GM, BFU-E, and CFU-GEMM numbers in over 100 cord blood collections. We plated both low-density cells (obtained after density cut separation with Ficoll-Hypaque at $<1.077\ \mathrm{gm/cm^3}$), as well as unseparated cord blood (limited sampling of three collections at that time) (Broxmeyer et al. 1989a). The results of a larger sampling of unseparated cord blood progenitor cell numbers are shown in Table 1. This table includes results of all collections used in the clinical transplants as well as others saved for possible HLA-complete or one-haplotype-matched transplants. A similar progenitor cell assessment of numbers in the umbilical cord blood collected at the birth of a child affected with Fanconi's anemia, who had been misdiagnosed on the basis of chorionic villus sampling, demonstrated a profound deficiency in colony numbers (Auerbach et al. 1990). This was consistent with previously reported abnormalities in the growth of Fanconi's anemia cells in culture and highlights the usefulness of these assays to detect quantitative and qualitative deficiencies of hematopoietic progenitor cells in cord blood collections. Although it is time-consuming to assess progenitor cell numbers, it is advisable to do so for all collections intended for transplantation purposes, at least for the present and near future. It is also advisable to have separate small aliquots of cord blood stored for assessment of cell viability after thawing to confirm the healthy state of the cells to be used.

Since our original studies (Broxmeyer et al. 1989a), it has become apparent that other culture conditions and growth factors allow enhanced detection of

Table 1. Pre-freeze nucleated and hematopoietic progenitor cells from unfractionated human umbilical cord blood of single collections

Parameters	Mean $\pm$ 1 SEM (18 collections)	Range
Blood collected (ml)	141 ± 14	55 – 282
Nucleated cells ($n \times 10^8$)	16.74 ± 2.53	5.4 – 46.7
Progenitors ($n \times 10^5$) in agar cultures		
Day 14 CFU-GM (colonies)	3.96 ± 1.21	0.4 – 21.9
(colonies and clusters)	7.77 ± 2.61	0.8 – 49.3
Progenitors in methyl cellulose ($n \times 10^5$)		
Day 14 CFU-GM (colonies)	13.86 ± 5.81	0.7 – 108.5
Day 14 BFU-E (colonies)	11.53 ± 2.73	2.5 – 38.1
Day 14 CFU-GEMM (colonies)	1.98 ± 0.71	0.4 – 12.8

progenitor cell numbers. For example, the use of agarose, in place of agar, as a semisupport medium allows one to detect rhuM-CSF responsive human macrophage progenitors (CFU-M) and rhuIL-3-responsive CFU-GM not seen when either agar or methylcellulose were used (BROXMEYER et al. 1990c). Additionally, a newly engineered rhuGM-CSF/IL-3 fusion protein, more potent on a weight to weight basis than the combination of rhuGM-CSF and rhuIL-3, allows detection of an apparently earlier subset of CFU-GM, and epo-dependent BFU-E and CFU-GEMM (CURTIS et al. 1991). Also the variously identified c-*kit* ligands (MGF, SCF, KL) allow detection of even greater numbers of progenitors when used in combination with CSF (WILLIAMS et al. 1991; ZSEBO et al. 1991; HUANG et al. 1991; BROXMEYER et al. 1991a). When we compared cord blood cells grown with MGF (a c-*kit* ligand), the following become apparent: for CFU-GM, colonies plated with 100 units/ml rhuGM-CSF plus either 50 ng/ml recombinant murine (rmu) or rhuMGF were 4–24- and 3–18-fold increased compared, respectively, to cells plated with 10% v/v 5637CM or 100 units/ml rhuGM-CSF. For CFU-GEMM, colonies plated with 1 unit/ml epo plus 50 ng/ml rhuMGF plus 100 units/ml rhuIL-3 were 2–29- and 2–37-fold increased compared to cells plated with 1 unit/ml epo plus either 10% v/v 5637CM or 100 units/ml rhuIL- 3. Increases in both number and size of CFU-GM and CFU-GEMM colonies were apparent when cells were grown in presence of MGF (BROXMEYER et al. 1991c).

Interestingly, when normal human adult bone marrow cells are grown in the presence of epo plus MGF, there is an increased detection of BFU-E and CFU-GEMM colonies (BROXMEYER et al. 1991a); however, the increased number of CFU-GEMM colonies detected with cord blood cells grown in the presence of epo plus MGF plus IL-3 were coincident with drastically decreased numbers of BFU-E when compared with cells plated with epo plus IL-3. In fact the total number of BFU-E and CFU-GEMM colonies formed in the presence of epo plus MGF plus IL-3 very closely approximated the number of such colonies formed in the presence of epo plus MGF plus IL-3, except that with IL-3 there were mainly BFU-E colonies and with MGF plus IL-3 there were mainly CFU-GEMM colonies (BROXMEYER et al. 1991c). This suggests that the numbers of cord blood CFU-GM and CFU-GEMM that we reported previously (BROXMEYER et al. 1989a) were gross underestimates, and that most or all BFU-E found in cord blood may in fact be CFU-GEMM. MGF has also been used by us recently to demonstrate that hematopoietic progenitor cells have greater replating capacity than previously recognized (CAROW et al. 1991), and in this context, in an as yet very limited setting, it appears that cord blood CFU-GEMM may have greater capacity for giving rise to secondary colonies than CFU-GEMM found in the bone marrow. Thus far, secondary CFU-GEMM colonies have been detected with cord blood, but not with adult bone marrow, CFU-GEMM (CAROW et al. 1991).

3 Conclusion

The broadness of applicability for clinical uses of umbilical cord blood transplantation remains to be determined, but the four successful transplants thus far performed are encouraging. The studies were originally based on estimates of hematopoietic progenitor cell numbers in single cord blood collec tions (Broxmeyer et al. 1989a), but more optimized culture assays, in part due to the identification of early acting co-factors such as the c-*kit* ligand, make it clear that original numbers were grossly underestimated. It is our opinion that there are probably enough stem/progenitor cells in cord blood to allow for successful transplantation into an adult; thus far transplants have been done only in relatively young children (<7 years old). Attempts at stem/progenitor cell expansion are encouraging, but significant improvements are needed to obtain the large numbers of such cells needed for transplantation from a small starting sample. Results from future studies on cord blood clinical transplantation and laboratory expansion of stem/progenitor cells will no doubt complement each other.

Acknowledgements. This work was supported by Public Health Service Grant RO1 HL46549, R37 CA36464, and RO1 CA36740 from the National Institutes of Health and the National Cancer Institute.

References

Auerbach AD, Liu Q, Ghosh R, Pollack MS, Douglas GW, Broxmeyer HE (1990) Prenatal identification of potential donors for umbilical cord blood transplantation for Fanconi anemia. Transfusion 30: 632–687

Bodger MP (1987) Isolation of hemopoietic progenitor cells from human umbilical cord blood. Exp Hematol 15: 869–876

Broxmeyer HE (1982) Hematopoietic stem cells. In: Trubowitz Z, Davis E (eds) Human bone marrow vol 1. Boca Raton, pp 77–123

Broxmeyer HE (1983) Colony assays of hematopoietic progenitor cells and correlations to clinical situations. CRC Crit Rev Oncol Hematol 1: 227–257

Broxmeyer HE (1990) Interacting effects of cytokines on hematopoietic stem and progenitor cells. In: Mertelesmann R, Herrmann F (eds) Hematopoietic growth factors in clinical applications. Dekker, New York, pp 3–24

Broxmeyer HE (1991a) Self-renewal and migration of stem cells during embryonic and fetal hematopoiesis: important, but poorly understood events. Blood Cells 17: 282–286

Broxmeyer HE (1991b) Umbilical cord BMTs. Oncol Times 13(1): 2

Broxmeyer HE, Williams DE (1988) The production of myeloid blood cells and their regulation during health and disease. CRC Crit Rev Oncol Hematol 8: 173–226

Broxmeyer HE, Douglas GW, Hangoc G, Cooper S, Bard J, English D, Arny M, Thomas L, Boyse EA (1989a) Human umbilical cord blood as a potential source of transplantable hematopoietic stem/progenitor cells. Proc Natl Acad Sci USA 86: 3828–3832

Broxmeyer HE, Cooper S, Gabig T (1989b) The effects of oxidizing species derived from molecular oxygen on the proliferation in vitro of human granulocyte-macrophage progenitor cells. N Y Acad Sci 554: 177–184

Broxmeyer HE, Gluckman E, Auerbach A, Douglas GW, Friedman H, Cooper S, Hangoc G, Kurtzberg J, Bord J, Boyes EA (1990a) Human umbilical cord blood: a clinically useful source of transplantable hematopoietic stem/progenitor cells. Int J Cell Cloning 8: 76–91

Broxmeyer HE, Sherry B, Lu L, Cooper S, Oh KO, Tekamp-Olsen P, Kwon BS, Cerami A (1990b) Enhancing and suppressing effects of recombinant murine macrophage inflammatory proteins on colony formation in vitro by bone marrow myeloid progenitor cells. Blood 76: 1110–1116

Broxmeyer HE, Cooper S, Lu L, Miller ME, Langefeld CD, Ralph P (1990c) Enhanced stimulation of human bone marrow macrophage colony formation in vitro by recombinant human macrophage colony stimulating factor in agarose medium at low oxygen tension. Blood 76: 323–329

Broxmeyer HE, Copper S, Lu L, Hongoc G, Anderson D, Cosman D, Lyman SD, Williams DE (1991a) Effect of murine mast cell growth factor (*c-kit* proto-oncogene ligand) on colony formation by human marrow hematopoietic progenitor cells. Blood 77: 2142–2149

Broxmeyer HE, Kurtzberg J, Gluckman E, Auerbach AD, Douglas G, Cooper S, Falkenburg JHF, Bard J, Boyse EA (1991b) Umbilical cord blood hematopoietic stem and repopulating cells in human clinical transplantation. Blood Cells 17: 313–329

Broxmeyer HE, Cooper S, Hangoc G (1991c) Growth and responsiveness of human umbilical cord blood myeloid progenitors to mast cell growth factor (MGF, a *c-kit* ligand). Exp Hematol 19: 467 (abstract)

Carow C, Hangoc G, Cooper S, Williams DE, Broxmeyer HE (1991) Most cell growth factor (*c-kit* ligand) supports the growth of multipotential (CFU-GEMM) progenitor cells with a high replating potential. Blood 78: In press (Nov. 1)

Cooper S, Broxmeyer HE (1991) Clonogenic methods in vitro for the enumeration of granilocyte-macrophage progenitor cells (CFU-GM) in human bone marrow and mouse bone marrow and spleen. J Tissue Culture Methods 13: 77–82

Curtis BM, Williams DE, Broxmeyer HE, Dunn J, Farrah T, Jeffrey E, Clevenger W, deRoos P, Martin U, Friend D, Craig V, Gayle R, Price V, Cosman D, March CJ, Park L (1991) Enhanced hemopoietic activity of a human GM-CSF/IL-3 fusion protein. Prof Natl Acad Sci USA 88: 5809–5813

Douay L, Gorin NC, Mary JY, Lemarie E, Lopez M, Najman A, Stackhowiak J, Giarratara MC, Baillou C, Salmon C, Duhamel G (1986) Recovery of CFU-GM form cryopreserved marrow and in vivo evaluation after autologous bone marrow transplantation are predictive of engraftment. Exp Hematol 14: 358–365

English D, Cooper S, Douglas G, Broxmeyer HE (1991) Collection and processing cord blood for preservation and hematopoietic transplantation. In: Areman E, Acher RAS, Deeg HJ (eds) Bone marrow steem cells processing: Davis, Philadelphia

Faille A, Maraninchi D, Gluckman E, Devergie A, Balitrand N, Niketels F, Dresch C (1981) Granulocyte progenitor compartments after allogeneic bone marrow grafts. Scand J Haematol 26: 202–214

Fauser AA, Messner HA (1978) Granuloerythropoietic colonies in human bone marrow, peripheral blood and cord blood. Blood 52: 1243–1248

Ferrero D, Broxmeyer HE, Pagliardi GL, Venuta S, Lange B, Pessono S, Rovera G (1983) Antigentically distinct subpopulations of myeloid progenitor cells (CFU-GM) in human peripheral blood and bone marrow. Proc Natl Acad Sci USA 80: 4114–4118

Fleischhauer K, Kernan NA, O'Reilly RJ, Dupont B, Yang SY (1990) Bone marrow-allograft rejection by T-lymphocytes recognizing a single amino acid difference in HLA-B44. N Engl J Med 323: 1818–1822

Gabutti V, Foa R, Mussa F, Aglietta M (1975) Behavior of human haematopoietic stem cells in cord and neonatal blood. Haematologica (Roma) 60: 492

Geissler K, Geissler W, Hinterberger W, Lechner K, Wurnig P (1986) Circulating committed and pluripotential haemopoietic progenitor cells in infants. Acta Haematol 75: 18–22

Gluckman E, Broxmeyer HE, Auerbach AD, Friedman HS, Douglas GW, Devergie A, Esperou H, Thierry D, Socie G, Lehn P, Cooper S, English D, Kurtzberg J, Bard J, Boyse EA (1989) Hematopoietic reconstitution in a patient with Fanconi's anemia by means of umbilical cord blood from an HLA-identical sibling, Proc Natl Acad Sci USA 321: 1174–1178

Hasson MW, Lutton JD, Levere RD, Rieder RE, Cederquist LL (1979) In vitro culture of erythroid colonies from human fetal liver and umbilical cord blood. Br J Haematol 41: 477–484

Huang E, Nocka K, Beier DR, Chu TY, Buck J, Lahm HW, Wellner D, Leder P, Besmer P (1991) The hematopoietic growth factor KL is encoded by the SL locus and is the ligand of the *c-kit* receptor, the gene product of the W locus. Cell 63: 225–233

Jacobsen N, Broxmeyer HE, Grossbard E, Moore MAS (1979) Colony forming units in diffusion chambers (CFU-D) and colony forming units in agar culture (CFU-C) obtained from human bone marrow: a possible parent progeny relationship. Cell Tissue Kinet 12: 213–226

Knudtzon S (1974) In vitro growth of granulocyte colonies from circulating cells in human cord blood. Blood 43: 357–361

Leary AG, Ogawa M, Strauss LC, Civin CI (1984) Single cell origin of multilineage colonies in culture. J Clin Invest 74: 2193–2197

Lu L, Broxmeyer HE (1983) The selective influence of hemin and products of erythrocytes on colony formation of human erythroid (BFU-E) and multipotential (CFU-GEMM) progenitor cells in vitro. Exp Hematol 11: 721–729

Lynch DE, Knott LY, Rodek CH, Huehns ER (1982) Studies of circulating hemopoietic progenitor cells in human fetal blood. Blood 59: 976–979

Ma DDF, Varga DE, Biggs JC (1987) Donor marrow progenitors (CFU-mix, BFU-E and CFU-GM) and haemopoietic engraftment following HLA matched sibling bone marrow transplantation. Leuk Res 11: 141–147

McGlave PB (1991) An expanded role for cord blood transplantation. Blood Cells 17: 330–337

Morris G (1991) Umbilical cord blood used for BMT. Oncol Times 12(11): 1

Nakahata T, Ogawa M (1982) Hemopoietic colony-forming cells in umbilical cord blood with extensive capability to generate mono- and multipotential hemopoietic progenitors. J Clin Invest 70: 1324–1328

Parham P (1991) Making just the right match. Nature 350: 111–113

Prindull G, Prindull P, Meulen N (1978) Haematopoietic stem cells (CFUc) in hyman cord blood. Acta Paediatr Scand 67: 413–416

Salahuddin SZ, Markham PD, Ruscetti FW, Gallo RC (1981) Long-term suspension culture of human cord blood myeloid cells. Blood 58: 931–938

Smith S, Broxmeyer HE (1986) The influence of oxygen tension on the long term growth in vitro of haematopoietic progenitor cells from human cord blood. Br Haematol 63: 29–34

Spitzer G, Verma DS, Fisher R, Zander A, Vellekoop L, Litham J, McCredie B, Dicke KA (1980) The myeloid progenitor cell—its value in predicting hematopoietic recovery after autologous bone marrow transplantation. Blood 55: 317–323

Tavassoli M (1991) Embryonic and fetal hemopoiesis: an overview. Blood Cells 17: 269–281

Tchernia G, Mielot F, Coulombel L, Mohandas N (1981) Characteristics of circulating erythroid progenitor cells in human newborn blood. J Lab Clin Med 97: 322–331

Vainchenker W, Gruchard J, Breton-Gorius J (1979) Growth of human megakaryocyte colonies in culture from fetal, neonatal and adult peripheral blood cells: ultrastructural analysis. Blood Cells 5: 25–42

Wagner JE, Broxmeyer HE, Cooper S, Zehnbauer B, Emanuel P, Santos GW (1991) Cord blood transplantation after myeloablative therapy in a patient with juvenile chronic myelogenous leukemia. Exp Hematol 19: 585 (abstract)

Williams DE, Eisenman J, Baird A, Rauch C, Van Ness K, March CJ, Park LS, Martin U, Mochizuki DY, Boswell HS, Burgess GS, Cosman D, Lyman SD (1990) Identification of a ligand for the c-*kit* proto-oncogene. Cell 63: 167–174

Zsebo KM, Wypych J, McNiece IK, Lu HS, Smith KA, Karkare SB, Sachdev RK, Yuschenkoff VN, Birkett NC, Williams LR, Satyagal VN, Tung W, Bosselman RA, Mendiaz EA, Langley EK (1991) Identification, purification, and biological characterization of hematopoietic stem cell factor from buffalo rat liver-conditioned medium. Cell 63: 195–201

Biology and Clinical Application of Peripheral Blood Stem Cells

T. C. SHEA and J. R. MASON

1 Introduction

Hematopoietic stem cells have been identified in several different locations within the human body. During early fetal development the majority of these cells are located in the liver. In late gestation and the early neonatal period a gradual transfer to the bone marrow takes place. While this shift from the liver to the bone marrow is underway, a large number of these primitive cells can be found in the peripheral blood circulation (GEISSLER et al. 1986; CHRISTENSON 1987; BROXMEYER et al. 1989). After infancy, however, the majority of these blood-forming cells reside in the bone marrow at a frequency of approximately one in 1000 nucleated cells. Although the bone marrow continues to be the primary site for normal adolescent and adult hematopoiesis, primitive cells remain detectable in the peripheral blood throughout life at a concentration estimated to be 1/10–1/50 of that in the marrow (MCCARTHY and GOLDMAN 1984). The focus of this chapter will be on those hematopoietic progenitor cells that are normally present in the peripheral blood circulation. Their biologic characteristics, factors influencing their frequency in the circulating blood, and their potential role in clinical treatment programs will be reviewed.

Division of Hematology/Oncology and the Cancer Center, University of California, 225 Dickinson Street (H-811 K), San Diego, CA 92103, USA

Current Topics in Microbiology and Immunology, Vol. 177

2 Biologic Characteristics

Peripheral blood stem cells (PBSC) are primitive hematopoietic cells that circulate in the blood and are defined by their ability to form progenitor cell colonies in methylcellulose/agar culture systems (KURNICK and ROBINSON 1971; RUBIN and COWAN 1973; RICHMAN et al. 1976). A large portion of the cells identified by these colony assays are partially committed cells that lack the capacity for self-renewal. It is clear from transplantation studies, however, that pluripotent cells capable of providing long-term hematologic and lymphoid reconstitution are also present in these peripheral blood cells (KORBLING et al. 1986; JUTTNER et al. 1985; STIFF et al. 1987; KESSINGER et al. 1989).

Many, but not all, of the growth characteristics and nutritional requirements of this heterogeneous population of PBSC are similar to those of progenitor cells found in the bone marrow. For example, the in vitro response of PBSC to the addition of exogeneous granulocyte-macrophage colony-stimulating factor (GM-CSF) and interleukin 3 (IL-3) is generally comparable to that exhibited by marrow-derived cells. In contrast, PBSC, unlike marrow cells, are unable to support in vitro colony formation following incubation with G-CSF, IL-6, and M-CSF (CARACCIOLO et al. 1989a, b; SOUZA et al. 1986).

As with progenitor cells obtained from the bone marrow, there appears to be a hierarchy in PBSC development. IL-3 acts at a more primitive level than GM-CSF and G-CSF which exert influences on progressively more differentiated cells along the myeloid pathway (CARACCIOLO et al. 1989a). Incubation of PBSC in vitro with IL-3 promotes expansion of the number of cells that are able to respond to the more terminal effects of G-CSF and erythropoietin. The role of IL-6 in this sequence is less clear. In studies with bone marrow progenitors, IL-6 has been shown to act synergistically with IL-3 at a relatively primitive level to support blast cell and multilineage colony formation in vitro and to augment hematopoietic recovery following transplantation in a mouse model (IKEBUCHI et al. 1987; OKANO et al. 1989; LEARY et al. 1987). In other systems, including human PBSC models, its effect appears later in augmenting the expansion and terminal differentiation of neutrophils (CARACCIOLO et al. 1989b; LEARY et al. 1988).

Recent preclinical and clinical studies have utilized the presence of the CD34 cell surface antigen to identify a population of cells that is markedly enriched in its capacity to form progenitor cells in vitro (CIVIN et al. 1984, 1987; LU et al. 1987; ANDREWS et al. 1989; SIENA et al. 1989). This antigen is present on as many as 3% of the cells in the mononuclear fraction of otherwise unselected bone marrow cells, but is present at a frequency of less than 1% in the normal peripheral blood mononuclear cell population (SIENA et al. 1989, 1991; ANDREWS et al. 1989). Further segregation of bone marrow and blood mononuclear cells into those that express both the CD34 and HLA-DR antigens defines a cell that is capable of forming either primitive colony-forming unit (CFU) blasts (CD34 positive/HLA-DR negative), CFU-MIX (CD34 positive/HLA weakly DR positive), or CFU-GM (CD34 positive/HLA-DR strongly positive) colony growth (CIVIN et al. 1987; LU et al. 1987;

SIENA et al. 1989). The co-expression of CD34 and CD33 antigens identify cells capable of forming committed progenitor colonies such as CFU-GM and BFU-E, while the CD34-positive/CD33-negative cell is more primitive and forms blast or multilineage colonies (SIENA et al. 1989; BUHRING et al. 1989; ANDREWS et al. 1989).

Circulating CD34-positive/CD33-negative cells, in contrast to those cells which express both CD34 and CD33 antigens, have been reported to yield a progenitor cell colony growth fraction as high as 70% in the presence of fetal calf serum, IL-3, GM-CSF, and erythropoietin (GABBIANELLI et al. 1990). When these highly purified peripheral blood-derived CD34-positive/CD33-negative cells are cultured in the presence of the above cytokines and basic fibroblast growth factor, colony growth occurred from over 90% of these cells in 14-day cultures; 3% of these colonies were CFU blast colonies capable of self-renewal.

The studies reviewed above indicate the presence of a heterogeneous collection of circulating blood cells capable of giving rise to progenitor colony growth in vitro. It is clear that some of these cells can generate more primitive growth of mixed and self-renewing blast colonies and that others are limited to the establishment of more differentiated colonies such as CFU-GM or BFU-E. Perhaps the strongest evidence for this heterogeneity is the capacity of cells obtained from the peripheral blood to establish long-term engraftment following transplantation, while at the same time providing partially differentiated progenitor cells leading to early hematopoietic recovery. This may account for the rapid hematopoietic recovery observed by some investigators following reinfusion of PBSC in conjunction with autologous bone marrow (GIANNI et al. 1989; PETERS et al. 1990).

3 Clinical Applications

PBSC have been utilized as an alternative to bone marrow in patients undergoing high-dose chemo/radiotherapy for many years (EPSTEIN and SARPEL 1979; ABRAMS et al. 1980; JUTTNER et al. 1985; KORBLING et al. 1986; STIFF et al. 1987; KESSINGER et al. 1989). This source of stem cells has generally been used when acquisition of marrow from normal sites such as the pelvis is not feasible because of prior irradiation or involvement with tumor. The use of PBSC in patients with tumor in the bone marrow may serve as an alternative to purging in vitro with chemotherapy or antibodies. This has been shown by KESSINGER et al. (1989) who reported that eight of 11 patients with non-Hodgkin's lymphoma in the marrow were treated with high-dose chemo/radiotherapy and PBSC reinfusion and have remained progression free for 96–795 days following therapy. A major difficulty with this approach, however, has been the large number of leukopheresis sessions (seven to ten) and time (3–4h each) required to obtain the 3–5 $\times 10^5$ CFU-GM/kg recommended to ensure adequate engraftment (TO et al. 1986; LUSKY 1989).

Table 1. Fold increase in PBSC following chemotherapy or cytokine administration

Reference	Animal	Chemotherapy[a]	Cytokine	CFU[b]	CD-34
DeWys et al. (1970)	Mouse	Cy		9	
Richman et al. (1976)	Human	Cy/Adr/DTIC		4	
Abrams et al. (1981)	Canine	Cy		16	
Gianni et al. (1989)	Human	Cy		34	
To et al. (1990)	Human	Cy		14	
Ottman et al. (1990)	Human		IL-3	2	
Geissler et al. (1990)	Monkey		IL-3	13	
Geissler et al. (1990)	Monkey		GM-CSF	12	
Socinski et al. (1988)	Human		GM-CSF	18	
Herrmann et al. (1989)	Human		GM-CSF	8.1	
Haas et al. (1990)	Human		GM-CSF	8.5	
Duhrsen et al. (1988)	Human		G-CSF	10	
Molineux et al. (1990)	Mouse		G-CSF	300[c]	
Geissler et al. (1990)	Monkey		IL-3 + GM-CSF	63	
Socinski et al. (1988)	Human	Adr/IFF/DTIC	GM-CSF	62.5	
Gianni et al. (1989)	Human	Cy	GM-CSF	283	
Siena et al. (1989)	Human	Cy	GM-CSF		5[d]
Toki et al. (1990)	Human	CHOP	G-CSF	109	
Mason et al. (1991)	Human	Carboplatin	GM-CSF	60	
Summary					
Chemotherapy alone (median value)				14	
(range)				4–34	
Cytokine alone (median value)				12	
(range)				2–300	
Chemotherapy plus cytokine (median value)				86	
(range)				60–283	

[a] Chemotherapy: Cy, cyclophosphamide; Adr, adriamycin; DTIC, dacarbazine; IFF, ifosfamide; CHOP, cyclophosphamide, doxorubacin, vincristine, prednisone

[b] CFU represent a composite of CFU-GM, BFU-E, and CFU-GEMM colonies. Numbers represent either the mean or median fold relative increase in PBSC content compared to baseline levels

[c] Fold increase in capacity for marrow repopulation following lethal irradiation and PBSC transplantation

[d] Mean fold increase in CD34-expressing cells compared to cyclophosphamide-treated control

By measuring the number of CD34-expressing cells and utilizing methods to increase the yield of PBSC in a pheresis product, the collection and reinfusion of PBSC is significantly more simple and reliable than in years past (Siena et al. 1989, 1991). For example, the frequency of circulating progenitor cells is markedly increased during the period of hematopoietic recovery following chemotherapy administration (Richman et al. 1976; DeWys et al. 1970; Abrams et al. 1981; Gianni et al. 1989; To et al. 1990) (Table 1). Collection of progenitor cells during this period enhances PBSC yields. This improved yield has facilitated the utilization of PBSC harvesting as a means of limiting the hematopoietic toxicity of patients treated with high-dose chemotherapy for a variety of malignancies including small cell lung cancer (Stiff et al. 1987), leukemia (Juttner et al. 1985), lymphoma (Korbling et al. 1986; Kessinger et al. 1989), and childhood malignancies (Takaue et al. 1989). While there was some concern about the ability of these cells to provide stable engraftment, this does not appear to be a problem if more than 3×10^5 CSU-GM/kg are reinfused (To et al. 1986; Lusky 1989).

Studies during the last 3 years have shown that the administration of GM-CSF can promote a marked increase in circulating progenitor cell content (Table 1). An increase in PBSC was reported by Socinski et al. (1988) following 1 week of GM-CSF administered by continuous infusion to patients with untreated metastatic sarcoma. These authors noted an 18-fold increase in CFU-GM concentration and an eightfold increase in the number of BFU-E colonies per milliliter of blood. When this same continuous infusion of 5.5 μg 1kg per day of GM-CSF was administered following multi-agent chemotherapy with adriamycin, ifosfamide, and decarbazine, there was a 62.5-fold increase in circulating CFU-GM a content in the peripheral blood. Herrmann et al. (1989) found an eightfold increase in circulating CFU-GM progenitor cells following 14 days of 120–1000 μg M^2 per day of GM-CSF administered by i.v. bolus or continuous infusion. No changes were seen in the progenitor content of the bone marrow (Socinski et al. 1988; Hermann et al. 1989). Similar results have been reported by Gianni et al. (1989) after the administration of 7 g/M^2 cyclophosphamide to seven patients with previously untreated breast cancer or lymphoma. In their original report the concentration of PBSC following high-dose chemotherapy alone was 30 times higher than the prechemotherapy baseline concentration. When these same patients received a concurrent infusion of GM-CSF during this time of postchemotherapy hematologic recovery, the total yield of both CFU-GM and BFU- E colonies increased an additional nine times compared to chemotherapy alone. This combination of chemotherapy and growth factor resulted in more than a 200-fold increase in circulating progenitor cells compared to the prechemotherapy values.

There is less extensive, but similar data on the value of G-CSF in expanding the concentration of PBSC. A study by Molineux et al. (1990) has indicated that G-CSF-primed peripheral blood is 300 times more potent than comparable non-G-CSF-stimulated blood cells in facilitating engraftment and survival in lethally irradiated mice. These results were comparable to the engraftment achieved with transplantation of 10^5 normal bone marrow cells. Duhresen et al. (1988) reported a 10–20-fold increase in circulating progenitor colony-forming cells following both i.v. and s.c. administration of 1-60 μg/kg per day of G-CSF to a group of 30 patients with metastatic cancer. As with the studies utilizing GM-CSF, there was no evidence for expansion in the relative number of bone marrow progenitor cells (Duhrsen et al. 1988; Socinski et al. 1988; Herrmann et al. 1989). Toki et al. (1990) evaluated the effect of administering G-CSF following cyclophosphamide, adriamycin, vincristine, plus prednisone chemotherapy for large cell lymphoma and reported a 109-fold increase in PBSC concentration in four patients treated with 2 μg/kg per day s.c. G-CSF for 2 weeks following chemotherapy.

Siena et al. (1989) have demonstrated an increase in the number of circulating CD34-positive cells in patients who were receiving GM-CSF during the period of hematologic recovery following high-dose chemotherapy. The total number of cells expressing the CD34 antigen in a pheresis product correlated highly ($r = 0.89$) with the number of CFU (CFU-GM, BFU-E and CFU-GEMM). In contrast, the number of leukocytes and CD34-negative cells were only weakly correlated with the number of day 14 colonies. Most importantly, in six patients who underwent successful autologous transplantation with GM-CSF-primed PBSC infusion after high-dose therapy, the number of CD34-positive/CD33-positive cells reinfused predicted the duration of post-therapy myelosuppression more strongly than did the number of CD34-positive/CD34-negative, CD34-negative/CD33-positive cells, CFU-GM, or BFU-E colony-forming cells reinfused. Similar results were obtained by HAAS et al. (1990) with GM-CSF-mobilized PBSC following high-dose therapy. The technical ease, reproducibility, and rapid analysis provided by the CD34 labeling represents a substantial improvement over colony assays for identifying the progenitor cell content of a pheresis product (SIENA et al. 1989).

Several reports have now documented the value of reinfusing PBSC together with autologous marrow following myeloablative high-dose therapy (GIANNI et al. 1989; PETERS et al. 1990). This combined approach has been particularly useful when the pheresis product has been collected during cytokine administration and maximal PBSC expansion. Compared to reinfusing autologous bone marrow and GM-CSF infusion, GIANNI et al. (1989) have reported a reduction in the duration of neutropenia and thrombocytopenia following high-dose treatment from 11.4 to 9.1 ($p = 0.001$) and 16.9 to 10.7 ($p = 0.02$) days, respectively, after reinfusing cytokine-primed PBSC and autologous marrow into patients receiving GM-CSF. The reduction in duration of thrombocytopenia reported by GIANNI et al. (1989) is especially important in that cytokines alone have generally have minimal impact on platelet recovery after high-dose therapy (BRANDT et al. 1988). PETERS et al. (1990) developed a similar treatment program and found a reduction in severe neutropenia from 11.7 to 7.5 days of neutrophils less than 100/μl ($p = 0.005$) and a reduction in severe infections from 39% to 6%. The presumed explanation for this beneficial effect lies in the large number of partially committed progenitor cells present in the PBSC pheresis product. When these cells are reinfused, particularly into patients receiving concurrent GM-CSF therapy, they are rapidly expanded and driven to differentiate into mature cells which then appear in the circulation and shorten the period of absolute neutropenia traditionally observed following marrow ablative therapy. The combination of PBSC and autologous marrow provides the early hematopoietic recovery often observed following reinfusion of large numbers of partially committed progenitor cells, while at the same time ensuring stable long-term engraftment with autologous marrow. This impact on the period of absolute neutropenia is greater than that observed with the use of autologous bone marrow and GM-CSF or G-CSF without PBSC infusion, where the rate of neutrophil recovery is enhanced; but the period of absolute neutropenia is not affected (BRANDT et al. 1988).

A third application of PBSC reinfusion is in the use of sequential cycles of high-dose chemotherapy together with growth factors and repeated infusions of PBSC (SHEA et al. 1990). In this approach, very intensive (three to four times normal) but non-marrow-ablative doses of carboplatin were administered every 28–35 days and followed by daily doses of GM-CSF. PBSC were collected during the hematopoietic recovery phase at the end of each cycle of chemotherapy and were reinfused during the next cycle of treatment. As with the study by SIENA et al. (1989), the number of reinfused CD34-positive cells and the number of CFU-GM, BFU-E, and CFU-GEMM all correlated inversely with the duration of neutropenia and thrombocytopenia after chemotherapy (MASON et al. 1991; SIENA et al. 1991). In comparison to a treatment group that received GM-CSF without PBSC reinfusion, the addition of PBSC resulted in a significant reduction in the duration of neutropenia ($p = 0.027$), thrombocytopenia ($p = 0.001$), number of hospital days per treatment cycle ($p = 0.01$), frequency of platelet ($p = 0.01$) and RBC transfusions ($p = 0.03$), and the number of days requiring i.v. antibiotics ($p = 0.007$) per cycle (SHEA et al. 1990). One drawback to this approach is the decreased PBSC yield following chemotherapy cycles two and three (MASON et al. 1991), thereby limiting the number of cycles in which this approach can be effectively applied.

4 Future Directions

Studies are currently underway to evaluate the impact of administering cytokines such IL-3, IL-6, and IL-1 alone and in combination with GM-CSF or G-CSF on the biology and frequency of PBSC. Data from GEISSLER et al. (1990) in a primates suggest that IL-3 stimulates a primitive progenitor and, in the process, makes it more responsive to GM-CSF or G-CSF expansion. By sequentially treating monkeys with 7–10 days of IL-3 followed by brief periods of GM-CSF exposure, a 63-fold rise in progenitor cell content was seen. Human studies of sequential administration of IL-3 followed by GM-CSF are currently underway and should result in a greater expansion of PBSC than has been shown with current trials using either GM-CSF or G-CSF alone. The roles of IL-1 and IL-6 in this process will also become clearer as clinical experience with them expands. Additional studies with agents such as the hybrid GM-CSF/IL-3 fusion molecule (WILLIAMS et al. 1990) and the stem cell factor isolated by ZSEBO et al. (1990) and ANDERSON et al. (1990) will likely by underway within the year. While the potential role for these compounds in stimulating hematopoietic progenitor cell expansion is very promising, these newer agents will need to be explored cautiously as the possibility of progenitor cell stimulation and exhaustion may become a more serious problem than has been seen with either G-CSF or GM- CSF.

With the availability of compounds to increase the number of PBSC and the use of an accurate and rapid assay of progenitor cell content, the acquisition of

PBSC is now considerably easier and more reliable than in the past. As a result, the potential application of these cells in clinical practice is also increasing. This approach may, for example, be suitable for allogeneic donors (J. Armitage, personal communication) who do not wish to undergo the risk of general anesthesia and the biopsy site discomfort associated with a bone marrow harvest. As described earlier in the paper by KESSINGER et al. (1989), there may also be a role for collection of these cells as an alternative to marrow harvesting and purging in patients undergoing autologous transplantation for diseases such as Hodgkin's and non-Hodgkin's lymphoma. These clinical avenues will in turn promote additional basic studies to identify the nutritional requirements, surface antigen characteristics, and capacity for self-renewal and proliferation of this heterogeneous cell population. These laboratory studies will be especially crucial in guiding the clinical use of agents which have the potential for altering the equilibrium between the quiescent and actively cycling hematopoietic stem cells (ZSEBO et al. 1990; WILLIAMS et al. 1990; ANDERSON et al. 1990). The interface between such basic studies and the utilization of these cells in more intensive and effective treatment programs represents an important area of investigation in the years ahead.

Acknowledgements. Dr. Shea is a recipient of a Faculty Development Award from the Pharmaceutical Manufacturer's Association Foundation. This work was supported in part by USPHS award CA-09290.

References

Abrams RA, Glaubiger D, Appelbaum FR, Deisseroth AB (1980) Result of attempted hematopoietic reconstitution using isologous, peripheral blood mononuclear cells: a case report. Blood 56: 516–520

Abrams RA, McCormack K, Bowles C, Deisseroth AB (1981) Cyclophosphamide treatment expands the circulating hematopoietic stem cell pool in dogs. J Clin Invest 67: 1392–1399

Anderson DM, Lyman SD, Baird A, Wignall JM, Eisenman J, Rauch C, March CJ, Boswell HS, Gimpel SD, Cosman D, Williams DE (1990) Molecular cloning of mast cell growth factor, a hematopoietin that is active in both membrane bound and soluble forms. Cell 63: 235–243

Andrews RG, Singer JW, Bernstein ID (1989) Precursors of colony-forming cells in humans can be distinguished from colony-forming cells by expression of the CD33 and CD34 antigens and light scatter properties. J Exp Med 169: 1721

Brandt SJ, Peters WP, Atwater SK, Kurtzberg J, Borowitz MJ, Jone RB, Shpall EJ, Bast RC Jr, Gilbert CJ, Oette DH (1988) Effect of recombinant human granulocyte-macrophage colony-stimulating factor on hematopoietic reconstitution after high-dose chemotherapy and autologous bone marrow transplantation. N Engl J Med 318: 869–876

Broxmeyer HE, Douglas GW, Hangoc G et al. (1989) Human umbilical cord blood as a potential source of transplantable hematopoietic stem/progenitor cells. Proc Natl Acad Sci USA 86: 3828–3832

Buhring H-J, Asenbauer B, Katrilaka K, Hummel G, Busch FW (1989) Sequential expression of CD34 and CD33 antigens on myeloid colony-forming cells. Eur J Haematol 42: 143–149

Caracciolo D, Clark SC, Rovera G (1989a) Human interleukin-6 supports granulocytic differentiation of hematopoietic progenitor cells and acts synergistically with GM-CSF. Blood 73: 666–670

Caracciolo D, Clark S, Rovera G (1989b) Differential activity of recombinant colony-stimulating factors in supporting proliferation of human peripheral blood and bone marrow myeloid progenitors in culture. Br J Haematol 72: 306–311

Christenson RD (1987) Circulating pluripotent hematopoietic progenitor cells in neonates. J Pediatr 110: 622

Civin CI, Strauss LC, Brovall C, Fackler MJ, Schwartz JF, Shaper JH (1984) Antigenic analysis of hematopoiesis. III. A hematopoietic progenitor cell surface antigen defined by a monoclonal antibody raised against KG-la cells. J Immunol 133: 157

Civin CI, Banquerigo ML, Strauss LC, Loken MR (1987) Antigenic analysis of hematopoiesis. VI. Flow cytometry characterization of My-10-positive progenitor cells in normal human bone marrow. Exp Hematol 15: 10

DeWys W, Goldin A, Mantel M (1970) Hematopoietic recovery after large doses of cyclophosphamide: correlation of proliferative state with sensitivity. Cancer Res 30: 1692–1697

Duhrsen U, Villeval J-L, Boyd J, Kannourakis G, Morstyn G, Metcalf D (1988) Effects of recombinant human granulocyte colony-stimulating factor on hematopoietic progenitor cells in cancer patients. Blood 72: 2074–2081

Epstein RB, Sarpel SC (1979) Processing of peripheral blood stem cells for transplantation. Exp Hematol 7: 125–134

Gabbianelli M, Sargiacomo M, Pelosi E, Testa U, Isacchi G, Peschle C (1990) "Pure" human hematopoietic progenitors: permissive action of basic fibroblast growth factor. Science 249: 1561–1564

Geissler K, Geissler W, Hinterberger W, Lechner K, Wurnig P (1986) Circulating committed and pluripotent haemopoietic progenitor cells in infants. Acta Haematol 75: 18

Geissler K, Valent P, Mayer P, Liehl E, Hinterberger W, Lechner K, Bettelheim P (1990) Recombinant human interleukin-3 expands the pool of circulating hematopoietic progenitor cells in primates—synergism with recombinant human granulocyte/macrophage colony-stimulating factor. Blood 75: 2305–2310

Gianni AM, Siena S, Bregni M, Tarella C, Stern AC, Pileri A, Bonadonna G (1989) Granulocyte-macrophage colony-stimulating factor to harvest circulating haemopoietic stem cells for autotransplantation. The Lancet, September 9: 580–584

Haas R, Ho AD, Bredthauer U, Cayeux S, Egerer G, Knauf W, Hunstein W (1990) Successful autologous transplantation of blood stem cells mobilized with recombinant human granulocyte-macrophage colony-stimulating factor. Exp Hematol 18: 94–98

Herrmann F, Schulz G, Lindemann A, Meyenburg W, Oster W, Krumwieh D, Mertelsmann R (1989) Hematopoietic responses in patients with advanced malignancy treated with recombinant human granulocyte-macrophage colony- stimulating factor. J Clin Oncol 7: 159–167

Ikebuchi K, Wong GG, Clark SC, Ihle JN, Hirai Y, Ogawa M (1987) Interleukin-6 enhancement of interleukin 3-dependent proliferation of multipotential hemopoietic progenitors. Proc Natl Acad Sci USA 84: 9035–9030

Juttner CA, To LB, Haylock DN, Branford A, Kimber RJ (1985) Circulating autologous stem cells collected in very early remission from acute non-lymphoblastic leukaemia produce prompt but incomplete haemopoietic reconstitution after high dose melphalan or supralethal chemoradiotherapy. Br J Haematol 61: 739–740

Kessinger A, Armitage JO, Smith DM, Landmark JD, Bierman PJ, Weisenburger DD (1989) High-dose therapy and autologous peripheral blood stem cell transplantation for patients with lymphoma. Blood 74: 1260–1265

Korbling M, Dorken B, Ho AD, Pezzuto A, Hunstein W, Fliedner TM (1986) Autologous transplantation of blood-derived hemopoietic stem cells after myeloablative therapy in a patient with Burkitt's lymphoma. Blood 67: 529–532

Kurnick J, Robinson W (1971) Colony growth of human peripheral white blood cells in vitro. Blood 37: 136–141

Leary AG, Yang Y-C, Clark SC, Gasson JC, Golde DW, Ogawa M (1987) Recombinant gibbon interleukin 3 supports formation of human multilineage colonies and blast cell colonies in culture: comparison with recombinant human granulocyte-macrophage colony-stimulating factor. Blood 70: 1343

Leary AG, Ikebuchi K, Hirai Y, Wong G, Yang YC, Clark SC, Ogawa M (1988) Syngergism between interleukin 6 and interleukin 3 supporting proliferation of human hemopoietic stem cells: Comparison with interleukin 1a. Blood 71: 1759–1763

Lu L, Walker D, Broxmeyer HE, Hoffman R, Hu W, Walker E (1987) Characterization of adult human marrow hematopoietic progenitors highly enriched by two-color cell sorting with MY10 and major histocompatibility class II monoclonal antibodies. J Immunol 139: 1823

Lusky LC (1989) Hematopoietic reconstitution using progenitors recovered from blood. Transfusion 29: 552

Mason JR, Mullen M, Bessent E, Taetle R, Shea TC (1991) Peripheral blood stem cells (PBSC) in patients receiving multiple cycles of high dose carboplatin (CBDCA) and granulocyte-macrophage colony stimulating factor (GM-CSF; Leucomax, Sandoz/Schering). Proc Asco 10: 108

McCarthy DM, Goldman JM (1924) Transfusion of circulating stem cells. CRC Crit Rev Clin Lab Sci 20: 1–24

Molineux G, Pojda Z, Hampson IN, Lord BI, Dexter TM (1990) Transplantation potential of peripheral blood stem cells induced by granulocyte colony-stimulating factor. Blood 76: 2153–2158

Okano A, Suzuki C, Takatsuki F, Akiyama Y, Koike K, Ozawa K, Hirano T, Kishimoto T, Nakahata T, Asano S (1989) In vitro expansion of the murine pluripotent hemopoietic stem cell population in response to interleukin 3 and interleukin 6: application to bone marrow transplantation. Transplantation 48: 495–498

Ottmann OG, Ganser A, Seipelt G, Eder M, Schulz G, Hoelzer D (1990) Effects of recombinant human interleukin-3 on human hematopoietic progenitor and precursor cells in vivo. Blood 76: 1494–1502

Peters WP, Kurtzberg J, Kirkpatrick G, Atwater S, Gilbert C, Borowitz M, Shpall E, Jones R, Ross M, Affronti M, Coniglio D, Mathias B, Oetter D (1990) GM-CSF primed peripheral blood progenitor cells (PBSC) coupled with autologous bone marrow transplantation (ABMT) will eliminate absolute leukopenia following high dose chemotherapy (HDC). Proc ASCO 9: 178

Richman CM, Weiner RS, Yankee RA (1976) Increase in circulating stem cells following chemotherapy in man. Blood 47: 1031–1039

Rubin S, Cowan D (1973) Assay of granulocyte progenitor cells in human peripheral blood. Exp Hematol 1: 127–131

Shea TC, Mason JR, Storniolo AM, Newton B, Breslin M, Mullen M, Ward D, Taetle R (1990) Beneficial effect from sequential harvesting and reinfusing of peripheral blood stem cells (PBSC) in conjunction with rHu GM-CSF (Schering/Sandoz) and high-dose carboplatin (CBDCA). Blood 76 [Suppl 1]: 651

Siena S, Bregni M, Brando B, Ravagnani F, Bonadonna G, Gianni AM (1989) Circulation of $CD34^+$ hematopoietic stem cells in the peripheral blood of high-dose cyclophosphamide-treated patients: Enhancement by intravenous recombinant human granulocyte-macrophage colony-stimulating factor. Blood 74: 1905–1914

Siena S, Bregni M, Brando B, Belli N, Ravagnani F, Gandola L, Stern AC, Lansdorp PM, Bonadonna G, Gianni AM (1991) Flow cytometry for clinical estimation of circulating hematopoietic progenitors for autologous transplantation in cancer patients. Blood 77: 400–409

Socinski MA, Cannistra SA, Elias A, Antman K, Schnipper L, Griffin JD (1988) Granulocyte-macrophage colony stimulating factor expands the circulating haematopoietic progenitor cell compartment in man. Lancet 28: 1194–1198

Souza LM, Boone TC, Gabrilove J, Lai PH, Zsebo KM, Muirdock DC, Chazin VR, Bruszewski J, Lu H, Chen KK, Barendt J, Platzer E, Moore MAS, Mertelsmann R, Welte K (1986) Recombinant human granulocyte colony-stimulating factor effects on normal and leukemic myeloid cells. Science 232: 61–75

Stiff PJ, Koester AR, Eagleton LE, Hindman T, Braud E (1987) Autologous stem cell transplantation using peripheral blood stem cells. Transplantation 44: 585–588

Takaue Y, Watanabe T, Kawano Y, Koyama T, Abe T, Suzue T, Satoh J, Shimokawa T, Ninomiya T, Kosaka M, Shimizu E, Ogura T, Kuroda Y (1989) Isolation and storage of peripheral blood hematopoietic stem cells for autotransplantation into children with cancer. Blood 74: 1245–1251

To LB, Dyson PG, Juttner CA (1986) Cell-dose effect in circulating stem-cell autografting. Lancet ii: 404–405

To LB, Shepperd KM, Haylock DN, Dyson PG, Charles P, Thorp DL, Dale BM, Dart GW, Roberst MM, Sage RE, Juttner CA (1990) Single high doses of cyclophosphamide enable the collection of high numbers of hemopoietic stem cells from the peripheral blood. Exp Hematol 18: 442

Toki H, Shimokawa T, Okabe K, Ishimitsu T (1990) Recombinant human granulocyte colony-stimulating factor (rG-CSF) amplifies the number of peripheral blood stem cells (PBSC) of lymphoma patients on chemotherapy. Proc ASCO 9: 188

Williams DE et al. (1990) Enhanced biological activity of a GM-CSF/IL-3 fusion protein. Exp Hematol 18: 256a

Zsebo KM, Wypych J, McNiece IK, Lu HS, Smith KA, Karkare SB, Sachdev RK, Yuschenkoff VN, Birkett NC, Williams LR, Satyagal VN, Tung W, Bosselman RA, Mendiaz EA, Langley KE (1990) Identification, purification, and biological characterization of hematopoietic stem cell factor from buffalo rat liver-conditioned medium. Cell 63: 195–201

Gene Transfer

Gene Transfer into Hematopoietic Stem Cells: Prospects for Human Gene Therapy

M. P. W. EINERHAND and D. VALERIO

1 Introduction

Advances in the knowledge of the molecular basis of inherited disease have made it theoretically feasible to develop therapeutic applications based on gene transfer. A respectable number of the genes responsible for congenital disorders has been cloned and characterized in recent years. In congruence with these discoveries, methods to introduce foreign sequences into cells are becoming more and more sophisticated. From these developments we can expect that the technical expertise to cure inborn errors of metabolism by so-called somatic cell gene therapy will soon become available. This therapy is based on the ability to introduce an intact version of a gene into the genetically crippled cells of a patient thereby restoring their function. The ethical considerations of gene therapy have been widely discussed, and the consensus has emerged that the genetic modification of a patient's somatic cells with the purpose of correcting severe genetic disorders is ethically acceptable (ANDERSON 1984; BELMONT and CASKEY 1986; FRIEDMANN 1983; MOTULSKY 1983; WALTERS 1986; WILLIAMSON 1982).

Gene Therapy Department, Institute of Applied Radiobiology and Immunology-TNO, P.O. Box 5815, NL-2280 HV Rijswijk, The Netherlands

Current Topics in Microbiology and Immunology, Vol. 177

The preferred method of gene therapy would involve the correction of a genetic defect through gene targeting. This would allow the replacement of defective sequences in the target tissue by added correct sequences. This goal can be achieved by homologous recombination. Homologous recombination has been shown to occur in mammalian cells (LIN et al. 1985; THOMAS et al. 1986). However, the mechanism by which the process proceeds in these cells is still obscure, and at this time the efficiency of the procedure is too low to be of use in somatic cell gene therapy protocols (THOMAS et al. 1986). Therefore, most of the research effort is focussed on gene addition. A functional copy of the affected gene is added to the genetic material of the cells. The defective copies are not removed. Thus, dominant inherited disorders cannot be treated with this procedure.

The principle of gene addition as a therapeutic tool was tested in murine models for inherited disease. In these studies, a correct copy of the defective gene was injected into fertilized oocytes resulting in transgenic mice carrying the introduced sequences in all cells. These experiments proved, for a variety of murine hereditary disease, that the altered phenotype could be reversed through gene addition i.e. in: ornithine transcarbamylase deficiency, β-thalassemia, growth hormone deficiency type I mucopolysaccharidosis VII, and hypothalamic gonadotropin-releasing hormone deficiency (CAVARD et al. 1988; CONSTANTINI et al. 1986; HAMMER et al. 1984; KYLE et al. 1990; MASON et al. 1986). Such studies will become possible for many other hereditary disorders when the appropriate mouse models have been developed. The latter can be created with the powerful mouse embryo-derived stem cell (ES) technology. Via standard molecular biological techniques, any desired mutation can be introduced into a cloned DNA sequence. This sequence can then be used to replace the sequence in the genome of ES cells via homologous recombination (gene targeting). Germ-line chimeras, carrying the mutated sequences, can be created by micro-injection of the mutant embryo-derived stem cells into normal mouse blastocysts. Interbreeding of heterozygous siblings will yield animals homozygous for the desired mutation (reviewed in CAPECCHI 1989).

Tissues that are considered to become suitable targets for gene therapy include epithelial cells, hepatocytes, keratinocytes, skin fibroblasts, and even cells derived from the central nervous system (BELMONT and CASKEY 1986). Currently, however, most progress has been achieved with cells derived from the hematopoietic system. These advances must be attributed to the relative ease with which bone marrow can be manipulated in vitro and to the extensive knowledge that has been obtained about its transplantation. In this chapter we will limit ourselves to a discussion of recent advances in the development of gene therapy protocols for the hematopoietic system.

Before clinical trails based on gene therapy are attempted, the procedure must have proven to be efficacious and safe. The relevant data should be obtained in animal species which are closely related to the human species. For bone marrow transplantations two animal models exist that fulfill this requirement (i.e., bone marrow transplantation for non-human primates and dogs). Both these models are being used for gene therapy research (reviewed in VAN BEUSECHEM and VALERIO, in press; F.G. SCHUENING, this volume). However, due to ethical, financial, and practical limitations attached to the use of these animals, most of the preparatory studies are executed in mice. The success of a gene therapy procedure depends on the availability of appropriate gene transfer systems. The gene delivery system should insure:

1. Efficient transfer of genetic information into relatively large cell populations
2. Stability of the introduced genes which should preferably be integrated in the host cell genome
3. Appropriate expression patterns of the newly introduced gene(s)
4. Acceptable safety margins for both the patient and its environment, to be evaluated in risk/benefit assessments

Currently, retrovirus-mediated gene transfer is the only technique that fulfills the above requirements. Moreover, improvements in the design of recombinant retroviral vectors and packaging cells has made the method safe and broadly applicable. Consequently, most of the current approaches toward gene therapy make use of retroviral vectors. The next section will deal with their design as useful tools in gene therapy procedures.

2 Retrovirus-Mediated Gene Transfer

In principle, all retroviruses can be used as vector systems. However, most retroviral vectors used in gene therapy studies are based on murine leukemia viruses (MuLV). MuLV genomes contain three genes designated *gag*, *pol*, and *env*. The *gag* gene encodes for proteins involved in viral encapsidation and assembly. The *pol* gene products direct synthesis of the viral DNA and its integration into the host chromosomal DNA. The env gene product is responsible for virus absorption to the target cells. This protein is located at the surface of the virions and is required for the recognition of specific cell surface receptors during the adhesion and penetration phases of viral infection (WEISS et al. 1984).

The protein-coding sequences do not have to be included in the retroviral vector. They can be provided *in trans* by virus-producing cell lines. Retroviral vectors must, however, carry the *cis*-acting sequences required for packaging of the RNA genome, its reverse transcription into a DNA genome, and the subsequent integration into the chromosomal DNA of the target cell. These sequences are located in the long terminal repeats (LTR) of the provirus and in short sequences flanking the LTR. They make up the minimal requirements for retroviral vectors. The space for foreign DNA, approximately 8 kb, is created by deleting the internal sequences that encode for the viral proteins. Inserted DNA can contain one or more genes of which the transcription is regulated either by the viral promoter in the 5′-LTR or by additional promoter elements (reviewed in TEMIN 1986).

With the knowledge of the location of the packaging signal, packaging cell lines for defective viruses have been made (reviewed in MILLER 1990; VALERIO in press). These cells express all viral proteins, but do not produce infectious viruses themselves. When the retroviral vector is transfected into packaging cells, its RNA is packaged by the viral proteins provided by the cells. Since the vector does not contain the viral genes, it can only go through one round of replication in the absence of viral proteins. It is therefore called replication defective. For safety reasons, this is considered to be a prerequisite in gene therapy procedures. Depending on the packaging cell line used, replication-defective retroviruses can be generated with an ecotropic or an amphotropic host range (respectively, infectious to mouse and rat cells, or to a wide variety of different mammalian cells including human cells).

Initially, packaging cell lines were made by simply deleting the packaging signal from a replication-competent retrovirus (MANN et al. 1983; MILLER et al. 1985). With these cell lines, helper-free virus production is possible. However, they often start to produce helper viruses after a while. This is presumably due to a recombination event between the vector, which supplies the packaging signal, and the crippled helper genome. More recently a number of packaging cell lines have been made in which the regions of homology between the vector and the helper genome(s) were drastically reduced (reviewed in MILLER 1990; VALERIO in press). The emergence of replication-competent retroviruses in these cell lines now requires a minimum of three recombinations events (DANOS and MULLIGAN 1988; MARKOWITZ et al. 1988a, b). To date, this has not been reported to occur. Thus, retroviral vector systems are available which provide an efficient and safe method for the transfer of genes in gene therapy protocols.

3 Retroviral Gene Transfer into Primary Hematopoietic Cells

3.1 Infection of Pluripotential Hematopoietic Stem Cells

Mature cell types circulating in the blood have a limited life span and need to be replenished continuously. This is achieved by a maturation process from more primitive cell types. As a result, a hierarchy of cells exists ranging from specialized mature cell types to pluripotential hematopoietic stem cells (PHSC) which reside in the bone marrow. Aside from the ability to generate committed progenitor cells for all hematopoietic lineages, the PHSC have extensive self-renewing capacity. A lasting genetic modification of the hematopoietic system can therefore be accomplished by efficient gene transfer into PHSC.

The successful infection of hematopoietic progenitor cells by recombinant retroviral vectors was first reported by JOYNER et al. (1983). They infected murine bone marrow with a Moloney MuLV (Mo-MuLV)-derived retroviral vector carrying the dominant selectable marker gene *neo*^{R}. As a result, 0.3% of the granulocyte-macrophage colony-forming units (CFU-GM) expressed *neo*^{R}-specific RNA and became resistant to the antibiotic G418. Subsequent studies revealed that also cells more primitive than committed progenitors could be infected. This was achieved with both eco- and amphotropic retroviral vectors (DICK et al. 1985; MCIVOR et al.1987; OSBORNE et al. 1990; VAN BEUSECHEM et al.1990; WILLIAMS et al. 1984). Three different assays have been used to demonstrate the presence of infected PHSC in bone marrow grafts:

1. The grafts were able to repopulate lethally irradiated recipients for more than 7 weeks with retrovirally infected cells (DICK et al. 1985, LEMISCHKA et al. 1986).
2. The grafts contained infected pluripotential cells. This was inferred once transplantation of infected bone marrow had resulted in the repopulation of both myeloid (bone marrow) and lymphoid (spleen and thymus) lineages with cells carrying the same clonal marker[1] (DICK et al. 1985; LEMISCHKA et al. 1986).
3. The grafts contained pluripotential cells with extensive self-renewing capacity. This was proven by the persistence of the same clonal marker in several hematopoietic cell lineages in secondary and even tertiary recipients (KELLER et al. 1985; VAN BEUSECHEM et al. 1990).

These results convincingly demonstrated the infectability of PHSC. More recently, using approaches similar to those described above, refined questions regarding the function and capabilities of transplanted PHSC were resolved (CAPEL et al. 1990; JORDAN and LEMISCHKA 1990; JORDAN et al. 1990; KELLER and SNODGRASS 1990; SNODGRASS and KELLER 1987). It is now clear that PHSC can repopulate two or more hematopoietic lineages. However, the contribution of a stem cell clone to the various lineages may be subject to changes in time (CAPEL et al. 1990; JORDAN and LEMISCHKA 1990; KELLER and SNODGRASS 1990). In addition, the extent to which a clone contributes to a lineage can show marked variation between different parts of the animal (JORDAN and LEMISCHKA 1990). The existence of so-called restricted stem cells which only repopulate a single lineage has also been demonstrated; however, to date there is no proof that these stem cells are indeed restricted in their differentiation. It might well be that a restricted repopulation is just one of the many ways in which repopulation by a PHSC can occur.

The most striking fact that emerged from these studies was that only a limited number of PHSC contributed to the expansion and the maintenance of a complete hematopoietic system post engraftment. The data confirmed the

---

[1] Retroviruses integrate apparently at random in the host chromosomal DNA, although slight preferences exist (SHIH et al. 1988; ROHDEWOHLD et al. 1987). The integration site is therefore unique to each infection and as a consequence unique to the infected cell and its progeny. Since independent integration events can be analyzed by Southern blotting, they can be used as clonal markers

results obtained in experiments using radiation-induced chromosomal aberrations as lineage markers (ABRAMSON et al. 1977; WU et al. 1968). These experiments led to the hypothesis that most of the transplanted stem cells remain in a quiescent state following transplantation, resulting in the so-called clonal succession theory. In this model, the life time of stem cells maintaining hematopoiesis is limited. Upon exhaustion, their function is taken over by a restricted number of formerly quiescent PHSC. Supportive evidence for this theory has been obtained in experiments designed to activate the resting PHSC in transplanted mice. This was achieved by either injecting 5-fluorouracil (5-FU), a cytostatic drug that is predominantly toxic to cycling cells, or by a retransplantation of the bone marrow from a first recipient into a secondary recipient. New clones appeared with both methods (KELLER and SNODGRASS 1990). However, the results were different when no additional stress was inflicted upon the system. In this case, fluctuations in stem cell proliferation are primarily observed within the first 4–6 months post engraftment. Only 7% of the investigated stem cell clones demonstrated a dynamic behavior after this period (JORDAN and LEMISCHKA 1990). Moreover, more than 50% of the stem cell clones present after stabilization of the hematopoietic system were also contributing to the reconstitution of the hematopietic system immediately following their transplantation. These experiments suggest that the hematopoietic system is not maintained by a clonal succession of PHSC, but rather by a few clones with extensive self-renewing capacity.

The mono- or oligoclonal reconstitution patterns make it difficult to derive quantitative information. It prevents an accurate assessment of the number of infected PHSC prior to transplantation, and it does not tell us what the toxicity of the procedure was on this cell type. In order to try and solve this problem, we have turned to an experimental design in which cells with extensive self-renewing capacity are identified by an assay which measures marrow-repopulating ability (MRA). Following infection of bone marrow cells, they are transplanted into lethally irradiated mice. Thirteen days post engraftment, the bone marrow of these mice is harvested and subsequently assayed for a primitive cell type. We use the class of primitive multipotent progenitors that, upon transplantation, form macroscopic colonies on the spleens of lethally irradiated recipients (CFU-S; TILL and MCCULLOCH 1961). Recent publications on the purification of PHSC indicate that the functional and phenotypic characteristics of this MRA-CFU-S closely resemble the cell type considered to be the PHSC (BERTONCELLO et al. 1985; HODGSON and BRADLEY 1979; PLOEMACHER and BRONS 1989). Since MRA-CFU-S can be titrated by means of this assay (unpublished results), the efficiency and/or toxicity of the infection procedure can be quantified in a primitive stem cell with extensive self-renewing capacity. We have recently discovered that clonal expansion of primitive hematopoietic cells occurs shortly after transplantation. By examining the bone marrow of mice 13 days following transplantation of an infected graft, we found that already at that time point, dominant clones made up more than 90% of CFU-S in the marrow (M.P.W. Einerhand et al., in preparation).

3.2 Factors Affecting the Survival and Infection of PHSC

The for the clinician most preferable and for the patient least compromising method of gene transfer into the hematopoietic system would be the direct administration of the genetic material into the blood stream. In this ideal situation, the information would be packaged in such a way as to warrant a proper targeting to the PHSC. Although this direct approach is promising for organs such as the liver and the muscle (DUBENSKI et al. 1984; WOLFF et al. 1990), the tools to target sequences to the PHSC via the blood stream are as yet not available. Therefore, gene transfer methods for these cells have concentrated on the manipulation of explanted bone marrow in vitro. This approach has been very successful. As described in the previous paragraph, the PHSC can be infected in vitro and subsequently transplanted without loss of pluripotentiality. However, the infection of hematopoietic progenitors was not very efficient in the first experiments (JOYNER et al. 1983; WILLIAMS et al. 1984). This efficiency was boosted as a result of the availability of virus-producing cell lines with high virus titers on the one hand and the improved manipulation and culture of PHSC in vitro on the other. Most of the initial studies were performed on CFU-S. Currently it is possible to infect up to 100% of the CFU-S with retroviral vectors (EGLITIS et al. 1985; LEMISCHKA et al. 1986).

3.2.1 Factors Shown to Affect the Efficiency with Which CFU-S Are Infected

3.2.1.1 Virus Titer and Type of Exposure

Maximal infection of CFU-S was obtained with virus titers greater than 2×10^5 CFU/ml (EGLITIS et al. 1985) or 10^6 CFU/ml (LEMISCHKA et al. 1986). These results were obtained when bone marrow was cocultivated with virus-producing cell lines. Infections with virus supernatants from the same cell lines resulted in a less efficient transduction of CFU-S (EGLITIS et al. 1985; KELLER, this volume).

3.2.1.2 Factors that Affect the Cycling Status of the Recipient Cells

Only a minority of the primitive cells in the bone marrow are replicating (10% of the CFU-S; ROSS et al. 1982). Since retroviruses preferentially infect actively replicating cells (WEISS et al. 1984), methods were devised to trigger the resting stem cells into cycle. These methods can be divided into two categories: source of cells and growth factors.

3.2.1.2.1 Source of Cells. Instead of using the bone marrow from normal mice, marrow from donors recovering from a treatment with 5-FU was investigated. As already mentioned, actively replicating cells are killed by 5-FU, the non-cycling primitive stem cells are relatively resistant to this treatment. Post-5-FU bone marrow is enriched for primitive stem cells (HODGSON and BRADLEY 1979), and, more importantly, the fraction of replicating CFU-S increases from 10% in untreated animals to 100% in 5-FU-treated mice (VAN ZANT 1984). LEMISCHKA et al. 1986) found that bone marrow taken 2 days following 5-FU injection was more susceptible to infection than 5-FU marrow taken a day earlier. BOWTELL et al. (1987) extended these studies and compared the infectability of bone marrow that was harvested 2–8 day post 5-FU. In this study, bone marrow taken 4 days post 5-FU treatment was optimal for the infection of CFU-S.

3.2.1.2.2 Growth Factors. Parallel to the experiments described above, studies were undertaken to stimulate the replication of stem cells during the co-cultivation with the virus-producing cell lines. Ideally, cocultivations should be executed in the presence of growth factors that can stimulate the proliferation of PHSC without inducing their differentiation (for review see METCALF 1989). PHSC can survive and become infected with retroviral vectors during cocultivation in the absence of externally added hematopoietic growth factors (WILLIAMS et al. 1984). However, the infectability of CFU-S was greatly enhanced by the addition of interleukin 3 (IL-3) to the medium (EGLITIS et al. 1985; LEMISCHKA et al. 1986; LIM et al. 1989). IL-3 is now routinely included in the coculture medium. Most laboratories include IL-1α in the coculture medium since it stimulates the proliferation of CFU-S in short-term cultures in a synergistic fashion with IL-3 (BELMONT et al. 1988; JORDAN and LEMISCHKA 1990; JORDAN et al. 1990; MOORE et al. 1990; VALERIO et al. 1989; VAN BEUSECHEM et al. 1990). Although the infection of CFU-S has been enhanced by the use of these growth factors, their numbers do not increase during coculture. Since also only 25% of the seeded MRA-CFU-S can be recovered after cocultivation in the presence of IL-1α and IL-3 (M.P.W. EINERHAND et al., in preparation) the conditions presently used might not be optimal for the replication and infection of these cells. Growth factors which can be considered plausible candidates for further enhancement of the procedure are: IL-6, Leukemia inhibitory factor (LIF), and stem cell factor (SCF) also known as mast cell growth factor (MGF).

IL-6 in combination with IL-3 enhances the formation of blast cell colonies (an in vitro-assay for multipotent hematopoietic cells with self-renewing capacity), as compared to IL-3 alone (IKEBUCHI et al. 1987). BODINE et al. (1989) found that upon co-cultivation in the presence of only IL-3, 30–37% of the CFU-S became infected, whereas 40–80% became infected when both IL-3 and IL-6 had been added to the co-cultures. However, the authors overlooked to study the endogenous production of IL-6 during such co-cultures. It is now clear that large amounts of IL-6 are produced by retrovirus packaging cells and bone marrow cells during co-cultivation (EINERHAND et al. 1991). Given the reported variability in CFU-S transduction (BODINE et al. 1989) it is questionable whether indeed the effect of the added IL-6 was measured in this study.

LIF is known for its ability to sustain the proliferation of embryonic stem cells without affecting their developmental potential. LIF also supports the proliferation of the IL-3-dependent cell line DA-1, implying that LIF might act on PHSC (MOREAU et al. 1988). Indications for a role in early hematopoiesis were found when it was shown that murine LIF enhanced the infection efficiency of CFU-S in a dose-dependent fashion during cocultivation with a virus-producing cell line (FLETCHER et al. 1990). It was not evaluated, however, whether this was a direct effect of LIF on CFU-S or an indirect effect, mediated through LIF-responsive accessory cell that, upon stimulation with LIF, produced additional growth factors.

The growth factor SCF (MGF) was recently cloned (HUANG et al. 1990; WILLIAMS et al. 1990; ZSEBO et al. 1990b) and found to be the ligand for the c-*kit* tyrosine kinase receptor (HUANG et al. 1990; ZSEBO et al. 1990a, b). Mice harboring mutations in the gene encoding SCF (Steel, Sl mice) or the receptor c-*kit* (W mice) display a similar pleiotropic phenotype with abnormalities in hematopoiesis, melanogenesis, and gametogenesis. In the hematopoietic system, mutations in c-*kit* affect the stem cell compartment, erythroid precursors, and tissue mast cells (reviewed in RUSSEL 1979). Bone marrow derived from W/W^v mice is unable to form macroscopic spleen colonies in normal lethally irradiated recipients and therefore contains no functional CFU-S (MCCULLOCH et al. 1964). In addition, normal syngeneic PHSC have a permanent and selective advantage in W/W^v hosts (ABRAMSON et al. 1977). This suggest that, under the assumption that SCF is the only c-*kit* ligand, a functional SCF/c-*kit* signal transduction pathway is essential for early hematopoiesis in vivo. In vitro, SCF is crucial to the survival of CFU-S on stromal cell lines. In addition, SCF induces colony formation in normal, lineage-negative, and 5-FU-treated bone marrow (ZSEBO et al. 1990a). These results imply that SCF might improve the survival and possibly the infection efficiency of CFU-S during coculture of bone marrow with virus-producing cell lines. To test this notion it should first be established whether virus-producing cells already synthesize (sufficient amounts of) SCF. Data obtained in binding studies with a soluble form of c-*kit* demonstrated the presence of membrane-bound SCF on Balb/c 3T3 fibroblasts (FLANAGAN and LEDER 1990). These cells are basically identical to the NIH/3T3 fibroblasts which are most commonly used to generate virus-producing cell lines.

The efforts to maximize the gene transfer procedure have resulted in a high gene transfer efficiency into CFU-S. With the knowledge at hand to date, upto 100% of the CFU-S can be infected. A frequently used protocol utilizes 5-FU-preteated bone marrow which is cocultivated with a high-titer virus-producing cell line in the presence of IL-1α and IL-3. Efforts to further optimize the gene transfer procedure will be directed at reducing the variability of the procedure and at enhancing the survival of primitive hematopoietic cells during the infection process. Given the natural function of SCF, expectations for its use in retrovirus-mediated gene transfer into PHSC should be high. A combination of SCF with LIF and/or IL-6 might even allow the replacement of coculture procedures, with their inherent variability, by infections with virus supernatants, which can be stored and tested prior to their use in the infection of stem cells (see also G. KELLER, this volume).

4 Gene Therapy for Severe Combined Immunodeficiency Caused by Adenosine Deaminase Deficiency

4.1 General Considerations

From the preceding sections it is obvious that the hematopoietic system can be genetically modified through the use of retroviral vector technology. In this section we will discuss the application of this technology in the development of gene therapy protocols for an inherited form of severe combined immunodeficiency (SCID) which is caused by the lack of sufficient functional adenosine deaminase (ADA). This disease, ADA^-SCID, is characterized by the absence of T and B cells, probably due to the accumulation of deoxyadenosine which is one of the ADA substrates (reviewed in KREDICH and HERSHFIELD 1983). Although ADA^-SCID is a rare disease (found in fewer than 1:100 000 newborns; Hirschhorm 1977) it is a particularly good model disease for initial attempts at gene therapy. The reasons for this are:

1. ADA deficiency is severely debilitating and lethal, if untreated. Effective treatment is possible by a bone marrow transplantation with bone marrow from HLA-identical siblings (see also TOURAINE, this volume). These are available only in one out of three cases. HLA-mismatched parental donors have also been used. However, in about 50% of these cases, treatment fails due to graft failure or rejection (FISCHER et al. 1986). For these patients, autologous transplantation of "cured" bone marrow would obviously be preferable.
2. It can be expected that the genetically cured blood cells will have a selective advantage over unaltered cells. This can be deduced from the chimeric reconstitution patterns of ADA^- SCID patients upon transplantation. In most of these cases, donor cell types can only be found in the lymphoid compartment, while all other hematopoietic lineages remain of the recipient type (DOOREN and VOSSEN 1985).
3. A precise and cell-type specific control of ADA gene expression is probably not required for a curative effect. As a member of the so-called house-keeping genes, the ADA gene is constitutively expressed in all cell lineages. In addition, large variations in ADA levels have been found among healthy individuals (HERBSCHLEB-VOOGT 1983; THOMPSON and SEEGMILLER 1980). With this in mind, it is expected that a slight under or over expression of the gene can be tolerated in the genetically altered cells.
4. Retroviral transduction of the hADA-cDNA into cultured B and T cells derived from an ADA-deficient patient reduced their sensitivity to deoxyadenosine (KANTOFF et al. 1986; BRAAKMAN, in press).

Most of the gene therapy research efforts for ADA^- SCID are directed toward the introduction of the hADA-coding sequences into the PHSC of the patients.

Reimplantation and outgrowth of these cells should result in a lasting restoration of the patient's lymphoid system. More recently, an alternative approach became expedient, based on the genetic modification of circulating T cells. The merits of both these strategies and recent advances are discussed below.

4.2 T Cell Gene Therapy for ADA$^-$SCID

Recently, some ADA$^-$SCID patients have been treated experimentally with purified bovine ADA enzyme coupled to polyethylene glycol (PEG) (HERSHFIELD 1987; LEVY et al. 1988). Although this form of treatment has been successful in some respects, a major problem was encountered when some of the patients developed ADA-inactivating antibodies (HERSHFIELD and FINKELBERG 1989; A. FISHER, personal communication). One of the advantages of this treatment is, however, that it offers a new possibility to perform gene therapy. PEG-ADA-treated patients usually have circulating T cells. These can be isolated and genetically cured by infection with recombinant ADA viruses (BRAAKMAN et al. in press; M. BLAESE, personal communication). Reinfusion of such corrected cells into the patient might have a therapeutic effect either directly, when enough (different) T cells have been cured to restore some immunologic functions, or indirectly by diminishing the concentration of the toxic adenine nucleosides in the patient's serum, as an improved form of enzyme therapy. Recently, such a protocol was approved by the authorities in the United States (GERSHON 1990) to be performed by the same group that had already studied the fate of retrovirally marked tumor-infiltrating lymphocytes (TIL) in patients (KASID et al. 1990; ROSENBERG et al. 1990). The first two patients are presently being treated according to this protocol. However, it is unlikely that this procedure will result in the outgrowth of a lasting, completely cured, and educated immune system. In our view, the transfer of a functional gene into PHSC, which can repopulate a complete hematopoietic system therefore remains one of the most important goals to be pursued for gene therapy in general and for the treatment of ADA$^-$SCID in particular.

4.3 Developments in Bone Marrow Gene Therapy for ADA$^-$SCID

Although the PHSC has formed the main target in ADA$^-$SCID gene therapy research for almost a decade now, the progress in the development of an effective protocol has been slow. Initially, vectors were made in which the hADA gene was inserted together with a dominant selectable markers gene. Efficient transfer of these hADA-gene-containing retroviruses into hematopoietic stem cells (HSC) has been reported (MCIVOR et al. 1987; VALERIO et al. 1986; WILLIAMS et al. 1986). However, these vectors were not capable of directing sustained expression of the hADA gene in vivo, although they were potent expressors in NIH/3T3 fibroblasts. This lack of expression in the hematopoietic system was also observed with other genes (reviewed in WILLIAMS 1990). To overcome this problem, most investigators introduced additional promoters into the viral transcription unit (BELMONT et al. 1988; BOWTELL et al. 1988; GUILD et al. 1988; LIM et al. 1987; MAGLI et al. 1987; MCIVOR et al. 1987). Although this has led to improved expression patterns, the outcome of such alterations is unpredictable. What was remarkable was that all studies that reported high and long-term hADA expression in vivo used the simple "one gene only" type of construct in which only the hADA gene was present (BELMONT et al. 1988; LIM et al. 1987, 1989; VAN BEUSECHEM et al. 1990; WILSON et al. 1990).

The unpredictable nature of retroviral gene expression in the hematopoietic system in vivo has hampered the development of retroviral vectors for gene therapy purposes. These studies are time consuming and form the bottleneck in gene therapy research. Attempts to set up reliable in vitro models for retroviral gene expression have failed thus far. Expression of the introduced hADA gene was detected in several hematopoietic cell lines (including the multipotent FDCP-mix cell line), whereas, using the same vector, expression was not detected in vivo in CFU-S-derived spleen colonies (WILLIAMS et al. 1986, 1988). Considering the discrepancies between the expression capability of retroviral vectors in cell lines and in primary hematopoietic cells, we have developed an in vitro model which utilizes primary murine HSC. For these experiments it was necessary to infect purified HSC. These can be obtained in relatively large numbers (up to 1×10^6 cells) from the bone marrow by a modification of the purification protocol described by BERTONCELLO et al. (1989) (EINERHAND et al. submitted). Purified stem cells were infected with retroviruses and cloned at a single cell per well in defined serum-free medium containing IL-1α and IL-3. Virus integration and expression were monitored in the progeny of the stem cells. Initial results obtained in this in vitro model gave further evidence for the presence of a retroviral expression block in primary HSC infected with Mo-MuLV.

Since the enhancer sequences in the LTR of retroviruses are known to be responsible for expression specificity (WEISS et al. 1984), our efforts to obtain expression in hematopoietic cells in vivo have been directed toward alteration of this element. We have recently described (VALERIO et al. 1989) the construction of a recombinant retrovirus in which a marker gene was placed under the transcriptional control of a hybrid LTR designated ΔMo + PyF101 (LINNEY et al. 1984). In this construct the Mo-MuLV enhancer was replaced by the enhancer of a mutant polyoma virus (PyF101) that was selected to grow in embryonal carcinoma (EC) cells. We showed that this vector directs gene expression in EC cells as well as in HSC. To test the applicapability of this LTR in gene therapy protocols, we placed the hADA-cDNA in a vector under the transcriptional control of the ΔMo + PyF101- LTR. Transplantation of bone marrow which was infected with this vector resulted in detectable levels of hADA upon reconstitution of the hematopoietic system. Retransplantation of the bone marrow of the first recipient into a secondary recipient revealed that hADA expression persisted in the myeloid, lymphoid, and erythroid lineage (VAN BEUSECHEM et al. 1990). For a quantification of the number of independent HSC expressing hADA, we turned to the previously described MRA-CFU-S assay. Preliminary data obtained with this assay indicate that approximately 50% of the infected MRA-CFU-S-derived spleen colonies express hADA. These characteristics render the ΔMo + PyF101-LTR quite useful for the design of vectors to be employed in future gene therapy protocols. An additional advantage of such vectors is that they are derived from a nonpathogenic retrovirus. The leukemogenic potential of Mo-MuLV is suppressed when the Mo-MuLV LTR is replaced by the ΔMo + PyF101-LTR (Davis et al. 1985).

5 Gene Therapy for Hemoglobin Disorders

Although ADA$^-$SCID forms the spearhead of gene therapy research, it is not a common inherited disease. More widespread, with millions of affected persons throughout the world and therefore of more clinical importance, are the hemoglobin abnormalities (for review see Wintrobe et al. 1981). Specifically patients suffering from β^0-thalassemia might be amenable to gene therapy. In the early days of gene therapy research, globin abnormalities were discarded as easy targets for gene therapy. This was largely due to the extremely complicated expression patterns of globin-like genes (Anderson 1984). Globin synthesis is highly requlated during development and confined to the cells of the erythroid lineage. Furthermore, the expression of the α- and β-globin-like chains is regulated such that they are maintained at a 1:1 ratio in the cell. Such careful control of gene expression was not feasible with the knowledge at hand in those days. However, current developments have revived the interest in gene therapy as an approach to treat this class of inherited disease. Using retroviral vectors containing a genomic human β-globin sequence, regulated expression was obtained upon transplantation of infected bone marrow (Bender et al. 1989; Dzierzak et al. 1988; Karlsson et al. 1988; Li et al. 1990). The expression levels were low, however, only 0.1%–5% as compared to the endogenous mouse β-globin expression. Recently, the sequences responsible for coordinated and high β-globin expression, independent of its site of integration, were found and localized to a region of 6.5 kb (Talbot et al. 1989). This region was originally termed "locus-activating region" or "dominant control region." It was renamed locus control region (LCR) at the biannual hemoglobin switching conference (Orkin 1990). Inclusion of a part of the LCR in a retroviral vector resulted in expression of the human β-globin gene in a murine erythro leukemia cell line (MEL) at a level of 35%–132% as compared to endogenous mouse β-globin RNA (Novak et al. 1990). These encouraging results provide a new basis for the development of gene therapy protocols for globin abnormalities.

LCR sequences are almost certainly not confined to the human β-globin locus. Sequences that direct position-independent, high-level, and cell type-specific expression were also located downstream of the human CD2 gene

(GREAVES et al. 1989) and in sequences flanking the chicken lysozyme gene (BONIFER et al. 1990). A more accurate definition of the sequences responsible for the action of the LCR is likely to warrant their inclusion into retroviral vectors. The thus created "mini-locus" containing retrovirus vectors could prove crucial in the development of gene therapy protocols for globin disorders and many other inherited disorders.

6 Conclusion

Gene transfer into hematopoietic cells has become a vital technology in biologic and clinical investigations. This became feasible through the development of retroviral vector technology which permits the highly efficient and safe transfer of foreign genes into cells that are otherwise difficult to transfect. In the clinic, these vectors are currently employed to track the fate of genetically marked tumor-infiltrating T lymphocytes in vivo. In addition, they are used to examine the feasibility of gene therapy for the treatment of inherited hematopoietic disorders. The first gene therapy protocol has recently been approved for patients suffering from ADA deficiency. Circulating T cells of these patients are harvested, cured in vitro through the addition of a functional copy of the hADA gene, and subsequently transplanted back into the patient. The first two patients are currently being treated. Although this treatment might be beneficial to the patients, one cannot expect a lasting cure with this method. Regular infusion of the corrected cells will most likely be necessary to replenish the supply of ADA enzyme in the circulation. Correction of the genetic defect in PHSC would not have this drawback.

For ADA deficiency the latter approach was initially hampered by the inefficient expression of the hADA gene in the hematopoietic system following gene transfer into PHSC. These problems have now been overcome, and long-term, high hADA expression has been obtained with a number of different vectors. The discovery of the so-called LCR, which direct position-independent, tissue-specific, and physiologic levels of gene expression, might prove important. It should be investigated whether these can be used to create retroviruses carrying mini-loci. Specifically, for genes with complicated expression patterns, such as β-globin, these LCRs might prove to be crucial to the development of vectors for gene therapy.

Now that the hurdles involved in the application of bone marrow gene therapy for ADA deficiency have been solved in mice, it must be investigated whether the technology can be translated to humans. For this purpose, we and others are currently testing our procedures in preclinical bone marrow transplantation models (i.e., the nonhuman primate model and the dog model). Preliminary data are encouraging and seem to indicate that results similar to those in the mouse model can be obtained in these relevant large animal models (VAN BEUSECHEM, submitted). We therefore, anticipate that the clinical application of a bone marrow gene therapy protocol can be expected in the coming years, most likely before the turn of the millennium!

References

Abramson S, Miller RG, Phillips RA (1977) The identification in adult bone marrow of pluripotent and restricted stem cells of the myeloid and lymphoid systems. J Exp Med 145: 1567–1579

Anderson WF (1984) Prospects for human gene therapy. Science 226: 401–409

Belmont JW, Caskey CT (1986) Developments leading to human gene therapy. In: Kucherlapati R(ed) Gene transfer. Plenum, New York

Belmont JW, MacGregor GR, Wager-Smith K, Fletcher FA, Moore KA, Hawkins D, Villalon D, Chang SMW, Caskey CT (1988) Expression of human adenosine deaminase in murine hematopoietic cells. Mol Cell Biol 8: 5116–5125

Bender MA, Gelinas RE, Miller AD (1989) A majority of mice show long-term expression of a human β-globin gene after retrovirus transfer into hematopoietic stem cells. Mol Cell Biol 9: 1426–1434

Bertoncello I, Hodgson GS, Bradley TR (1985) Multiparameter analysis of transplantable hemopoietic stem cells homing to marrow and spleen on the basis of rhodamine-123 fluorescence. Exp Hematol 13: 999–1006

Bertoncello I, Bradley TR, Hodgson GS (1989) The concentration and resolution of primitive hemopoietic cells from normal mouse bone marrow by negative selection using monoclonal antibodies and dynabead monodisperse magnetic microspheres. Exp Hematol 17: 171–176

Bodine DM, Karlsson S, Nienhuis AW (1989) Combination of interleukins 3 and 6 preserves stem cell function in culture and enhances retrovirus-mediated gene transfer into hematopoietic stem cells. Proc Natl Acad Sci USA 86: 8897–8901

Bonifer C, Vidal M, Grosveld F, Sippel AE (1990) Tissue specific and position independent expression of the complete gene domain for chicken lysozyme in transgenic mice. EMBO J 9: 2843–2848

Bowtell DDL, Johnson GR, Kelso A, Cory S (1987) Expression of genes transferred to haematopoietic stem cells by recombinant retroviruses. Mol Biol Med 4: 229–250

Bowtell DDL, Cory S, Johnson GR, Gonda TJ (1988) Comparison of expression in hemopoietic cells by retroviral vectors carrying two genes. J Virol 62: 2464–2473

Braakman E, Van Beusechem VW, Van Krimpen BA, Fischer A, Bolhuis RLH, Valerio D, Genetic correction of cultured T cells from an adenosine deaminase deficient patient: characteristics of non-transduced and transduced T cells. Eur J Immunol (in press)

Capecchi MR (1989) The new mouse genetics: altering the genome by gene targetting. Trends Genet 5: 70–76

Capel B, Hawley RG, Mintz B (1990) Long- and short-lived murine hematopoietic stem cell clones individually identified with retroviral integration markers. Blood 75: 2267–2270

Cavard C, Grimber G, Dubois N, Chasse JF, Bennoun M, Minet-Thuriaux M, Kamoun P, Briand P (1988) Correction of mouse ornithine transcarbamylase deficiency by gene transfer into the germ line. Nucl Acids Res 16: 2099–2110

Constantini F, Chada K, Magram J (1986) Correction of murine β- thalassemia by gene transfer into the germ line. Science 223: 1192–1194

Danos O, Mulligan RC (1988) Safe and efficient generation of recombinant retroviruses with amphotropic and ecotropic host ranges. Proc Natl Acad Sci USA 85: 6460–6464

Davis B, Linney E, Fan H (1985) Suppression of leukaemia virus pathogenicity by polyoma virus enhancers. Nature 314: 550–553

Dick JE, Magli MC, Huszar D, Phillips RA, Bernstein A (1985) Introduction of a selectable gene into primitive stem cells capable of long-term reconstitution of the hematopoietic system of W/W^v mice. Cell 42: 71–79

Dooren LJ, Vossen JM (1985) Severe Combined Immunodeficiency: Reconstitution of the immune system following bone marrow transplantation. In: Van Bekkum DW, Löwenberg B (eds) Bone marrow transplantation 3. Dekker, New York

Dubenski TW, Campbell BA, Villarreal LP (1984) Direct transfection of viral and plasmid DNA into the liver or spleen of mice. Proc Natl Acad Sci USA 81: 7529–7533

Dzierzak EA, Papayannopoulou T, Mulligan RC (1988) Lineage-specific expression of a human β-globin gene in murine bone marrow transplant recipients reconstituted with retrovirus-transduced stem cells. Nature 311: 35–41

Eglitis MA, Kantoff P, Gilboa E, Anderson WF (1985) Gene expression in mice after high efficiency retroviral-mediated gene transfer. Science 230: 1395–1398

Einerhand MPW, Bakx TA, Valerio D (1991) Interleukin-6 production by retrovirus packaging cells and cultured bone marrow cells. Hum Gene Ther 2: 301–306

Einerhand MPW, Bakx TA, Visser JWM, Valerio D, Retrovirus infection and culture of purified murine hemopoietic progenitor cells. (submitted)

Fischer A, Landais P, Friedrich W, Morgan G, Gerritsen B, Fasth A, Porta F, Griscelli C, Goldman SF, Levinsky R (1990) European experience of bone marrow transplantation for severe combined immunodeficiency. Lancet 336: 850–854

Flanagan JG, Leder P (1990) The *kit* ligand: a cell surface molecule altered in steel mutant fibroblasts. Cell 63: 185–194

Fletcher FA, Williams DE, Maliszewski C, Anderson D, Rives M, Belmont JW (1990) Murine leukemia inhibitory factor enhances retroviral-vector infection efficiency of hematopoietic progenitors. Blood 76: 1098–1103

Friedmann T (1983) Gene therapy: fact and fiction. Cold Spring Harbor Laboratory, New York

Gershon D (1990) First experiment approved. Nature 346: 402

Greaves DR, Wilson FD, Lang G, Kioussis D (1989) Human CD2 3′-flanking sequences confer high-level, T cell-specific, position-independent gene expression in transgenic mice. Cell 56: 979–986

Guild BC, Finer MH, Housman DE, Mulligan RC (1988) Development of retrovirus vectors useful for expressing genes in cultured murine embryonal cells and hematopoietic cells in vivo. J Virol 62: 3795–3801

Hammer RE, Palmiter RD, Brinster RL (1984) Partial correction of murine hereditary growth disorder by germ-line incorporation of a new gene. Nature 311: 65–67

Herbschleb-Voogt E (1983) Expression of human adenosine deaminase and its complexing protein in health and disease. PhD thesis, State University of Leiden

Hershfield MS, Finkelberg Z (1989) ADA deficiency treatment. Science 246: 1375

Hershfield MS, Buckley RH, Greenberg ML, Melton AL, Schiff R, Hatem C, Kurtzberg J, Markert ML, Kobayashi RH, Kobayashi AL, Abuchowski A (1987) Treatment of adenosine deaminase deficiency with polyethylene glycol-modified adenosine deaminase. N Engl J Med 316: 589–596

Hirschhorn R (1977) Defects of purine metabolism in immunodeficiency diseases. In: Schwartz RS (ed) Progress in clinical immunology, vol 3. Grune and Stratton, New York

Hodgson GS, Bradley TR (1979) Properties of haematopoietic stem cells surviving 5-fluorouracil treatment: evidence for a pre-CFU-S cell? Nature 281: 381–382

Huang E, Nocka K, Beier DR, Chu TY, Buck J, Lahm HW, Wellner D, Leder P, Besmer P (1990) The hematopoietic growth factor KL is encoded by the Sl locus and is the ligand for the c-*kit* receptor, the gene product of the W locus. Cell 63: 225–233

Ikebuchi K, Wong GG, Clark SC, Ihle JN, Hirai Y, Ogawa M (1987) Interleukin 6 enhancement of interleukin 3-dependent proliferation of multipotential hemopoietic progenitors. Proc Natl Acad Sci USA 84: 9035–9039

Jordan GT, Lemischka IR (1990) Clonal and systemic analysis of long-term hematopoiesis in the mouse. Genes Dev 4: 220–232

Jordan GT, McKearn JP, Lemischka IR (1990) Cellular and developmental properties of fetal hematopoietic stem cells. Cell 61: 953–963

Joyner A, Keller G, Phillips RA, Bernstein A (1983) Retrovirus transfer of a bacterial gene into mouse haematopoietic progenitor cells. Nature 305: 556–558

Kantoff PW, Kohn DB, Mitsuya H, Armentano D, Sieberg M, Zwiebel JA, Eglitis MA, McLachlin JR, Wiginton DA, Hutton JJ, Horowitz SD, Gilboa E, Blaese RM, Anderson WF (1986) Correction of adenosine deaminase deficiency in cultured human T and B cells by retrovirus-mediated gene transfer. Proc Natl Acad Sci USA 83: 6563–6567

Karlsson S, Bodine DM, Perry L, Papayannopoulou T, Nienhuis AW (1988) Expression of the human β-globin gene following retroviral-mediated transfer into multipotential hematopoietic progenitors of mice, Proc Natl Acad Sci USA 85: 6062–6066

Kasid A, Morecki S, Aebersold P, Cornetta K, Culver K, Freeman S, Director E, Lotze MT, Blaese RM, Anderson WF, Rosenberg SA (1990) Human gene transfer: characterization of human tumor-infiltrating lymphocytes as vehicles for retroviral-mediated gene transfer in man. Proc Natl Acad Sci USA 87: 473–477

Keller G, Snodgrass R (1990) Life span of multipotential hematopoietic stem cells in vivo. J Exp Med 171: 1407–1418

Keller G, Paige C, Gilboa E, Wagner EF (1985) Expression of a foreign gene in myeloid and lymphoid cells derived from multipotent haematopoietic precursors. Nature 318: 149–154

Kredich NM, Hershfield MS (1983) Immunodeficiency diseases caused by adenosine deaminase deficiency and purine nucleoside phosphorylase deficiency. In: Stanbury JB et al. (eds) The metabolic basis of inherited disease, 5th edn. McGraw-Hill, New York

Kyle JW, Birkenmeier EH, Gwynn B, Vogler C, Hoppe PC, Hoffmann JW, Sly WS (1990) Correction of murine mucopolyssacharidosis VII by a human β-glucuronidase transgene. Proc Natl Acad Sci USA 87: 3914–3918

Lemischka IR, Raulet DH, Mulligan RC (1986) Developmental potential and dynamic behavior of hematopoietic stem cells. Cell 45: 917–927

Levy Y, Hershfield MS, Fernandez-Meija C, Polmar SH, Scudiery D, Berger M, Sorensen RU (1988) Adenosine deaminase deficiency with late onset of recurrent infections: response to treatment with polyethylene glycol-modified adenosine deaminase. J Pediatr 113: 312–317

Li CL, Dwarki VJ, Verma IM (1990) Expression of human α-globin and mouse/human hybrid β-globin genes in murine hemopoietic stem cells transduced by recombinant retroviruses. Proc Natl Acad Sci USA 87: 4349–4353

Lim B, Williams DA, Orkin SH (1987) Retrovirus-mediated gene transfer of human adenosine deaminase: expression of functional enzyme in murine hematopoietic stem cells in vivo. Mol Cell Biol 7: 3459–3465

Lim B, Apperley JF, Orkin SH, Williams DA (1989) Long-term expression of human adenosine deaminase in mice transplanted with retrovirus-infected hematopoietic stem cells. Proc Natl Acad Sci USA 86: 8892–8896

Lin FL, Sperle K, Sternberg N (1985) Recombination in mouse L cells between DNA introduced into the cells and homologous chromosomal sequences. Proc Natl Acad Sci USA 82: 1391–1395

Linney E, Davis B, Overhauser J, Chao E, Fan H (1984) Non-function of a Moloney murine leukaemia virus regulatory sequence in F9 embryonal carcinoma cells. Nature 308: 470–472

Magli MC, Dick JE, Huszar D, Bernstein A, Phillips RA (1987) Modulation of gene expression in multiple hematopoietic cell lineages following retroviral vector gene transfer. Proc Natl Acad Sci USA 84: 789–793

Mann R, Mulligan RC, Baltimore D (1983) Construction of a retrovirus packaging mutant and its use to produce helper-free replication defective retrovirus. Cell 33: 153–159

Markowitz D, Goff S, Bank A (1988a) Construction and use of a safe and efficient amphotropic packaging cell line. Virology 167: 400–406

Markowitz D, Goff S, Bank A (1988b) A safe packaging line for gene transfer: separating viral genes on two different plasmids. J Virol 62: 1120–1124

Mason AJ, Pitts SL, Nikolics K, Szonyi E, Wilcox JN, Seeburg PH, Stewart TA (1986) The hypogonal mouse: reproductive functions restored by gene therapy. Science 234: 1372–1378

McCulloch EA, Siminovitch L, Till JE (1964) Spleen-colony formation in anemic mice of genotype W/W^v. Science 144: 844–846

McIvor RS, Johnson MJ, Miller AD, Pitts S, Williams SR, Valerio D, Martin DW Jr, Verma IM (1987) Human purine nucleoside phosphorylase and adenosine deaminase: gene transfer into cultured cells and murine hematopoietic stem cells by using recombinant amphotropic retroviruses. Mol Cell Biol 7: 838–846

Metcalf D (1989) The molecular control of cell division, differentiation commitment and maturation in haemopoietic cells. Nature 339: 27–30

Miller AD (1990) Retrovirus packaging cells. Hum Gene Ther 1: 5–14

Miller AD, Law MF, Verma IM (1985) Generation of helper-free amphotropic retroviruses that transduce a dominant-acting methotrexate-resistant dihydrofolate reductase gene. Mol Cell Biol 5: 431–437

Moore KA, Fletcher FA, Villalon DK, Utter AE, Belmont JW (1990) Human adenosine deaminase expression in mice. Blood 75: 2085–2092

Moreau JF, Donaldson DD, Bennett F, Witek-Giannotti J, Clark SC, Wong GG (1988) Leukaemia inhibitory factor is identical to the myeloid growth factor human interleukin for DA cells. Nature 336: 690–692

Motulsky AG (1983) Impact of genetic manipulation on society and medicine. Science 219: 135–140

Novak U, Harris EAS, Forrester W, Groudine M, Gelinas R (1990) High-level β-globin expression after retroviral transfer of locus activating region-containing human β-globin gene derivatives into murine erythroleukemia cells. Proc Natl Acad Sci USA 87: 3386–3390

Orkin SH (1990) Globin gene regulation and switching: circa 1990. Cell 63: 665–672

Osborne WRA, Hock RA, Kaleko M, Miller AD (1990) Long-term expression of human adenosine deaminase in mice after transplantation of bone marrow infected with amphotropic retroviral vectors. Hum Gene Ther 1: 31–41

Ploemacher RE, Brons NHC (1989) Separation of CFU-S from primitive cells responsible for reconstitution of the bone marrow hemopoietic stem cell compartment following irradiation: evidence for a pre-CFU-S cell. Exp Hematol 17: 263–266

Rohdewohld H, Weiher H, Reik W, Jaenisch R, Breindl M (1987) Retrovirus integration and chromatin structure: moloney murine leukemia proviral integration sites map near DNAse I-hypersensitive sites. J Virol 61: 336–343

Rosenberg SA, Aebersold P, Cornetta K, Kasid A, Morgan RA, Moen R, Karson EM, Lotze MT, Yang JC, Topalian SL, Merino MJ, Culver K, Miller AD, Blaese RM, Anderson WF (1990) Gene transfer into humans—immunotherapy of patients with advanced melanoma, using tumor-infiltrating lymphocytes modified by retroviral gene transduction. N Engl J Med 323: 570–578

Ross EAM, Anderson N, Micklem HS (1982) Serial depletion and regeneration of the murine hematopoietic system. Implications for hematopoietic organization and the study of cellular aging. J Exp Med 155: 432–444

Russel ES (1979) Hereditary anemias of the mouse; a review for geneticists. Adv Genet. 20: 357–459

Shih CC, Stoye JP, Coffin JM (1988) Highly preferred targets for retrovirus integration. Cell 53: 531–537

Snodgrass R, Keller G (1987) Clonal fluctuation within the haematopoietic system of mice reconstituted with retrovirus-infected stem cells. EMBO J 6: 3955–3960

Talbot D, Collis P, Antoniou M, Vidal M, Grosveld F, Greaves DR (1989) A dominant control region from the human β-globin locus conferring integration site independent gene expression. Nature 338: 352–355

Temin HM (1986) Retrovirus vectors for gene transfer: efficient integration into and expression of exogenous DNA in vertebrate cell genomes. In: Kucherlapati R (ed) Gene transfer. Plenum, New York

Thomas KR, Folger KR, Capecchi MR (1986) High frequency targetting of genes to specific sites in the mammalian genome. Cell 44: 419–428

Thompson LF, Seegmiller JE (1980) Adenosine deaminase deficiency and severe combined immunodeficiency disease. In: Meister A (ed) Advances in enzymology Vol 51, Academic, New York

Till JE, McCulloch EA (1961) A direct measurement of the radiation sensitivity of normal mouse bone marrow cells. Radiat Res 14: 213–222

Valerio D, Retrovirus vectors for gene therapy procedures. In: Grosveld F, Kollias G (eds) Transgenic mice in biology and medicine. Academic, New York. (in press)

Valerio D, Visser TP, Wagemaker G, Van der Eb AJ, Van Bekkum DW (1986) The introduction of human ADA sequences into mouse hematopoietic cells. In: Vossen J, Griscelli C (eds) Progress in immunodeficiency research and therapy: Proceedings, 2nd Meeting of the European Group for Immunodeficiency (t6iD) Congress Centre Roldue, Verkrader, 11–14 June 1986, vol 2. Elsevier, Amsterdam

Valerio D, Einerhand MPW, Wamsley PM, Bakx TA, Li CL, Verma IM (1989) Retrovirus-mediated gene transfer into embryonal carcinoma cells and hemopoietic stem cells: expression from a hybrid long terminal repeat. Gene 84: 419–427

Van Beusechem VW, Kukler A, Einerhand MPW, Bakx TA, Van der Eb AJ, Van Bekkum DW, Valerio D (1990) Expression of human adenosine deaminase in mice transplanted with hemopoietic stem cells infected with amphotropic retroviruses. J Exp Med 172: 729–736

Van Beusechem VW, Valerio D, Prospects for human gene therapy. In: Murray JAH (ed) Transgenesis-applications of gene transfer. Open University Press, London (in press)

Van Beusechem VW, Kuhler A, Heidt PJ, Valerio D, Preclinical testing of bone marrow gene therapy procedures in non-human primates. (submitted)

van Zant G (1984) Studies of hematopoietic stem cells spared by 5-fluorouracil. J Exp Med 159: 679–690

Walters LR (1986) The ethics of human gene therapy. Nature 320: 225–227

Weiss R (1984) Experimental biology and assay of RNA tumor viruses. In: Weiss R, Teich N, Varmus H, Coffin J (eds) RNA tumor viruses. Cold Spring Harbor Laboratory, New York

Williams DA (1990) Expression of introduced genetic sequences in hematopoietic cells following retroviral-mediated gene transfer. Hum Gene Ther 1: 229–239

Williams DA, Lemischka IR, Nathan DG, Mulligar RC (1984) Introduction of new genetic material into pluripotent haematopoietic stem cells of the mouse. Nature 310: 476–480

Williams DA, Orkin SH, Mulligan RC (1986) Retrovirus-mediated transfer of human adenosine deaminase gene sequences into cells in culture and into murine hematopoietic cells in vivo. Proc Natl Acad Sci USA 83: 2566–2570

Williams DA, Lim B, Spooncer E, Longtine J, Dexter TM (1988) Restriction of expression of an integrated recombinant retrovirus in primary but not immortalized murine hematopoietic stem cells. Blood 71: 1738–1743

Williams DE, Eisenman J, Baird A, Rauch C, van Ness K, March CJ, Park LS, Martin U, Mochizuki DY, Boswell HS, Burgess GS, Cosman D, Lyman SD (1990) Identification of a ligand for the c-*kit* proto-oncogene. Cell 63: 167–174

Williamson B (1982) Gene therapy. Nature 298: 416–418

Wilson JM, Danos O, Grossman M, Raulet DH, Mulligan RC (1990) Expression of human adenosine deaminase in mice reconstituted with retrovirus-transduced hematopoietic stem cells. Proc Natl Acad Sci USA 87: 439–443

Wintrobe MM, Lee GR, Boggs DR, Bithell TC, Foerster J, Athens JW, Lukens JN (1981) Clinical hematology, 8th edn. Lea and Febiger, Philadelphia

Wolff JA, Malone RW, Williams P, Chong W, Acsadi G, Jani A, Felgner PL (1990) Direct gene transfer into mouse muscle in vivo. Science 247: 1465–1468

Wu AM, Till JE, Siminovitch L, McCulloch EA (1968) Cytological evidence for a relationship between normal hematopoietic colony-forming cells and cells of the immune system. J Exp Med 127: 455–463

Zsebo KM, Williams DA, Geissler EN, Broudy VC, Martin FH, Atkins HL, Hsu RY, Birkett NC, Okino KH, Murdock DC, Jacobsen FW, Langley KE, Smith KA, Takeishi T, Cattanach BM, Galli SJ, Suggs SV (1990a) Stem cell factor is encoded at the Sl locus of the mouse and is the ligand for the c-*kit* tyrosine kinase receptor. Cell 63: 213–224

Zsebo KM, Wypych J, McNiece IK, Lu HS, Smith KA, Karkare SB, Sachdev RK, Yuschenkoff VN, Birkett NC, Williams LR, Satyagai VN, Tung W, Bosselman R, Mendiaz E, Langley K (1990b) Identification, purification, and biological characterization of hematopoietic stem cell factor from buffalo rat liver-conditioned medium. Cell 63: 195–201

Gene Transfer into Hematopoietic Stem Cells

F. G. SCHUENING

1 Introduction

Hematopoietic stem cells represent a preferred target for gene transfer. These cells are pluripotent and able to produce mature cells of all hematopoietic lineages including lymphocytes, erythrocytes, granulocytes, macrophages, mast cells, and platelets. They are also characterized by extensive self-renewal capacity thereby enabling them to provide long-term hematopoiesis. Pluripotent stem cells are best assayed in long-term in vivo repopulation studies, and they are extremely rare in normal marrow (approximately 10^{-4}–10^{-5}) (BOGGS et al. 1982; HARRISON et al. 1988). Gene transfer techniques therefore have to be highly efficient, and their possible effect needs to be documented by long-term repopulating assays. Currently, retroviruses represent the preferred method for gene transfer because they provide high efficiency of infection, stable integration of the provirus into the host genome, low copy number, and high levels of expression (VARMUS 1988).

2 Transfer of Selectable Marker Genes into Murine Hematopoietic Stem Cells

DICK et al. (1985) were the first to report insertion of a selectable marker gene into murine pluripotent stem cells capable of long-term production of myeloid and lymphoid progenitor cells using a helper virus-free vector. To obtain high gene

Clinical Research Division, Fred Hutchinson Cancer Research Center, 1124 Columbia Street, Seattle, WA 98104, USA

Current Topics in Microbiology and Immunology, Vol. 177

transfer efficiency, these investigators used donor animals treated with 5-fluorouracil to kill differentiated cells, thereby enriching for stem cells. This strategy also stimulated replication of normally nondividing early precursors and perhaps stem cells as shown by HODGSON and BRADLEY (1979) and VAN ZANT (1984). Replication of stem cells is necessary because retroviral vectors promote gene transfer only in replicating cells (MILLER et al. 1990). In these experiments, bone marrow was removed from both femurs of the animals 2 days after 5-fluorouracil treatment and infected by 24-h cocultivation with irradiated (15 Gy) fibroblasts which produced retroviral vectors containing the bacterial neomycin phosphotransferase gene (*neo*). After cocultivation in the presence of hematopoietic growth factors (conditioned medium from WEHI-3B (D) cells), marrow cells were exposed for 2 days to the selective agent G418 to kill uninfected cells thereby increasing the percentage of repopulating cells having the transferred gene. Using this infection protocol 60%–100% of in vitro colony-forming cells were G418 resistant, and 10/10 spleen colony forming units (CFU-S) contained one to three proviruses integrated at different chromosomal sites as compared to 1/11 CFU-S arising from nonpreselected marrow. To determine whether exogenous genes can be inserted and expressed in pluripotent hematopoietic stem cells, infected marrow cells (5×10^5) were injected into mutant W/W^v mice which were irradiated with a nonlethal dose (4 Gy) of total body irradiation (TBI) to promote both myeloid and lymphoid reconstitution. Cells from marrow, spleen, and thymus were obtained 11–17 weeks after transplantation and analyzed by Southern blot analysis for the presence of the *neo* provirus. Seven of 15 mice were at least partially reconstituted by stem cells harboring the *neo* gene indicating that the *neo* gene was stably inserted into about 50% of pluripotent stem cells. Some of these seven mice were shown to have the *neo* provirus in marrow, thymus, and spleen cells indicating that a pluripotent stem cell had been infected which gave rise to myeloid as well as lymphoid progeny. Others contained the *neo* provirus only in either lymphoid (spleen, thymus) or myeloid (marrow) cells suggesting the existence of more committed stem cells restricted to either the myeloid or lymphoid lineage. Expression of the *neo* gene in the progeny of the stem cells that had reconstituted the W/W^v mice was demonstrated by assaying marrow cells for G418-resistant in vitro colony-forming cells. All but one mouse showed G418-resistant granulocyte-macrophage CFU (CFU-GM). The mouse which did not show G418-resistant CFU-GM had been characterized by DNA analysis to contain *neo* provirus only in lymphoid cells, a finding which is consistent with the result of the functional assay. These observations demonstrating the insertion and expression of a selectable marker gene into murine pluripotent hematopoietic stem cells have been confirmed by others (KELLER et al. 1985; EGLITIS et al. 1985; LEMISCHKA et al. 1986; BOWTELL et al. 1987; SNODGRASS and KELLER 1987).

3 Transfer of Therapeutically Relevant Genes into Murine Hematopoietic Stem Cells

Deficiency of the enzyme adenosine deaminase (ADA) leads to severe combined immunodeficiency, a disorder that could potentially be corrected by transfer of the human ADA gene into pluripotent hematopoietic stem cells of ADA-deficient patients with subsequent autologous transplantation. LIM et al. (1989) were the first to show long-term expression of human ADA in mice up to 6 months after transplantation of marrow infected with helper virus-free vectors. Marrow cells were incubated for 48-h in conditioned medium from WEHI-3B cells as a source of hematopoietic growth factors (including interleukin 3) before 48-h cocultivation with vector-producing packaging cells. Without any further in vitro selection all 13 W/W^v mice expressed human ADA in peripheral blood cells 30 days after TBI (0.3 Gy) and marrow transplantation ($4–6 \times 10^6$ cells). In seven of 13 mice, human ADA expression decreased gradually over 1–4 months, while in six of 13 mice no decrease in the level of human ADA expression was observed until they were sacrificed between 16 and 26 weeks after marrow transplantation. Tissue obtained from these mice at the time of sacrifice was analyzed by in situ enzyme analysis, and expression of human ADA was detected in lymphoid, myeloid, and erythroid cells. Marrow cells from some of these mice were injected into secondary transplant recipients which were shown to express human ADA in peripheral blood cells for at least 6 weeks after transplantation. No helper virus was detected in any of the long-term reconstituted animals. These results suggest successful infection of pluripotent hematopoietic stem cells by human ADA-containing retroviral vectors and subsequent long-term stable gene expression in the progeny of these infected pluripotent stem cells. Presence and long-term expression of the human ADA gene in murine pluripotent hematopoietic stem cells have been confirmed by others (WILSON et al. 1990; KALEKO et al. 1990; OSBORNE et al. 1990; MOORE et al. 1990).

Successful long-term expression of another therapeutically relevant gene, the human β-globin gene, has also been shown in mice. DZIERZAK et al. (1988) infected marrow cells of female mice with retroviral vectors containing the human β-globin gene and *neo* gene by 48-h cocultivation on virus-producing cells and subsequently transplanted the infected marrow cells ($5 \times 10^4–4 \times 10^6$) into lethally irradiated (11 Gy) syngeneic male mice. Peripheral blood cells, obtained 1–2 months later, contained proviral DNA sequences in 8 of 104 animals. Animals were sacrificed 4–9 months after transplantation. DNA analysis of spleen and marrow cells with a Y-chromosome-specific probe showed engraftment with donor-derived cells in all animals, although the extent of donor engraftment varied from 40%–98%. Analysis with a β-globin probe as well as a *neo* probe showed an infection efficiency ranging from 2% to 100%. Eight of the 18 mice found to harbor peripheral cells containing proviral sequences showed expression of human β-globin 4–9 months after marrow transplantation. Expression was restricted to erythroid cells. Human β-globin mRNA varied between only 0.04% and 0.56% of endogenous mouse β-globin mRNA levels while it is estimated that levels of 10% will be needed for the treatment of β-thalassemia. Other reports have also described low expression of human β-globin gene following retroviral-mediated transfer into murine pluripotent stem cells (BODINE et al. 1989; BENDER et al. 1989). Recent experiments have identified sequences approximately 50 Kb upstream and 20 Kb downstream from the structural β-globin gene which, when juxtaposed to a minimal β-globin-containing DNA segment, can strongly enhance β-globin expression independent of the viral integration site (GROSVELD et al. 1987; TALBOT et al. 1989). These sequences were not included in the vectors used in the studies described above, and it appears reasonable that their inclusion might increase β-globin expression. However, such inclusion is technically difficult because of the large size of the sequences, and it leads to inhibition of virus production (NOVAK et al. 1990).

4 Gene Transfer into Hematopoietic Stem Cells of Large Outbred Animals

Gene transfer into hematopoietic stem cells and long-term gene expression in large outbred animals is more difficult to achieve than in mice. One major difference is that, while the number of marrow donors is basically unlimited in the inbred murine model, in outbred animals marrow is only available from one animal, the marrow recipient. This would also be the case in possible human applications. The amount of marrow cells available for retroviral infection is therefore limited. Methods which have been shown in the murine model to increase the efficiency of gene transfer, such as exposure of the marrow after infection to the selective agent to kill uninfected cells, are more difficult to apply to outbred animals. These procedures can lead in the outbred model to a considerable loss of stem cells to the point that the number of transplanted stem cells is too small to allow hematopoietic recovery after TBI and marrow transplantation, and the animals die of infections or bleeding complications secondary to marrow aplasia. It is therefore important to develop infection procedures for large outbred animals which allow a high infection rate of pluripotent hematopoietic stem cells without causing considerable stem cell loss.

We have established a canine model for studying retrovirally mediated gene transfer into canine hematopoietic progenitor cells. Successful transfer and expression of either the *neo* gene or a mutant dihydrofolate reductase (DHFR*) gene into canine CFU-GM has been shown using amphotropic helper-free retroviral vectors (KWOK et al. 1986). The average rate of gene expression in CFU-GM after cocultivation for 24-h with virus-producing packaging cells was approximately 10%, which is 3–10-fold higher than that obtained following exposure of hematopoietic cells to virus-containing medium. Subsequently, infection of canine hematopoietic stem cells by 24-h cocultivation was investigated by transplanting infected autologous marrow into the otherwise lethally irradiated (5 Gy TBI) marrow donor and by studying the reconstituted marrow for the presence and expression of retroviral vectors (STEAD et al. 1988). Five dogs received marrow infected with DHFR* virus and one dog *neo*-virus-infected marow. All dogs engrafted. Three of the five dogs which were given DHFR*-virus-infected marrow received methotrexate as in vivo selection after transplantation. Two of the three dogs died due to gastrointestinal toxicity of the methotrexate. The surviving dog showed 0.03%–0.1% methotrexate-resistant CFU-GM colonies 3 and 5 weeks post-transplant suggesting that only committed hematopoietic progenitor cells had been infected.

In an attempt to improve the efficiency of gene transfer into canine hematopoietic precursor cells, marrow was cocultured for 24-h with *neo* or DHFR*-virus-producing packaging cells as before and then kept for 6 days in a long-term marrow culture system which was fed three times with virus-containing medium (SCHUENING et al. 1989). Subsequently marrow cells were harvested, cultured in CFU-GM assay with and without selective drug and, on average, gave rise to 40%–50% drug-resistant colonies as compared to about 10% after cocultivation only. Possible factors contributing to the improved infection rate are prolonged exposure of marrow cells to retroviral vectors and the stimulation of cell division by repeated feeding of long-term cultures (TOKSÖZ et al. 1980). Subsequent experiments investigated whether combining cocultivation with long-term marrow culture also leads to infection of hematopoietic stem cells (SCHUENING et al. 1991). This was done by transplanting autologous marrow cells infected by 24-h cocultivation and additional incubation in a virus-containing long-term marrow culture system for 4 days. Infected marrow was reinfused into the otherwise lethally irradiated (9.2 Gy TBI) marrow donor, and reconstituted marrow was examined for the presence and expression of retroviral vectors. Four dogs received marrow (0.7–1.0 × 10^8 cells/kg) infected with retroviral vectors which contained both the *neo* gene and the human ADA gene. Two of these dogs engrafted and showed intermittently between 1% and 11% G418-resistant CFU-GM colonies for, so far, over 1.7 years after marrow transplantation. Culture results were confirmed by polymerase chain reaction (PCR) demonstrating the presence of the *neo* gene in marrow cells, so far, up to 1.7 years after transplantation. *Neo* gene was also found by PCR in peripheral blood and lymph node lymphocytes obtained, so far, up to 1.8 years post-transplant. Repeated analysis of peripheral blood and marrow cells for human ADA gene expression by starch gel electrophoresis was negative.

In a different experimental approach, marrow, obtained 7 days after cycloposphamide treatment (40 mg/kg once i.v.) at the time of the peripheral blood neutrophil nadir, was cocultivated for 24 h on vector- producing cells and reinfused (0.06–0.18 × 10^8 cells/kg) into the otherwise lethally irradiated (9.2 Gy TBI) marrow donor. One of three dogs engrafted and showed 2%–6% G418-resistant CFU-GM colonies intermittently for, so far, over 1.5 years similarly to the two surviving dogs of the previous experiment. PCR analysis of marrow cells and lymph node lymphocytes obtained, so far, up to 1.7 years after marrow transplantation demonstrated the presence of the *neo* gene. Similarly to the previous results, no expression of human ADA was detected. Dilution experiments indicated that in the three surviving dogs the fraction of marrow cells, peripheral blood, and lymph node lymphocytes containing the *neo* gene was ≈1%–10% of the total. Peripheral blood samples of all three dogs were free of helper virus more than 1.5 years after transplantation, and no long-term side effects were observed. These data demonstrate for the first time successful infection of pluripotent hematopoietic stem cells in a large outbred adult animal model. Intermittent gene expression is probably due to the small number of stem cells infected, which randomly produce committed progenitor cells for limited periods of time. Alternatively, a small number of infected stem cells may be randomly distributed in the marrow cavity, and marrow is only occasionally aspirated from these spots where infected stem cells are located. Although those results are encouraging, they also indicate that efficiency of stem cell infection and gene expression needs to be further improved.

Studies in nonhuman primates have demonstrated short-term expression of transferred genes, but long-term expression has not been found so far. KANTOFF et al. (1987) transplanted autologous marrow infected with a vector containing the *neo* and the human ADA gene into lethally irradiated nonhuman primates. Southern blot analyses of peripheral blood and marrow-derived mononuclear cells for detection of proviral DNA have been repeatedly negative. Peripheral blood mononuclear cells of two monkeys showed low levels of human ADA several times during the early posttransplant period up to day 120, but no human ADA was detected in either animal after day 160 after transplantation. In situ hybridization for vector-encoded mRNA in peripheral blood monocuclear cells of one of the two animals obtained on day 127 post-transplant was positive in 0.8% of the cells counted. BODINE et al. (1990) used very high- titer viruses generated by vector amplification during cocultivation of different packaging cell lines to infect marrow from rhesus monkeys. The packaging cell lines also produced helper virus. Proviral DNA was detected in ≈1% of the peripheral blood cells of all three animals and in marrow and spleen cells of one animal. Observation time was between 32 and 99 days after transplantation of the infected marrow. This short observation time and the presence of helper virus do not allow definitive conclusions with regard to possible stem cell infection.

An alternative approach, in utero gene transfer, has been described by KANTOFF et al. (1989). Fetal sheep hematopoietic cells were obtained 7 weeks before birth by exchange transfusion, infected in vitro with retroviral vectors containing the *neo* gene, and returned to the donor fetus. Persistently infected cells were demonstrated after birth by detection of *neo* gene in peripheral blood, marrow, spleen, and thymus cells from some animals 1 week after birth, detection of neomycin phosphotransferase in marrow of one animal 6 weeks after birth, and by the intermittent presence of G418-resistant hematopoietic colony forming cells in marrow for more than 2 years after birth. However, the same infection protocol was not successful when infected marrow cells from an adult sheep were transplanted into a heterologous recipient fetal lamb. This difference is probably due to the high replication rate of fetal hematopoietic cells which makes them more susceptible to infection by retroviral vectors.

5 Conclusion

Successful gene transfer into murine pluripotent hematopoietic stem cells has been shown by using both selectable marker genes as well as therapeutically relevant genes. Long-term expression of human ADA in mice has been demonstrated achieving ADA levels in blood which seem to be sufficient for treating ADA deficiency. In contrast, human β-globin expression was only 1%–2% of that of murine β-globin which is probably too low for the treatment of β-thalassemia. Current experiments are trying to construct different retroviral vectors containing regulatory elements which strongly enhance β-globin expression in erythroid cells to remedy this limitation. Gene transfer into hematopoietic stem cells of large outbred animals is more diffcult to achieve than in mice mainly because the number of marrow cells available for retroviral infection is limited to those harvested from one animal, the marrow recipient. It is, therefore, important to develop infection procedures which lead to a high infection rate of pluripotent hematopoietic stem cells without causing considerable stem cell loss. Experiments in dogs have shown infection of pluripotent hematopoietic stem cells by 24-h cocultivation on vector-producing cells combined with subsequent 4-day incubation in a virus-containing long-term marrow culture system as well as by 24-h cocultivation of marrow obtained after cyclophosphamide pretreatment. Stem cell infection was documented by demonstration of drug-resistant marrow derived CFU-GM colonies for more than 1.5 years after transplantation of the infected marrow and by PCR results confirming the presence of the transferred *neo* gene in marrow cells, peripheral blood, and lymph node lymphocytes more than 1.5 years after transplantation. These results are encouraging insofar as for the first time successful gene transfer into pluripotent hematopoietic stem cells in a large outbred adult animal model has been shown without any long-term side effects. On the other hand, the obsevation that the transferred gene was expressed only intermittently suggests that the efficiency of stem cell infection and gene expression needs to be further improved before possible application in humans.

Acknowledgements. This work was supported by grants CA15704, CA18105, CA18221, and CA31787 awarded by the National Cancer Institute and HL36444 awarded by the National Heart, Lung, and Blood Institute, Department of Health and Human Services.

References

Bender MA, Gelinas RE, Miller AD (1989) A majority of mice show long-term expression of a human β-globin gene after retroviral transfer into hematopoietic stem cells. Mol Cell Biol 9: 1426–1434

Bodine DM, Karlsson S, Nienhuis AW (1989) Combination of interleukins 3 and 6 preserves stem cell function in culture and enhances retrovirus-mediated gene transfer into hematopoietic stem cells. Proc Natl Acad Sci USA 86: 8897–8901

Bodine DM, McDonagh KT, Brandt SJ, Ney PA, Agricola B, Byrne E, Nienhuis AW (1990) Development of a high-titer retrovirus producer cell line capable of gene transfer into rhesus monkey hematopoietic stem cells. Proc Natl Acad Sci USA 87: 3738–3742

Boggs DR, Boggs SS, Saxe DF, Gress LA, Canfield DR (1982) Hematopoietic stem cells with high proliferative potential. Assay of their concentration in marrow by the frequency and duration of cure of W/W^V mice. J Clin Invest 70: 242–253

Bowtell DDL, Johnson GR, Kelso A, Cory S (1987) Expression of genes transferred to haemopoietic stem cells by recombinant retroviruses. Mol Biol Med 4: 229–250

Dick JE, Magli MC, Huszar D, Phillips RA, Bernstein A (1985) Introduction of a selectable gene into primitive stem cells capable of long-term reconstitution of the hemopoietic system of W/W^V mice. Cell 42: 71–79

Dzierzak EA, Papayannopoulou T, Mulligan RC (1988) Lineage-specific expression of a human β-globin gene in murine bone marrow transplant recipients reconstituted with retrovirus-transduced stem cells. Nature 331: 35–41

Eglitis MA, Kantoff P, Gilboa E, Anderson WF (1985) Gene expression in mice after high efficiency retroviral-mediated gene transfer. Science 230: 1395–1398

Grosveld F, Van Assendelft GB, Greaves DR, Kollias G (1987) Position-independent, high-level expression of the human β-globin gene in transgenic mice. Cell 51: 975–985

Harrison DE, Astle CM, Lerner C (1988) Number and continuous proliferative pattern of transplanted primitive immunohematopoietic stem cells. Proc Natl Acad Sci USA 85: 822–826

Hodgson GS, Bradley TR (1979) Properties of haematopoietic stem cells surviving 5-fluorouracil treatment: evidence for a pre-CFU-S cell? Nature 281: 381–382

Kaleko M, Garcia JV, Osborne WRA, Miller AD (1990) Expression of human adenosine deaminase in mice after transplantation of genetically-modified bone marrow. Blood 75: 1733–1741

Kantoff PW, Flake AW, Eglitis MA, Scharf S, Bond S, Gilboa E, Erlich H, Harrison MR, Zanjani ED, Anderson WF (1989) In utero gene transfer and expression: a sheep transplantation model. Blood 73: 1066–1073

Kantoff PW, Gillio AP, Mclachlin JR, Bordignon C, Eglitis MA, Kernan NA, Moen RC, Kohn DB, Yu S-F, Karson E, Karlsson S, Zwiebel JA, Gilboa E, Blaese RM, Nienhuis A, O' Reilly RJ, Anderson WF (1987) Expression of human adenosine deaminase in nonhuman primates after retrovirus-mediated gene transfer. J Exp Med 166: 219–234

Keller G, Paige C, Gilboa E, Wagner EF (1985) Expression of a foreign gene in myeloid and lymphoid cells derived from multipotent hematopoietic precursors. Nature 318: 149–154

Kwok WW, Schuening F, Stead RB, Miller AD (1986) Retroviral transfer of genes into canine hemopoietic progenitor cells in culture: a model for human gene therapy. Proc Natl Acad Sci USA 83: 4552–4555

Lemischka IR, Raulet DH, Mulligan RC (1986) Developmental potential and dynamic behavior of hematopoietic stem cells. Cell 45: 917–927

Lim B, Apperley JF, Orkin SH, Williams DA (1989) Long-term expression of human adenosine deaminase in mice transplanted with retrovirus-infected hematopoietic stem cells. Proc Natl Acad Sci USA 86: 8892–8896

Miller DG, Adam MA, Miller AD (1990) Gene transfer by retrovirus vectors occurs only in cells that are actively replicating at the time of infection. Mol Cell Biol 10: 4239–4242

Moore KA, Fletcher FA, Villalon DK, Utter AE, Belmont JW (1990) Human adenosine deaminase expression in mice. Blood 75: 2085–2092

Novak U, Harris EAS, Forrester W, Groudine M, Gelinas R (1990) High-level β-globin expression after retroviral transfer of locus activation region-containing human β-globin gene derivatives into murine erythroleukemia cells. Proc Natl Acad Sci USA 87: 3386–3390

Osborne WRA, Hock RA, Kaleko M, Miller AD (1990) Long-term expression of human adenosine in mice after transplantation of bone marrow infected with amphotropic retroviral vectors. Hum Gene Ther 1: 31–41

Schuening FG, Storb R, Stead RB, Goehle S, Nash R, Miller AD (1989) Improved retroviral transfer of genes into canine hematopoietic progenitor cells kept in long-term marrow culture. Blood 74: 152–155

Schuening FG, Kawahara K, Miller AD, To R, Goehle S, Stewart D, Mullally K, Graham TC, Appelbaum FR, Hackman R, Osborne WRA, Storb R (1991) Retrovirus-mediated gene transduction into long-term repopulating marrow cells of dogs. Blood (in press)

Snodgrass R, Keller G (1987) Clonal fluctuation within the heamatopoietic system of mice reconstituted with retrovirus-infected stem cells. EMBO J 6: 3955–3960

Stead RB, Kwok WW, Storb R, Miller AD (1988) Canine model for gene therapy: Inefficient gene expression in dogs reconstituted with autologous marrow infected with retroviral vectors. Blood 71: 742–747

Talbot D, Collis P, Antoniou M, Vidal M, Grosveld F, Greaves DR (1989) A dominant control region from the human β-globin locus conferring integration site-independent gene expression. Nature 338: 352–355

Toksöz D, Dexter TM, Lord BI, Wright EG, Lajtha LG (1980) The regulation of hemopoiesis in long-term bone marrow cultures. II. Stimulation and inhibition of stem cell proliferation. Blood 55: 931–936

van Zant G (1984) Studies of hematopoietic stem cells spared by 5-fluorouracil. Exp Med 159: 679–690

Varmus H (1988) Retroviruses. Science 240: 1427–1435

Wilson JM, Danos O, Grossman M, Raulet DH, Mulligan RC (1990) Expression of human adenosine deaminase in mice reconstituted with retrovirus-transduced hematopoietic stem cells. Proc Natl Acad Sci USA 87: 439–443

Subject Index

Current Topics in Microbiology and Immunology

Volumes published since 1987 (and still available)

Vol. 133: **Oldstone, Michael B. (Ed.):** Arenaviruses. Genes, Proteins, and Expression. 1987. 39 figs. VII, 116 pp. ISBN 3-540-17246-7

Vol. 134: **Oldstone, Michael B. (Ed.):** Arenaviruses. Biology and Imminotherapy. 1987. 33 figs. VII, 242 pp. ISBN 3-540-14322-6

Vol. 135: **Paige, Christopher J.; Gisler, Roland H. (Ed.):** Differentiation of B Lymphocytes. 1987. 25 figs. IX, 150 pp. ISBN 3-540-17470-2

Vol. 136: **Hobom, Gerd; Rott, Rudolf (Ed.):** The Molecular Biology of Bacterial Virus Systems. 1988. 20 figs. VII, 90 pp. ISBN 3-540-18513-5

Vol. 137: **Mock, Beverly; Potter, Michael (Ed.):** Genetics of Immunological Diseases. 1988. 88 figs. XI, 335 pp. ISBN 3-540-19253-0

Vol. 138: **Goebel, Werner (Ed.):** Intracellular Bacteria. 1988. 18 figs. IX, 179 pp. ISBN 3-540-50001-4

Vol. 139: **Clarke, Adrienne E.; Wilson, Ian A. (Ed.):** Carbohydrate-Protein Interaction. 1988. 35 figs. IX, 152 pp. ISBN 3-540-19378-2

Vol. 140: **Podack, Eckhard R. (Ed.):** Cytotoxic Effector Mechanisms. 1989. 24 figs. VIII, 126 pp. ISBN 3-540-50057-X

Vol. 141: **Potter, Michael; Melchers, Fritz (Ed.):** Mechanisms in B-Cell Neoplasia 1988. Workshop at the National Cancer Institute, National Institutes of Health, Bethesda, MD, USA, March 23–25, 1988. 1988. 122 figs. XIV, 340 pp. ISBN 3-540-50212-2

Vol. 142: **Schüpach, Jörg:** Human Retrovirology. Facts and Concepts. 1989. 24 figs. 115 pp. ISBN 3-540-50455-9

Vol. 143: **Haase, Ashley T.; Oldstone Michael B. A. (Ed.):** In Situ Hybridization 1989. 22 figs. XII, 90 pp. ISBN 3-540-50761-2

Vol. 144: **Knippers, Rolf; Levine, A. J. (Ed.):** Transforming. Proteins of DNA Tumor Viruses. 1989. 85 figs. XIV, 300 pp. ISBN 3-540-50909-7

Vol. 145: **Oldstone, Michael B. A. (Ed.):** Molecular Mimicry. Cross-Reactivity between Microbes and Host Proteins as a Cause of Autoimmunity. 1989. 28 figs. VII, 141 pp. ISBN 3-540-50929-1

Vol. 146: **Mestecky, Jiri; McGhee, Jerry (Ed.):** New Strategies for Oral Immunization. International Symposium at the University of Alabama at Birmingham and Molecular Engineering Associates, Inc. Birmingham, AL, USA, March 21–22, 1988. 1989. 22 figs. IX, 237 pp. ISBN 3-540-50841-4

Vol. 147: **Vogt, Peter K. (Ed.):** Oncogenes. Selected Reviews. 1989. 8 figs. VII, 172 pp. ISBN 3-540-51050-8

Vol. 148: **Vogt, Peter K. (Ed.):** Oncogenes and Retroviruses. Selected Reviews. 1989. XII, 134 pp. ISBN 3-540-51051-6

Vol. 149: **Shen-Ong, Grace L. C.; Potter, Michael; Copeland, Neal G. (Ed.):** Mechanisms in Myeloid Tumorigenesis. Workshop at the National Cancer Institute, National Institutes of Health, Bethesda, MD, USA, March 22, 1988. 1989. 42 figs. X, 172 pp. ISBN 3-540-50968-2

Vol. 150: **Jann, Klaus; Jann, Barbara (Ed.):** Bacterial Capsules. 1989. 33 figs. XII, 176 pp. ISBN 3-540-51049-4

Vol. 151: **Jann, Klaus; Jann, Barbara (Ed.):** Bacterial Adhesins. 1990. 23 figs. XII, 192 pp. ISBN 3-540-51052-4

Vol. 152: **Bosma, Melvin J.; Phillips, Robert A.; Schuler, Walter (Ed.):** The Scid Mouse. Characterization and Potential Uses. EMBO Workshop held at the Basel Institute for Immunology, Basel, Switzerland, February 20–22, 1989. 1989. 72 figs. XII, 263 pp. ISBN 3-540-51512-7

Vol. 153: **Lambris, John D. (Ed.):** The Third Component of Complement. Chemistry and Biology. 1989. 38 figs. X, 251 pp. ISBN 3-540-51513-5

Vol. 154: **McDougall, James K. (Ed.):** Cytomegaloviruses. 1990. 58 figs. IX, 286 pp. ISBN 3-540-51514-3

Vol. 155: **Kaufmann, Stefan H. E. (Ed.):** T-Cell Paradigms in Parasitic and Bacterial Infections. 1990. 24 figs. IX, 162 pp. ISBN 3-540-51515-1

Vol. 156: **Dyrberg, Thomas (Ed.):** The Role of Viruses and the Immune System in Diabetes Mellitus. 1990. 15 figs. XI, 142 pp. ISBN 3-540-51918-1

Vol. 157: **Swanstrom, Ronald; Vogt, Peter K. (Ed.):** Retroviruses. Strategies of Replication. 1990. 40 figs. XII, 260 pp. ISBN 3-540-51895-9

Vol. 158: **Muzyczka, Nicholas (Ed.):** Viral Expression Vectors. 1992. 20 figs. IX, 176 pp. ISBN 3-540-52431-2

Vol. 159: **Gray, David; Sprent, Jonathan (Ed.):** Immunological Memory. 1990. 38 figs. XII, 156 pp. ISBN 3-540-51921-1

Vol. 160: **Oldstone, Michael B. A.; Koprowski, Hilary (Ed.):** Retrovirus Infections of the Nervous System. 1990. 16 figs. XII, 176 pp. ISBN 3-540-51939-4

Vol. 161: **Racaniello, Vincent R. (Ed.):** Picornaviruses. 1990. 12 figs. X, 194 pp. ISBN 3-540-52429-0

Vol. 162: **Roy, Polly; Gorman, Barry M. (Ed.):** Bluetongue Viruses. 1990. 37 figs. X, 200 pp. ISBN 3-540-51922-X

Vol. 163: **Turner, Peter C.; Moyer, Richard W. (Ed.):** Poxviruses. 1990. 23 figs. X, 210 pp. ISBN 3-540-52430-4

Vol. 164: **Bækkeskov, Steinnun; Hansen, Bruno (Ed.):** Human Diabetes. 1990. 9 figs. X, 198 pp. ISBN 3-540-52652-8

Vol. 165: **Bothwell, Mark (Ed.):** Neuronal Growth Factors. 1991. 14 figs. IX, 173 pp. ISBN 3-540-52654-4

Vol. 166: **Potter, Michael; Melchers, Fritz (Ed.):** Mechanisms in B-Cell Neoplasia 1990. 143 figs. XIX, 380 pp. ISBN 3-540-52886-5

Vol. 167: **Kaufmann, Stefan H. E. (Ed.):** Heat Shock Proteins and Immune Response. 1991. 18 figs. IX, 214 pp. ISBN 3-540-52857-1

Vol. 168: **Mason, William S.; Seeger, Christoph (Ed.):** Hepadnaviruses. Molecular Biology and Pathogenesis. 1991. 21 figs. X, 206 pp. ISBN 3-540-53060-6

Vol. 169: **Kolakofsky, Daniel (Ed.):** Bunyaviridae. 1991. 34 figs. X, 256 pp. ISBN 3-540-53061-4

Vol. 170: **Compans, Richard W. (Ed.):** Protein Traffic in Eukaryotic Cells. Selected Reviews. 1991. 14 figs. X, 186 pp. ISBN 3-540-53631-0

Vol. 171: **Kung, Hsing-Jien; Vogt, Peter K. (Eds.):** Retroviral Insertion and Oncogene Activation. 1991. 18 figs. X, 179 pp. ISBN 3-540-53857-7

Vol. 172: **Chesebro, Bruce W. (Ed.):** Transmissible Spongiform Encephalopathies. 1991. 48 figs. X, 288 pp. ISBN 3-540-53883-6

Vol. 173: **Pfeffer, Klaus; Heeg, Klaus; Wagner, Hermann; Riethmüller, Gert (Ed.):** Function and Specificity of γ/δ T Cells. 1991. 41 figs. XII, 296 pp. ISBN 3-540-53781-3

Vol. 174: **Fleischer, Bernhard; Sjögren, Hans Olov (Eds.):** Superantigens. 1991. 13 figs. IX, 137 pp. ISBN 3-540-54205-1

Vol. 175: **Aktories, Klaus (Ed.):** ADP-Ribosylating Toxins. 1992. 23 figs. IX, 148 pp. ISBN 3-540-54598-0

Vol. 176: **Holland, John J. (Ed.):** Genetic Diversity of RNA Viruses. 34 figs. IX, 226 pp. ISBN 3-540-54652-9